MW01134564

Müller's Lab

Laura Otis

Müller's Lab

2007

OXFORD
UNIVERSITY PRESS

Oxford University Press, Inc., publishes works that further
Oxford University's objective of excellence
in research, scholarship, and education.

Oxford New York
Auckland Cape Town Dar es Salaam Hong Kong Karachi
Kuala Lumpur Madrid Melbourne Mexico City Nairobi
New Delhi Shanghai Taipei Toronto

With offices in
Argentina Austria Brazil Chile Czech Republic France Greece
Guatemala Hungary Italy Japan Poland Portugal Singapore
South Korea Switzerland Thailand Turkey Ukraine Vietnam

Published by Oxford University Press, Inc.
198 Madison Avenue, New York, New York 10016

www.oup.com

Library of Congress Cataloging-in-Publication Data
Otis, Laura, 1961–
Müller's lab / Laura Otis.
p. cm.
Includes bibliographical references.
ISBN: 978-0-19-530697-2
1. Müller, Joh., 1801–1858. 2. Anatomists—Germany—Biography.
3. Physiologists—Germany—Biography. 4. Life sciences—Germany
—History—19th century. I. Title.
[DNLM: 1. Müller, Joh., 1801–1858. 2. Anatomy—history
—Germany. 3. Physiology—history—Germany. 4. History, 19th Century
—Germany. 5. Mentors—Germany. QT 11 GG4 O88m 2007]
QM16.M85O85 2007
611'.0092—dc22 2006026796

For all the people who have spent their lives exploring the wonders of the human body.

Acknowledgments

First, I would like to extend my deepest thanks to the John D. and Catherine T. MacArthur Foundation and the Alexander von Humboldt Stiftung for the generous fellowships that supported me while I was writing *Müller's Lab*. I am also grateful to Hofstra University for granting me general leaves in 2000–2002 and 2003–2004 and to Emory University for allowing me a research leave in 2005–2006 so that I could do the research and writing for this book.

I am extremely thankful to the Max Planck Institute for the History of Science in Berlin, which has hosted me for four years and offered me a research paradise, including inspiring interactions with colleagues, a superb library, and the best office I have ever had. I offer my warmest thanks to Michael Hagner and Helmut Müller-Sievers, who brought me to the Institute, and to Hans-Jörg Rheinberger, who has allowed me to work in Department III under his guidance. I am grateful to Ellen Garske, Matthias Schwerdt, and the other Institute librarians for their indefatigable work in locating the sources for this book. And I extend my warmest thanks to department secretary Antje Radeck, whose generosity and good humor made me feel welcome in Berlin.

In addition, I must thank the many archivists who have granted me access to hand-written documents and, when possible, made copies for me to keep. I am indebted to Dr. Wolfgang Knobloch at the Archiv der Berlin-Brandenburgische Akademie der Wissenschaften for allowing me to examine Virchow's and Helmholtz's papers, and I am grateful to the archivists of the Geheimes Staatsarchiv Preussischer Kulturbesitz for letting me see the files on Müller's Anatomical Museum. I thank the staff of the Handschriftenabteilung of the Staatsbibliothek zu Berlin, Haus 2, for giving me access to Emil du Bois-Reymond's letters. I offer my deepest thanks to Dr. Thomas Bach of the Ernst-Haeckel-Haus in Jena for helping me to locate and date Ernst Haeckel's fragmented autobiographical sketches and for transcribing his challenging handwritten descriptions of Müller. In Heidelberg, I am indebted to Universitätsarchiv Director Werner Moritz and to Sabine Breihofer for welcoming me and showing me some of Jakob Henle's correspondence. I am also grateful to André Tourneux and Raphael de Smedt of the Bibliothèque Royale Albert I in Brussels for granting me stack access so that I could examine the books from Müller's library.

Many scholars have helped to shape *Müller's Lab* with their knowledge and intelligence, but I am most indebted to those who have read and re-read the chapters, offering valuable advice and much-needed wake-up calls. Foremost among these are Michael Hagner, Larry Holmes, and Hans-Jörg Rheinberger. Like other historians of science, I deeply regret Holmes's untimely death and will miss his fine judgment and perceptive insights. I thank my readers at Oxford University Press, one of whom, Peter Vinten-Johansen, offered copious suggestions about how I could present material more clearly.

I am also grateful to all the generous scholars at the Max Planck Institute and elsewhere who have led me to essential sources of information: David Cahan, for reliable histories of nineteenth-century Germany; Tom Couser and Craig Howes, for analyses of life writing; Sven Dierig and Henning Schmidgen, for designing the Virtual Laboratory and showing me how to use it as a resource; Gabriel Finkelstein, for the du Bois-Reymond–Bence Jones correspondence, which helped me to see these scientists in a whole new light; Sander Gilman, for information about *Burschenschaft* practices and Jewish culture in nineteenth-century Germany; Christoph Gradmann, for the Henle-Magnus letters; Volker Hess, for histories of medicine in Berlin; Lynn Nyhart, for studies of academic competition; Ohad Parnes, for Friedrich Bidder's biography and some other vital accounts of Müller's and Henle's teaching; Henning Schmidgen, for the Fonds Müller at the Bibliothèque Royale; Thomas Schnalke, for a rundown of the best Virchow biographies and letters; and Julia Voss, for Albert Gunther's valuable portrayal of Müller's last years.

Since I was trained as a neuroscientist and literary scholar, not as a historian, I am indebted to all the people who have enlightened me on how the history of science works. For all the epistemological conversations from which I learned what it means to investigate past scientists, I thank David Cahan, Raine Daston, Gabriel Finkelstein, Peter Galison, Michael Hagner, and Hans-Jörg Rheinberger.

Finally, a substantial part of writing *Müller's Lab* consisted of translating nineteenth-century German into English, which was a joy despite the time it consumed. As a creative writer and comparative literature scholar, I loved the challenge of these translations, but I must emphasize that I did not do them alone. I am profoundly grateful to the many colleagues here in Berlin who explained what strange words meant and untangled labyrinthine sentences: Cornelius Borck, Christina Brandt, Soraya Chadarevian, Sven Dierig, Uljana Feest, Josephine Fenger, Peter Geimer, Sabine Hoehler, Christoph Hoffmann, Manfred Laublicher, Andreas Mayer, Staffan Müller-Wille, Antje Radeck, Henning Schmidgen, Iris Schroeder, Friedrich Steinle, and Julia Voss. I thank all of these scholars for teaching me German as a culture and state of mind as well as a language.

Contents

Introduction: The Lab That Never Was xi

1 Müller's Net 3

2 Cells and Selves: The Training of Jakob Henle and Theodor Schwann 42

3 Emil du Bois-Reymond as a Scientific and Literary Creator 76

4 Physiological Bonds: The Training of Hermann von Helmholtz 111

5 Rudolf Virchow's Scientific Politics 132

6 Banned from the Academy: The Mentoring of Robert Remak 165

7 Ernst Haeckel's Evolving Narratives 190

Afterword: Remembering and Dismembering a Scientist 224

Abbreviations 237

Notes 239

Bibliography 293

Index 305

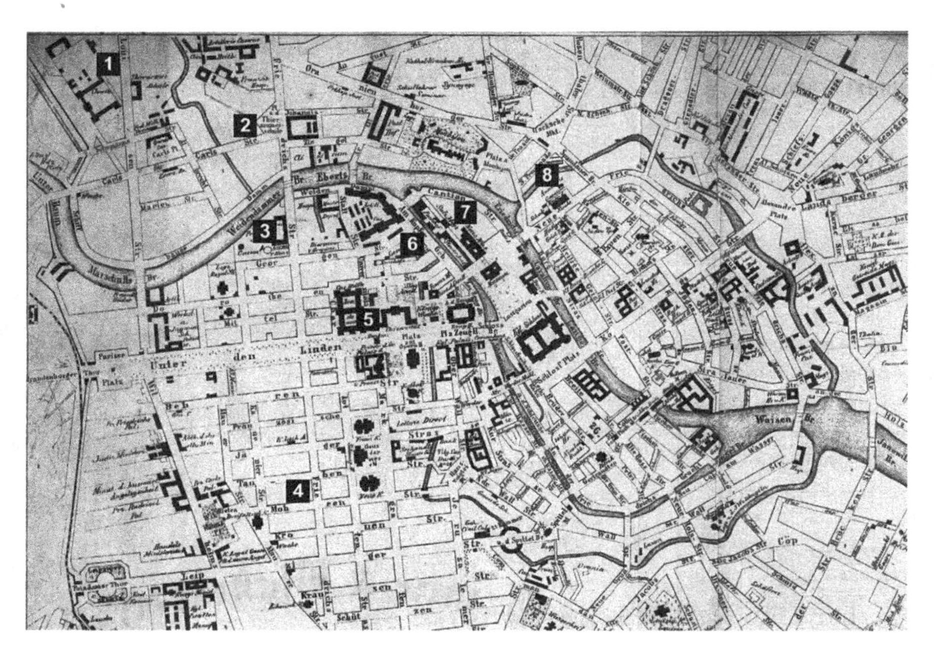

Map of Berlin in 1857. Source: Berlin: Heintze and Blanckertz, 1857, SBPK, Haus 1, Kartenabteilung, printed with kind permission of the Bildarchiv Preussischer Kulturbesitz. The numbers identify the buildings listed below:

1. The Charité Hospital
2. Carlstrasse 20–21, home of du Bois-Reymond 1845–1860
3. The Friedrichs-Wilhelms-Institut (Pépinière), where Helmholtz and Virchow studied medicine
4. The "Hotel Hilgendorf," home of Henle and Schwann 1835–1839
5. The Berlin University building, home of Müller's Anatomical Museum
6. Müller's first home (1833–1840), am Kupfergraben 4a
7. Müller's second home (1840–1858), Cantianstrasse 5
8. The Anatomical Institute (Anatomical Theater)

Introduction
The Lab That Never Was

Historians wish to alter views of the past
or they would not be writing.
—Peter Loewenberg

Johannes Müller never had a lab. He and his students performed experiments that laid the foundations of modern biology, but they never worked in a designated space or functioned as a unified group. Improvising, they did science anywhere and everywhere they could: in tiny rooms adjacent to the medical students' dissecting hall, in the window nooks of the Anatomical Museum, in shacks along the North Sea coast, and in a run-down guest house on Friedrichstrasse. The whole notion of Müller's lab is a myth, traceable to remarks by his students and elaborated by generations of historians.[1] This book bears the name of a nonexistent lab because it explores fiction as well as science, fiction in the sense of history that is actively made. We know Müller today only through the accounts of his students, each of whom described him in a different way. This study will compare seven great scientists' depictions of their teacher, exploring the many factors that can shape histories of science past.[2] It is not a biography but a labography, the story of eight individuals whose intersecting lives allowed modern biology to emerge.

A cobbler's son, Müller became the most respected anatomist and physiologist in the mid-nineteenth-century German-speaking world. In Europe, his *Handbook of Human Physiology* was the definitive source for much of the nineteenth century. Müller acquired this reputation for good reasons. As a young investigator, he studied nervous and sensory systems in a wide range of animals and devised the law of specific sense energies, the idea that each sensory organ can respond to stimuli in only one way (i.e., as light, sound, or pain). In a classic series of experiments, he used frogs to show that the dorsal and ventral roots of spinal nerves carry mainly sensory and motor information, respectively. Müller also discovered the arteries responsible for penile erections and showed that tumors are made of cells. As a comparative anatomist and early marine biologist, he designed a net still used to scoop up plankton and described the development of countless organisms. Between 1822 and 1858, he established methods that transformed the experimental and descriptive life sciences.

Despite, or perhaps because of, his ceaseless activity, Müller was a troubled scientist. A workaholic, he could scan tissue under his microscope for ten hours at a stretch, drinking strong coffee and sleeping only with difficulty. Over the course of his life, he suffered at least five depressive episodes, and according to one student, he committed suicide at fifty-six. The son of the Swedish anatomist Anders Retzius, Müller's best friend, called Müller "the great investigator with the serious, often melancholy face and flashing eyes, which no one who once saw them could ever forget."[3]

From 1833 to 1858, when Müller taught in Berlin, this charismatic physiologist attracted brilliant students. Any physician or biologist who has studied *Gray's Anatomy* will know the bodily structures that bear their names. Fluids in the kidney's renal tubules pass through the loop of Henle. In the peripheral nervous system, Schwann cells insulate axons with layers of fatty membrane. In sympathetic nerves, signals pass along the fine, gray fibers of Remak. Although less well known today than his pupils, Müller trained the microscopists and experimentalists who mapped the human body at the cellular level.

Like their teacher, Müller's students made crucial discoveries while facing considerable challenges. Jakob Henle (1809–1885), Müller's first and favorite student, taught anatomy to the bacteriologist Robert Koch and preceded Louis Pasteur in claiming that diseases are caused by living organisms. He was also arrested for participating in a political fraternity, and could continue his microscopic studies only when Müller got him out of jail. Theodor Schwann (1810–1882), a devout Catholic, showed that all animals are made of cells, but gradually stopped doing research after a religious crisis in 1838. Emil du Bois-Reymond (1818–1896) collaborated with instrument makers to design super-sensitive galvanometers that could detect the electrical activity of nerves and muscles. Hermann von Helmholtz (1821–1894) measured the velocity of nerve impulses and proposed the currently accepted model of color vision. Rudolf Virchow (1821–1902) convinced physicians to think about disease at the cellular level, and in his political career he drove through legislation that vastly improved public health. As an elected representative in the Prussian Parliament, he so enraged Chancellor Otto von Bismarck with his opposition to military spending that Bismarck challenged him to a duel. Robert Remak (1815–1865), the first unbaptized Jew to teach on the Berlin Medical Faculty, discovered that nerve cell bodies and nerve fibers are connected and described the three germ layers of human embryos. Because he refused to renounce his religion, he obtained an academic appointment only at the end of his life, due largely to Müller's unflagging efforts. Ernst Haeckel (1834–1919), a philosopher as well as an embryologist, became famous for his controversial theory that ontogeny (individual development) recapitulates phylogeny (the development of a species or race). When one considers how greatly these students' interests and career choices differed, it is hardly surprising that their portraits of Müller varied. When describing what Müller was like as an adviser, each student tells a different tale.

Why did each scientist see a different Johannes Müller? When they wrote about him, why did they represent him as they did? As a literary-historical case study, this book will juxtapose seven students' accounts of an inspiring teacher in order to show how social conditions, institutional conflicts, scientific interests, and personal relationships can affect the writing of history. The case of Müller and his students is not simply a new instance of three blind men and an elephant. Their tales do not differ merely because they were groping a very large scientist and each felt only a small piece. Ultimately, readers must decide for themselves why these students experienced Müller so differently, but as they do so, I urge them to ask three questions to which I will return as I discuss their science.

First, *What kind of relationship did each student have with Müller, and how did his social background affect his dealings with his mentor?* Helmholtz and Virchow came from families that could not afford a university education, so that they learned science on military medical scholarships. Their clinical obligations restricted the amount of time that they could spend with Müller, so that they interacted with him much less than his university students, Henle and du Bois-Reymond. With the exceptions of Virchow and Remak, all of the pupils examined in this book were from a higher social class than Müller. Consequently, they came to science and medicine with higher career expectations, and thus were less patient about working for years as the poorly paid (or unpaid) assistants of a star scientist. The social and institutional conditions under which they worked aggravated conflicts based on scientific interests. As Müller's focus shifted from physiology of the nervous system and sensory organs to comparative anatomy and marine life, he disappointed the students drawn by his early physiological experiments. Schwann and du Bois-Reymond, who had wanted to build on his work of the late 1820s, found themselves sorting fish in his museum. Because Prussian universities of the 1830s and 1840s strictly limited the numbers of academic positions available, ambitious students could become professors only if they took jobs in remote cities (if Müller recommended them) or if he moved or died. Forced to sacrifice either the advanced technology and scientific community of Berlin or the possibility of academic advancement, most chose temporary exile, while others reluctantly stayed on as Müller's satellites. These decisions affected their relationships with him and the tales that they subsequently wrote.

As readers judge their stories, the second question I hope they will ask is *How did each student's relationship with Müller affect the science he eventually did?* If one believes the pupils' retrospective accounts, they founded modern biology largely in spite of their adviser, not because of him. None of the seven liked to think of his work as the product of influence.

The issue of scientific originality resonates most strongly in the third question, the one most essential to this study: *How did each student's relationship with Müller and the work that the younger scientist did affect the way that he represented Müller in the historical record?* As his own science developed, how did he touch up the portrait of his teacher? Like most people, Müller changed with time, so that his scientific ideas and his treatment of young students in 1854, when Haeckel knew him, were not what they had been twenty-five years earlier, when he was training Henle. And as the students' careers advanced, most of them adjusted their memories of Müller to fit their evolving ideas about cells, tumors, and the nature of life.[4] In describing Müller, they were not passive recorders but active creators.

Accurately reconstructing the lives of Müller and his students means admitting that one is working with fictions. In making this claim, I do not wish to imply that science is just language, or stories, or texts. It is not a mere narrative or way of writing. Science is a frustrating, challenging, gritty experience and a courageous attempt to learn how the world works. Müller and his students really lived, and every student of medicine today benefits from the knowledge that they produced. For historians, of course, the problem is access. We can learn about how Müller's students built this knowledge only through their own accounts, and each of them had his own agenda in writing science history.[5] Johannes Müller's pupils provide the only firsthand descriptions of his work, and their letters and memorial addresses depict their teacher in ways that serve their own interests. None can be read as a transparent window yielding a view of past science. Instead, their conflicting accounts of Müller reflect their understandings of themselves as scientists.

The students' descriptions are self-serving—but as opposed to what? It would be naïve to expect to find unbiased accounts of Müller's science, including those of Müller himself. L. S. Jacyna has remarked that efforts to discover the "real" scientists behind written descriptions are as futile as nineteenth-century attempts to uncover the "real" Jesus.[6] Every existing narrative is already an interpretation.

How do you learn about the history of science when you can't trust anything you read? It is one thing to admit that the sources are unreliable, but quite another to act on it. I have acted by both using and exposing oft-quoted accounts, telling good-sounding stories only to undermine them. Since there is no trustworthy master narrative of Müller's life, I suspect that one can approach the truth only by hearing his story from as many perspectives as possible. There is a truth about Johannes Müller in the sense that there was a lived reality. Someone was born on 14 July 1801, experimented with frogs, and gave the lectures recorded in volumes of notes in state archives. Mathematically, each student's account might be regarded as a curve that intersects this lived reality at several points. By studying these curves sequentially, then superimposing them in their minds' eyes, readers will see the shape of the Müller function even if they can't plot all of its points.[7]

To show what it was like to learn science from Müller, I will present each student's portrait of him in turn. I have tried to write the history of science as William Faulkner might have written it, offering multiple viewpoints of some rather grim conditions. One by one, I will explore each student's perspective, emphasizing the dynamic interplay of politics, material environs, human emotion, and human intelligence. Readers are invited to meet each successive pupil and assess his version of Müller's science in the context of his own experiences and achievements. Like detectives, they must then reconstruct the "lab" of this troubled physiologist for themselves. In this respect, they will find demands placed upon them more akin to those facing readers of fiction than those of science history. Based on a series of conflicting clues, they must form a mental image of the situation that generated these traces. It is something that both scientists and avid fiction readers do very well.

To some degree, anyone who has ever written a biography has weighed conflicting evidence from multiple sources. My aim, however, is to focus more on the differences than the similarities. Consequently, this book bears some resemblance to postmodern biographies of the 1990s, which abandoned any attempt to chronicle a life with a fluent, unified narrative. Rejecting the notion that a human being is a coherent whole whose life can be captured in words, writers of these new biographies offered readers multiple perspectives and discontinuities, destabilizing their own authority as narrators.[8] My goals are quite different. I provide multiple perspectives on Müller's science not because I think the truth about it can never be learned, but because I believe we can approach that truth most closely if we hear about it from scientists with different points of view. This book is not a biography but a study of how history is written, a project with literary as well as scientific roots.

Having worked in biology labs for eight years and taught English for twenty, I have applied literary techniques to the history of science for the benefit of scientists, historians, and English teachers alike. This study is meant to complement those of perseverant historians who have invested years locating and analyzing scientists' unpublished writings. It is my hope that the interpretive strategies of historians, literary scholars, and scientists can be combined to produce knowledge about the way that science is done. By "literary techniques" I mean nothing obscure, nothing that biologists and historians don't use every day. For instance, I examine the students' motives for writing and their consciousness of their readers. In their narratives about Müller, I study the ways that they depict changes in their ideas, either as gradual shifts or as sharp, dramatic breaks. I call attention not just to their choices of words but to their favorite words, terms that they apply to surprisingly different situations and that suggest links between them. If an English teacher were handed Emil du Bois-Reymond's letters to his friend Eduard Hallmann, s/he might look for recurring metaphors (such as comparing science to war) and narrative tactics (such as describing his "progress" with Müller as an eighteenth-century libertine reported the progress of his seductions). The results can be amusing, but they are much more than that. They reveal the ways that scientists conceive of themselves, their work, and their predecessors.

In exploring the personal style of each scientist in this book, I have paid close attention to each one as a writer. It is a myth that scientists do not write well. Two decades as a teacher and scholar have shown me not just the logical coherence of experimenters' arguments but the creativity of their expressive forms as well. When Müller's students wrote about their mentor, they were acutely aware of their audiences and the ways that their words would be received. In Choderlos de Laclos's *Dangerous Liaisons*, Madame de Merteuil warns Cécile, "You must see that when you write to someone, it is for him and not for you: you must therefore try to tell him less what you think than what will please him most."[9] Only naïve writers compose to pour out their feelings, she implies. Worldly ones write to influence others and to shape the record to their advantage.

From Müller's thousands of students, I have chosen these seven not just for their crucial scientific work but also for the quality of their writing. Their comments on their adviser are sharp, witty, deeply revealing, and often hilarious. In weighing their accounts of Müller, I have considered their memorial addresses to him, their references to his work in their scientific writings, their recollections of him in public lectures and popular essays, and their descriptions of him in private letters. Their letters reveal the most about what they actually thought of him, since in their published works they had to use the formulas expected in the nineteenth-century scientific community. Yet even their correspondence offers no direct access to the writers' minds. If Madame de Merteuil is right, their letters can reveal only the thoughts that they wanted their readers to attribute to them. Even these private sketches of Müller should be judged only in the context of their public statements and the other students' portraits of their teacher.

In assessing the motives of these seven students describing their mentor, it is tempting to turn to psychoanalysis. The combined love and frustration of a group of male scientists for their adviser (in German, *Doktorvater*) certainly suggests Freud's primal horde, as described in *Totem and Taboo* (1913):

> One day the brothers who had been driven out came together, killed and devoured their father and so made an end of the patriarchal horde. United, they had the courage to do and succeeded in doing what would have been impossible for them individually. (Some cultural advance, perhaps, command over some new weapon, had given them a sense of superior strength.) Cannibal savages as they were, it goes without saying that they devoured their victim as well as killing him. The violent primal father had doubtless been the feared and envied model of each one of the company of brothers.[10]

At first, Freud's fantasy seems like an apt description of Müller's students, all of whom were eager to assert their own scientific identities. Du Bois-Reymond's memorial address for Müller is a kind of slaying and eating, and his "new weapon" might be a galvanometer. But for several reasons, Freud's myth falls short. First, Müller's students never banded together. Those with similar interests and technical challenges established friendships and exchanged ideas, such as Henle and Schwann in the 1830s and du Bois-Reymond and Helmholtz in the 1840s. But none of these alliances was motivated by an urge to overthrow Müller, whose laissez-faire style was anything but oppressive. The students' failure to unite was largely his doing, since he repeatedly ignored opportunities to introduce students pursuing related questions. Then there is the fact that Freud's most influential teacher in Vienna, the physiologist Ernst Brücke (1819–1892), learned to experiment from Johannes Müller. If I were to claim that an ancient patricide explains the dynamics of Müller's "lab," I would be leading readers in a circle. Probably, Freud got the idea of the primal horde by watching the relationships among Viennese scientists and their students.

Literary studies offer one perceptive adaptation of Freud's notions about fathers and sons that may help us understand the relationships among scientists. Drawing on Freud's theories of neurosis, the critic Harold Bloom has coined the phrase "anxiety of influence" to describe poets' quests for uniqueness. Bloom bases his theory of poetry not on *Totem and Taboo*, in which the father is murdered, but on an earlier essay, "The Family Romance of the Neurotic" (1909), in which he is merely rewritten. Since Müller's students seized control of his memory through writing, this insight of Freud's is much more relevant. According to Freud:

> The whole effort at replacing the real father by a superior one is only an expression of the child's longing for the happy, vanished days when his father seemed to him the noblest and strongest of men. . . . He is turning away from the father whom he knows today to the father in whom he believed in the earlier years of his childhood; and his fantasy is no more than the expression of a regret that those happy days have gone.[11]

Devised to explain the narratives of neurotics, Freud's notion of the family romance may shed light on students' stories about their scientific advisers. Not that science students are neurotic, of course—but like Freud's fantasizing children, they need to rewrite the roles of their advisers as they create the narratives of their own scientific identities. Freud himself generalizes the family romance to include childhood tales of infants replaced at birth, delivered up to strange parents who failed to understand them. It aptly describes du Bois-Reymond's memorial address for Müller, which emphasizes the physiologist of the late 1820s, and Haeckel's adoring portraits of Müller the anatomist, which depict a very different figure. Both describe the "happy days" with Müller they wish they had had.

Taking Freud's insight as a point of departure, Bloom argues that to forge their individual identities as artists, poets "swerve away" from their precursors' works, whose influence they perceive as threatening, and make "corrective movements" in their own lyrics. According to Bloom, the younger poet "read[s] the parent-poem as to retain its terms but to mean them in another sense, as though the precursor had failed to go far enough."[12] Although Bloom developed this idea by studying poets, it applies to scientists equally well. Like poets, scientists think creatively and compare their original visions to those of the previous generation. Anxiety of influence does not describe all adviser-student relationships, but it does provide a plausible account of how students' ideas diverge from their teachers.' In choosing projects, students almost always retain something of their advisers' assumptions. And sometimes—not always—their original contributions to science grow out of reactions to what they see as their advisers' mistakes.

As a work that examines scientists as historians, this book might be read as a psychohistorical study. Starting from the premise that "no phenomenon has an inherent meaning," psychohistorians analyze scholars' feelings about the material they are studying, weighing the unconscious motives not just of historical figures but of the people writing about them.[13] By examining historians' active involvement with the individuals they are describing, psychohistorians try to avoid the "distortion [that] arises from the failure to account for the observer in each act of knowledge."[14] Like psychoanalysts, historians often have to work with people's memories of past events, determining which elements represent actual happenings and which have been added over time.[15] Exploring Müller's relationships with his students and their emotional—as well as practical—reasons for depicting him as they did, this book shares some methodological ground with psychohistory. This study differs from most psychohistorical ones, however, in that it emphasizes scientific, institutional, and social factors over psychological ones in explaining why these seven relationships evolved as they did.[16]

In all human interactions, there is an unpredictable element. No combination of practical factors—scientific interests, social class, academic structures—can fully explain the attractions and repulsions among these eight scientists. There remains the murky issue of Müller's and the students' personalities, and of the dynamic interplay among them. How the personalities of past investigators affected their research is not a question historians can answer definitively, but an interpretive project like this one may provide some insights by aligning seven scientists' responses to one individual. Most of the students refer to Müller's imposing appearance, his magnetic looks, his joy in discovery, and his ability to inspire. Hopefully, this book will offer a glimpse into the way that personal relationships can drive science.

It is partly because of this unpredictable element that readers should consider all seven versions of Müller. As a professor at the Berlin University, he had thousands of students and a long series of assistants and helpers. I have chosen these seven because of their importance as scientists, their creativity as writers, and their strikingly different family backgrounds, career choices, and personalities. Because few readers will be familiar with all of these scientists' lives and works, I will spend much of each chapter describing each student's scientific experiences and achievements as context for his portrait of Müller. Since their careers took such different paths, no two chapters are alike, but each contains some information about the student's family relationships, scientific and medical training, major achievements, and interactions with Müller. This essential background will be followed by an assessment of the student's writing and an analysis of his comments on Müller. Where these comments are scarce, as in the case of Robert Remak, more attention will be paid to his citations of Müller in his scientific works. In cases in which the students created extensive narratives of their teacher's life, as did Emil du Bois-Reymond, I will offer a literary-style close reading. Even readers interested in a particular student will benefit from learning about all seven. Just as the curves of their lives intersected Müller's, they repeatedly crossed one another, and their remarks on each other's science are as illuminating as their comments on Müller's.

The opening chapter will familiarize readers with Müller's scientific career and mental challenges, and will begin to describe the academic, social, and political conditions under which he and his students worked. Students' comments in this chapter will be kept to a minimum so that readers can hear Müller's own voice in his letters and scientific papers, although his reflections on his own life should not be taken as more authoritative than the observations of his students.

By opening with my own narrative, I risk making readers think that this chapter is the definitive account of Müller's life. That is not my intention. Instead, I am offering a template to be adjusted and edited as readers meet the next seven versions of Johannes Müller.

The main purpose of this book is to deepen our knowledge of how history is written by juxtaposing seven scientists' accounts of their mentor. As a study of nineteenth-century German science, however, it examines the development of particular fields that will be of special interest to certain readers. Chapter 2, which describes the experiences of Jakob Henle and Theodor Schwann, offers information on the history of cell theory and microscopy. Chapters 3 and 4, on Emil du Bois-Reymond and Hermann von Helmholtz, illustrate the challenges facing the scientists who developed electrophysiology. Readers who would like to learn about medical training in the mid-nineteenth century will find useful information (and some rather horrifying descriptions) in chapter 5, on Rudolf Virchow. Chapter 6, on Robert Remak, depicts the obstacles facing an Orthodox Jewish scientist who was determined to teach and do research in Prussia but refused to convert to Christianity. Finally, chapter 7, on Ernst Haeckel, shows how a scientist from a new generation rewrote early nineteenth-century comparative anatomy so that it would support evolutionary theory.[17] Each chapter follows the path of a scientist who took Müller's teachings in a different direction, suggesting how his interactions with his teacher and his subsequent choices led him to conceive of Müller in a particular way.

While this analysis of Müller's "lab" cannot be generalized into a picture of typical laboratory science, it may shed some light on contemporary lab conflicts. Aligning seven scientists' tales of their personal relationships and everyday annoyances, it invites readers to consider how these factors can influence experiments. Combining the methods of history and literary studies, this book seeks to build knowledge as relevant to science as to the humanities. It makes no claim, however, to reduce science to writing. Johannes Müller and his students are not constructs of language, and they are not fictional characters; they are deceased scientists who learned how the human body works. Studying their motives for writing, their senses of audience, and their creations of personae can help us to understand their experiments. Their lives and their work were real, but we can gain access to them only by reading the stories they wrote to serve their own interests.

Müller's Lab

1

Müller's Net

Growing up in the Rhineland, 1801–1819

Johannes Müller was born in Koblenz on 14 July 1801, and his life reflects the shifts of nineteenth-century politics as a seismograph records the earth's motions. Since his untimely death in 1858, students and historians have been turning his life into contrasting stories. Müller's father was a cobbler who wanted his son to become a saddle maker, but even these simple facts, formulated in different ways, have become the starting points for contradicting tales. In his memorial address for Müller, Emil du Bois-Reymond called Müller's father "a shoemaker in good circumstances," whereas the more liberal Rudolf Virchow described Müller as "the shoemaker's son."[1] In his history of the Berlin University, Max Lenz represented Müller senior as "an honorable master shoemaker."[2] In the late twentieth century, the medical historian Johannes Steudel called Müller's father "a reputable master craftsman, the owner of a large leather-working shop," whereas the historian of science Nelly Tsouyopoulos depicted Müller as a "son of the working class [*Handwerkerklasse*]."[3] According to preference, even this basic information from Müller's students has permitted the creation of a different father, either a well-to-do artisan or a humble cobbler. Putting the students' statements into their own words, historians have told the stories they wanted to tell. Müller's social origin is highly significant, since his fear of poverty and respect for his Prussian mentors would give him a conservative outlook inimical to some of his students. But historians have represented it varyingly, and it is not easily defined in modern terms. Müller's father worked with his hands but earned enough to send his son to the Koblenz *Gymnasium*, an elite high school designed to prepare students for university study.

During Müller's youth, the troubled lands that would become Germany expended much of their energy fighting the French. In the early nineteenth century, the area that is now Germany consisted of numerous small states whose inhabitants defined themselves through loyalty to local rulers rather than to any larger, German-speaking culture. In the year of Müller's birth, 1801, Napoleon conquered the Rhineland, and until Müller's fourteenth year, this region was administered by France. Napoleon's three wars against Austria and its allies (1792–1806) finished off the anachronistic Holy Roman Empire, of which the German territories had been a part. Hoping to create a buffer between France and its more formidable enemy, Russia, Napoleon organized sixteen small German states into the Confederation of the Rhine, of which he served as Protector.[4] But Prussia, the most powerful German state apart from Austria, remained independent and in 1806 joined Austria in one last struggle against the French invaders. During the reign of Friedrich der Grosse (Frederick the Great, 1740–1786), Prussia had become a major military power, but two weak kings (Friedrich Wilhelm II, 1786–1797, and Friedrich Wilhelm III, 1797–1840) allowed its institutions to deteriorate.[5] Within a week, Napoleon defeated the Prussian army at Jena, and in the ensuing Peace of Tilsit forced Prussia to surrender much of its land, population, and army and to pay staggering reparations.[6]

As a result of this humiliation, Prussian officials who had been unable to enact much-needed reforms were given the power to modernize their feudal society. Aiming to create a system in which free, responsible individuals worked and fought for their country because they felt a personal investment in it—not because they were terrified of its rulers—they drafted decrees that would motivate talented citizens to work for the state.[7] An edict of 1810 freed the serfs, who could not marry, own land, or move to new estates without permission from their aristocratic landowners, the *Junkers*. In reality, the lives of these agricultural workers changed little, since they could not afford to buy land and the *Junkers* continued to control local governments.[8] In addition, the king's chief ministers undertook financial reforms that eliminated customs duties within Prussia, created a single income tax, and tried—without much result—to eliminate the *Junkers'* tax loopholes.[9] They also enacted military reforms, including a universal draft, modernized training, and promotion based on merit.[10] These changes would affect Müller, who served in the Prussian army before entering the university.

While Prussian bureaucrats were just starting to enact these reforms, Russia's surprising defeat of Napoleon made his overthrow a feasible goal. In 1812, Austrian, Russian, and Prussian troops vanquished the French army at Leipzig, and within months, Napoleon was driven out of German territory. In 1815, representatives of the prevailing states (Austria, England, Russia, and Prussia) met in Vienna to redesign Central Europe. The Austrian foreign minister, Klemens Wenzel Lothar von Metternich (1773–1859), served as chief power broker, trying to create an equilibrium that would restore order and maximize the influence of his own Habsburg Empire. Prussia hoped to annex Saxony, which contained the cities of Leipzig and Dresden, but seeing that this move would anger the smaller German states, Metternich suggested a more appropriate prize. So that

France would not invade German lands again, he offered Prussia extensive territories along the Rhine. As a result of these negotiations, Prussia doubled in size, acquiring regions that would become industrial powerhouses and gaining 5.5 million new citizens, among them fourteen-year-old Johannes Müller.[11]

Although the Prussian territories of 1815 look vaguely like modern Germany, the German lands of that year were far from united. With little interest in unification, they formed a Germanic Confederation of thirty-nine states, including Austria and Prussia, the kingdoms of Bavaria and Saxony, the duchy of Baden, and the free cities of Hamburg and Frankfurt.[12] Bound only by a federal Diet with little real power, these states varied considerably in their cultures and forms of government. Baden, in the southwest, had its own constitution and parliament, whereas Prussia remained authoritarian and bureaucratic.[13]

Still, a movement for a unified, German-speaking nation began to develop, especially at the universities. Between 1815 and 1819, politically active students organized themselves into *Burschenschaften*, fraternities that called for the unification of German-speaking lands.[14] At the time, most of those who supported German nationhood held liberal views, demanding freedom of expression, a universal draft, and equality before the law.[15] Creating a German state would mean reforming societies built on submission to local authorities. When a deranged fraternity member killed a conservative playwright in 1819, Metternich convinced the Prussian king and other Confederation rulers to outlaw *Burschenschaften*, but they continued to attract bright, idealistic students throughout the 1820s.[16] During his first two years at the Bonn University, Müller participated actively in these fraternities, leading a chapter that rallied the students behind the black, red, and gold colors of German unity.[17] After only a year, however, the group dissolved, and Müller abandoned his commitment to a liberal, unified Germany.

Industrialization, which increased exponentially during Müller's thirty-five years as a scientist, played a more crucial role in German unification.[18] When he began his physiological experiments in the 1820s, German investors still had to rely on British machines and engineers to realize their projects. Although the German territories had iron, coal, and willing workers, each of the thirty-nine states issued its own currency and levied its own taxes and tariffs, making it hard to import raw materials and distribute products. This situation changed in 1834 when the German states formed a customs union, making it much easier to move goods across state lines. In the late 1830s, engineers constructed the German territories' first railroad lines, which enabled the mass transport of fuel and factory-made goods. In the 1840s, Werner von Siemens (1816–1892) designed Prussia's telegraph network. Berlin, the city in which Müller and his students practiced science, became a major center for machine construction, home not just to Siemens's workshop but to August Borsig's (1804–1854) locomotive factory. Part of the tension that would arise between Müller and his students in the 1840s can be traced to their eagerness to use local machinists' discoveries in their experiments, while Müller preferred more traditional observations of natural forms.[19]

Prussia's reform of its educational system had the greatest impact on Müller and his students. From 1815 onward, as part of the program to create responsible, self-motivated citizens, Prussia's ministers redesigned its high schools and universities. Those nine-year-olds who showed aptitude (or whose parents believed they had aptitude) were sent to *Gymnasien*, elite high schools where they worked for nine years, developing their native gifts through exposure to culture.[20] The *Gymnasien* were to foster *Bildung*, a lifelong process of self-development inspired through the study of Greco-Roman scholarship and art.[21] Only after completing this course of study could a young man enter a university and, subsequently, a career in the civil service, law, medicine, or theology. Theoretically, these new Prussian universities fostered scholarship and knowledge for their own sake. The Berlin and Bonn universities, founded in 1810 and 1818, respectively, were to educate future scholars and civil servants by exposing them to the research of professors who were both scholars *and* civil servants.[22] At these two institutions, where Müller learned and taught medicine, professors were government employees, hired, promoted, and funded by the Prussian Cultural Ministry.[23] Viewing higher education as indispensable to strong statehood, the government maintained tight control over its faculty, intervening to help promising scholars—and to remove potential subversives.

In 1815, as this competitive system began to emerge, Müller the shoemaker's son was in the right place at the right time. At the Koblenz *Gymnasium*, his gifts for mathematics and classical languages impressed the educational reformer Johannes Schulze (1786–1869), who convinced Müller's father to send him to the new Bonn University instead of teaching him leatherworking.[24] Throughout his life, Müller received help from powerful patrons, each with his own motives for answering his requests for aid. Since 1818, Cultural Minister Karl von Altenstein (1770–1840) had been trying to make the Bonn University a center for scientific studies.[25] Appointed by the Prussian king, the cultural minister controlled all aspects of higher education, such as the appointment and promotion of university professors and the awarding of travel grants. In the 1810s and 1820s, Altenstein was doing his best to build schools that would train knowledgeable, loyal, Christian servants of the Prussian state. When Schulze told him about Müller, he was happy to help, for Müller's success as a scientist would show that in the new, reformed Prussia, a talented young man could rise through merit alone.[26] In 1819, Müller began studying medicine at the Bonn University.

Nature Philosophy and Experimental Physiology in Bonn, 1819–1832

At this time, the most respected approach to science in the German territories—and at the Bonn University in particular—was *Naturphilosophie* (nature philosophy), and the question of whether Müller developed his physiology out of or in opposition to this earlier science has raised considerable controversy.[27] According to his students, Müller either (1) freed physiology from its pernicious influence or (2) failed as a scientist because he retained its central ideas.[28] Du Bois-Reymond argued both views simultaneously. His claim becomes less contradictory, however, when one considers the number of different viewpoints encompassed by nature philosophy.

In the 1820s, nature philosophy had as many forms as it had practitioners, but its essential concepts emerged from the philosophy of Immanuel Kant (1724–1804).[29] In his *Critique of Judgment* (1790), Kant had proposed that living things differ from inorganic ones in that organic parts have their purpose and existence only in relation to the whole.[30] The parts of living organisms exhibited a unique *Zweckmässigkeit* (purposefulness, or appropriateness in achieving a goal) and found their form and function only in relation to one another. According to Kant, an animal differed fundamentally from a watch, since in a watch, no wheel could be regarded as the cause of any other wheel, and the purpose of the whole device was the will of a person outside of it. Living things had their own inherent purpose, irrespective of anything outside of themselves. Consequently, organized beings could never be fully explained or understood by reference to the mechanical principles of nature.

The nature philosopher Friedrich Wilhelm Joseph von Schelling (1775–1854) developed Kant's ideas into a system for acquiring reliable knowledge about living organisms. Aiming to "explain the ideal in terms of the real," Schelling used the latest findings in biological research to argue for a continuum between the world perceived and the human consciousness that perceived it.[31] Because Müller's students represented nature philosophy as empty theorizing, subsequent scientists have not realized how greatly Schelling respected physiological experiments. He studied Alexander von Humboldt's investigations of animal electricity and attempted his own laboratory work.[32] Like Kant, however, Schelling and his followers believed that experiments could yield only limited information. While they revealed mechanical laws, they could not show nature's overarching organization, which could be discovered only if practical and theoretical studies of nature were combined.

In the 1790s and 1810s, some insightful German scientists pursued these organizational secrets by comparing the forms of animals and searching for patterns.[33] Like Schelling, Johann Wolfgang von Goethe (1749–1832) used Kant's notion of inherent purposefulness as a point of departure for studying animal life. To make sense of diverse bodily forms, Goethe proposed that scientists seek an "anatomical type, a general image which would contain all possible animal forms and according to which one could assign each animal a place in a certain order."[34] This general type would never be represented by any existing animal. To discover it, anatomists had to study as many bodily forms as possible until they could imagine one general form that could give rise to them all. Although this method sounds like a quest for a common ancestor, Goethe did not think of his strategy in terms of evolution or descent. It was more like doing a jigsaw puzzle, with the emerging picture guiding the placement of the pieces and the placement of pieces revealing the big picture. Succeeding meant learning nature's intent and design.

To solve the puzzle, some investigators created elaborate analogies. Lorenz Oken (1779–1851) proposed that the "multitudinous [*manchfaltigen*] forms" of the animal kingdom represented the activities of the individual human organs.[35] In a work of 1808, he argued that the universe was a continuation of the human sensory system.[36] Oken's comparisons represent nature philosophy at its most theoretical, but he did German-speaking scientists tremendous good on a practical level. He founded the Society of German Scientists and Physicians, organizing its first meeting in 1822, and he created the journal *Isis*, in which Müller would publish his first article.[37]

While studying medicine, Müller began making minute observations of animals. His patrons, who admired nature philosophy, were pleased by his early achievements. He quickly fulfilled Altenstein's expectations, winning a contest with an experimental study of fetal respiration. Unfortunately, however, Müller's father died during his second semester, leaving him and his four younger brothers and sisters without financial support. For the next ten years, Müller struggled to survive economically, as his family and his Prussian patrons fought to keep his research alive. In 1822 he wrote a doctoral thesis on animal movement—not the movement of one particular animal, but of animals in general.[38] As Müller put it, he was exploring "the parabolic line along which life plays," and he found that "flexion and extension are the two poles and marks of life in motion."[39] In Schelling's nature philosophy, polarity and dualism played key roles in explaining how diverse structures had developed.[40] By scrutinizing the movements of many individual animals, Müller was trying to learn how polarity had allowed living matter to acquire such a staggering variety of forms. Oken liked the young physiologist's work and published his doctoral research in his new journal.

When Müller received his medical degree in December 1822, Bonn University curator Philipp Josef von Rehfues wrote to Berlin for a scholarship so that Müller could continue his studies elsewhere.[41] The young scientist wanted to go to Paris to work with his role model, Georges Cuvier (1769–1832), Europe's leading comparative anatomist. At the Natural History Museum in Paris, Cuvier had assembled the world's most comprehensive anatomical collection and had proposed a new classificatory scheme to describe the relationships among animals. Viewing life as a great network of species, he was seeking a system whose patterns could be revealed through careful anatomical observations. What Rehfues did instead had great consequences for Müller's scientific development. His Bonn patron sent him to the Berlin anatomist Carl Asmund Rudolphi (1771–1832), who encouraged him to do microscopic studies. Müller quickly became an adept microscopist, and when he passed his Prussian state medical exam and returned to Bonn in the winter of 1824, Rudolphi gave him his own Fraunhofer microscope to conduct his research.[42] With this instrument, he began studies that would transform anatomy and physiology.

In the fall of 1824, Müller's research won him a post as lecturer in anatomy and physiology at the Bonn University. In his inaugural lecture, "On Physiology's Need for a Philosophical View of Nature," Müller outlined a scientific strategy that he would use for much of his life, combining close observation of natural forms with philosophical theories about their interrelations.[43] In this methodological lecture, Müller attacked both unchecked theorizing and blind empiricism, advocating natural science based on observation, experiment, and philosophical quests for organization.

During his years in Bonn, Müller revealed crucial information about the visual, circulatory, endocrine, and reproductive systems. He rarely studied any function in one animal alone, preferring to compare the ways that different organisms solved physiological problems. His 1826 study, *On the Comparative Physiology of Vision in Men and Animals*, explained the mechanism of human binocular vision but also contained a long section on the structure of insect eyes.

In this book, Müller first expressed his law of specific sense energies. Through his comparative studies of nervous systems, he had realized that sensory receptors are not passive receivers of outer stimuli, since the same external event (a mechanical pinch, for example) affects different sensory organs in different ways and can be perceived as light, sound, or pain. As he later put it in his *Handbook of Human Physiology*, "Sensation is not the conduction to our consciousness of a quality or circumstance of external bodies, but the conduction to our consciousness of a quality or circumstance *of our nerves* brought about by an external cause."[44] The auditory nerve, for instance, has "the characteristic of perceiving shocks as tones."[45] Each nerve can respond to stimuli only in a specific way, so that our knowledge of the world reflects the structure of our nervous system.

To learn what the nerves and sensory organs contributed to perception, Müller began studying internally generated images.[46] Noticing that when he was falling asleep, he sometimes saw imaginary people and things, he tried to manipulate these figures in a series of rigorous self-experiments. He published the results in *On Fantasy Images* (*Ueber die phantastischen Gesichtserscheinungen*) in 1826, the same year his comparative visual study appeared. This more speculative work complemented the one of exact description, showing that the visual system is an active, not a passive, recorder of external events.

Figure 1.1 Johannes Müller in 1826, as painted by his friend Johann Heinrich Richter. From Wilhelm Haberling, *Johannes Müller: Das Leben des rheinischen Naturforschers* (Leipzig: Akademische Verlagsgesellschaft, 1924), plate III, 73.

As a researcher, Müller worked at a frenetic pace, exploring the sensory, nervous, lymphatic, circulatory, and reproductive systems, sometimes simultaneously. In Bonn he rose rapidly from lecturer to associate professor (1826) to full professor (1830). But in April 1827 he suffered an incapacitating, six-month depressive episode that began just two weeks before he married the musically gifted Nanny Zeiller.[47] He had just published two books, and he had been depriving himself of sleep to conduct a grueling course of self-experiments. According to du Bois-Reymond:

> Müller fell into a state of nervous irritability in which, among other things, he felt little shocks in his fingers any time that he overexerted his hand. Associated with this was a feeling of the greatest flaccidity, which made any bodily exertion impossible and even hindered his walking.[48]

In late July 1827, his former professor Dr. Philipp von Walther wrote to Cultural Minister von Altenstein—who was naturally concerned about his investment—vouching for Müller's health. For the past three and a half months, Walther explained, Müller had been suffering from "a kind of hypochondria" which Walther had seen before in young scholars: "[Müller] thought he was suffering from a disease of the spine which would lead to the total paralysis of his legs and end with his death."[49] In the mid-nineteenth century, "hypochondria" was a general term for depression, one that Müller's students used in their letters.[50] Walther's assurances to Altenstein and the students' frequent, ironic use of the term suggest that at the time, "hypochondria" was a culturally defined mental state rather than a pathological condition. Probably, Müller simply collapsed from exhaustion. By the fall of 1827, he had recovered and begun studies of the endocrine and reproductive systems.

Müller's decision to study genitalia soon after his marriage and breakdown suggests a strong personal investment in this work, but his motivation seems to have been mainly scientific. In February 1828, his daughter Maria was born; in October 1829, his son Max.[51] Surely Müller's own capacity as a reproductive animal must have been on his mind during this period, and he began his observations of genitalia in late 1828. But here the urge to tell a satisfying story about the relation between life and work confronts the urge to tell an accurate one. There is a better explanation.

In the summer of 1828, Rehfues got Müller a grant to attend the 1828 meeting of the Society of German Scientists and Physicians in Berlin, where he met the embryologists Karl Ernst von Baer (1792–1876) and Martin Rathke (1793–1860). After talking with these scientists, Müller became intrigued by the Wolffian bodies, structures that appeared, then disappeared, in developing embryos and that Rathke thought produced the kidneys. Other scientists disputed this hypothesis and believed that they gave rise to the sex organs.[52] In his usual style, Müller compared the Wolffian bodies in the embryos of many species, which led to his study of male erections throughout the animal kingdom and his 1835 discovery of the Ranken arteries that make them possible. Perhaps Müller's own reproduction stimulated his interest in animals' genitalia, but his scientific life was a more likely inspiration than his life at home.

At the 1828 meeting, Müller also met the Swedish anatomist Anders Retzius (1796–1860), who would become one of his closest friends, and the patron of sciences Alexander von Humboldt (1769–1859).[53] A leader of the eighteenth-century Berlin Enlightenment, Humboldt had begun his scientific career as a Prussian mining inspector, during which time he became interested in mineralogy, geology, and magnetic forces.[54] In the 1790s he had conducted valuable studies of animal electricity, and from 1799 to 1804 had traveled extensively through South America, recording geomagnetic forces, weather patterns, and the forms of previously unknown plants and animals. From his return to Europe until 1827, Humboldt had lived mainly in Paris, writing about his discoveries and sharing ideas with members of Europe's most advanced scientific community. But in 1827, Humboldt went home to Berlin. He had exhausted his private fortune and was hoping to convince the Prussian court to fund a new trip to Siberia, but he also had another goal in mind. Humboldt wanted to develop science in Berlin that would rival Parisian learning, and he saw in the Berlin University (organized by his brother Wilhelm in 1810) the opportunity to do so. Alexander von Humboldt advocated exact science based on close observation, experiments, and quantitative measurements, and he energetically supported any young man who performed such investigations. In 1828, Müller's science was exactly the kind that he wanted to promote.

Between 1828 and 1832, Müller did some of his finest physiological work. In 1830 he published extensive comparative studies of genitalia and secretory glands, showing that glands give off substances that control bodily functions. During these same years, he made one of his greatest contributions to physiology, the experimental demonstration that the dorsal roots of spinal nerves (those initially heading upward along the back) carry mainly sensory fibers, whereas the ventral ones (those initially heading downward toward the belly) carry mainly motor fibers. The Scottish physician Charles Bell (1774–1842) had proposed the idea in 1811 but had provided no experimental evidence. The French physiologist François Magendie (1783–1855) had conducted experiments testing the hypothesis in 1822, but his live dogs had been in such pain that his results were questionable.[55] In 1831 Müller thought of repeating Magendie's experiments in frogs, hardier animals in which the spinal cord could be more readily exposed.[56] Not only did he confirm Magendie's findings; he publicized a new system through which young experimenters could study the functions of muscles and nerves. In his article on spinal nerve roots, Müller defined the criteria for a good physiological experiment. Like a physical one, he wrote, "in any place, at any time, under the same conditions, it should present the same certain, unambiguous phenomena, so that it can always be confirmed." He urged all physiologists "to repeat the simple experiments that I have described."[57] Both because of his methods and because of the way that he presented them, Müller's study became a model for physiologists during the next two decades.

Knowing the value of his research, Müller gladly accepted the help of patrons such as Humboldt, Rehfues, and Altenstein. He wanted to become the most powerful anatomist and physiologist in Prussia, not to improve his social status but to shape the development of these fast-growing fields. Müller's students have all noted his keen ambition, but his desire for scientific clout must not be misread. His biographer Gottfried Koller claims patronizingly that his drive was understandable in a gifted man "stemming from very simple social circles," but Müller sought power so that he could promote exact science.[58] He built his reputation partly through skillful self-promotion, although with Müller, one must include his students in his definition of "self."

From the time that he met Altenstein, Müller actively cultivated their relationship, sending him copies of his articles and stressing the practical value of his achievements. When he made a discovery, he made sure to publish it quickly and to send it to influential people.[59] One can see the success of Müller's campaign from Altenstein's responses to his letters. Thanking Müller for two journal volumes containing his articles, the cultural minister wrote, "Certainly these [studies] . . . will have a very healthy influence on practical surgery, offering new evidence that precise physiological investigations can have direct medical applications."[60] Through careful lobbying, Müller convinced Prussian officials that physiology and anatomy were worthwhile pursuits.

In 1832, Müller's former mentor Rudolphi died, and the most prestigious Prussian chair of anatomy and physiology became available. Although Müller's Bonn professorship was a respectable job, the Prussian capital, Berlin, offered a far superior cultural and intellectual environment. By the early 1830s, German research in the life sciences was overtaking that of France and England, which had dominated up to that point. A key cause of this growth was the intense competition among German universities.[61] Unlike France and England, the German territories constituted no unified state and had no central university complex like those of Paris and Oxford-Cambridge.[62] German-language universities lay in different regions (Bonn, Berlin, and Königsberg, in Prussia; Heidelberg, in Baden; Würzburg, in Bavaria) and even in different countries (Zurich, in Switzerland; Vienna, in Austria; Dorpat, in Russia [now Estonia]). These institutions varied considerably in quality. Based on the number of scholars who moved because of a promotion, or who moved despite the lack of a promotion, Berlin and Vienna seem to have been the most attractive universities; Heidelberg and Würzburg, somewhat less desirable.[63] No one ever wanted to go to Dorpat.

Because professors and students could move freely from one university to another, institutions competed to attract the best scholars, luring them with higher salaries, more specialized teaching, and eventually research facilities.[64] As a result, scientists with offers could bargain for work space, assistantships, and the ability to define their own fields. Each offer set off chain reactions as positions were vacated and universities tried to match their competitors' innovations.[65] One important consequence of this decentralization was the creation of new scientific fields.[66] During Müller's lifetime as a scientist, anatomy and physiology emerged as disciplines, not just because of his students' research but also because of their demands that their chairs be defined as professorships of physiology or pathology. Once one university established a new chair, the others followed suit.[67] If they didn't, they risked losing their brightest students and scholars.

Even in this dynamic, open system, academic jobs were scarce, and the tensions between the powerful (*Ordinarien*, or full professors) and the less powerful (*Extraordinarien*, or associate professors, and *Privatdozenten*, or lecturers), affected the relationships between faculty and students.[68] Generally, a university appointed only one full professor per discipline, and younger scholars could advance only if that professor moved or died. Universities held full professors responsible for teaching everything in their subjects, and a substantial portion of their income came from student lecture fees.[69] Unless they actively wanted to delegate courses in order to leave more time for research, they became unhappy when innovative young lecturers lured their students away with microscopy labs or physiological experiments. Scientists who had just finished their doctoral research and hoped to teach at universities could survive only by cultivating relationships with full professors who would recommend them, find them funding, and—if their interests were sufficiently close—suggest exciting research projects.[70]

Müller's all-out effort to win the Berlin job illustrates these broader cultural and institutional trends. Interestingly, he found his greatest advocates on the Berlin philosophical faculty, though the appointment was on the medical faculty. The university's nature philosophers wanted an experimentalist, and on their own initiative, they wrote to the Cultural Ministry:

> Recently, a new path to higher goals has been broken in anatomy and physiology. Formerly, one was content to observe events [*Erscheinungen*]; now one tries to create new events through experiments. . . . We already have an outstanding observer; we would obtain an experimenter in [Friedrich] Tiedemann or Johann[es] Müller.[71]

But Christian Samuel Weiss (1780–1856), who ran the university's Mineralogical Museum, filed a dissenting minority report, wanting to know why the faculty had not considered the microscopist Jan Evangelista Purkinje (1787–1869, after whom the large cerebellar cells are named), or the embryologist Karl Ernst von Baer. "It's not physiological experiments that founded Professor Müller's literary reputation," he wrote, "but his much more observational anatomical-physiological work on glands, and among the aforementioned scientists there are several who have done more work in the experimental field of physiology than he has."[72] Even at this early stage of Müller's career, some scientists suspected his real love was anatomy.

Müller might have won the job even if he had left things to themselves, but instead he wrote directly to Altenstein, recommending himself for the chair. According to du Bois-Reymond, this was an unusual step, but subsequent research has shown that it was not uncommon.[73] Rudophi's assistant Friedrich Schlemm (1795–1859), an anatomist who would become Müller's colleague, also applied for the job, writing to the cultural minister, "I am more practiced with the knife; [Müller] knows better how to conduct microscopic investigations."[74] In his own letter, Müller emphasized the importance of physiological research for medical advancement and echoed Humboldt's desire to make Berlin a research center equal to Paris. A crucial part of the job, he argued, was the creation of a world-class anatomical museum where scholars could compare the structures of all known animals. "Now is the decisive moment," he wrote, "to put the increase of the collections and their contents, those magnificent fruits, into the hands of a director who understands not just how to tolerate talented people but how to attract, inspire, employ, and challenge them."[75] While he suggested Johann Friedrich Meckel (1781–1833) for the chair, Müller repeatedly invoked his own youthful energy. (Meckel died soon afterward.) At thirty-one, Müller was not the faculty's first choice. They favored the older, more established anatomist Friedrich Tiedemann (1781–1861), but he rejected the offer. Trusting the glowing reports from Bonn, Cultural Minister von Altenstein took a chance.[76] In the spring of 1833, he made Johannes Müller Berlin's professor of anatomy and physiology.

Berlin in the Mid-Nineteenth Century

In 1833, the Prussian capital did not offer the best environment in which to break scientific paths. In an 1834 guidebook, Leopold Freiherr von Zedlitz bragged:

> Beautiful, blooming, and populous, [Berlin] ranks among the finest European cities, bejeweled with many works of architecture and sculpture, rich in teaching and instructive institutes for the arts and sciences . . . it is the showplace of the mighty works of the Hohenzollern dynasty.[77]

Not all visitors got this impression. When the Bostonian Henry Adams (1838–1918) studied law during Müller's last year in Berlin (1858), he found that

> Berlin was a poor, keen-witted, provincial town, simple, dirty, uncivilized, and in most respects disgusting. . . . Overridden by military methods and bureaucratic pettiness, Prussia was only beginning to free her hands from internal bonds. Apart from discipline, activity scarcely existed. . . . The condition of Germany was a scandal and a nuisance to every earnest German, all whose energies were turned to reforming it from top to bottom.[78]

During the twenty-five years that Müller and his students worked in Berlin, they experimented in a rapidly industrializing city whose aggressive drive to match Paris in the arts and sciences contrasted with its oppressive politics and rough living conditions.

Since its emergence in the thirteenth century, Berlin had gained economic importance in the seventeenth century, when canals linked the Spree River bisecting the town to the Elbe and the port of Hamburg. In the eighteenth century, it had become a cultural center as the home city of the Hohenzollern kings. By 1834, 265,000 people lived in Berlin, including a garrison of over 16,000 soldiers.[79] About 8000 of these were Catholic; 5000, Jewish; and the rest, Protestant. In the year that Müller came to Berlin, 9378 children were born, every seventh one out of wedlock. Of the 8053 registered deaths, 445 were stillborn children, 1896 were children under one year, and 1791 were children between one and fourteen. Sixty-seven mothers died in childbirth, and sixty-three people committed suicide. Art and science meant little to most of Berlin's inhabitants, who were trying to keep warm, feed their children, and avoid disease.

The Berliners of the 1830s worked in shops and factories scattered throughout the city.[80] Of its 50,000 or so families, about 4050 earned their income from the Prussian civil service; 1350, from the military; and the rest, from a privately owned business such as those of the city's 1116 master shoemakers.[81] Already the city was developing an industrial zone. The iron foundry and growing machine industry were concentrated on Berlin's north side, near the Oranienburg Gate.

The 1835 *Academic Mentor*, a guide for students entering the Berlin University, reveals a great deal about Berliners' everyday life. In the confusing monetary system, 1 Prussian taler was equal to 30 *Silbergroschen*, and 1 *Friedrichsd'or* was worth 5 taler and 5 *Silbergroschen*. The author warned students that living in Berlin would cost twice as much as it did in a small university town like Bonn, but any student "strong enough to resist the attractions and pleasures" of the capital could get by on 150–200 taler a year. Many civil servants supported entire families on 300–800 taler.[82] As a full professor in Bonn, Müller had earned 800 taler; in Berlin, he received 2500 taler but spent a great deal of it on his anatomical museum.[83] For 3 taler a month, a student could rent a small room on an interior courtyard with a bed, a table, and some cane chairs; for 5 to 8 taler, he could get a "quite tolerable" room on the street with a "bed, canopy, wardrobe, secretary desk, dresser and the necessary chairs."[84] The students who assisted Müller for 10 taler a month as museum helpers most likely had the former kind. In gray Berlin, exterior apartments were more desirable because they got more light, but the guide warned students to avoid corner apartments on major intersections because of the "unbearable noise" between 4 A.M. and 10 P.M. With 150 night watchmen who whistled each hour, the damp, interior quarters had some advantages.[85]

Most lodgings offered "bed-making, turning-out, and dust-removal," but not indoor plumbing. A city ordinance forbidding "the pouring of water, much less of any other unclean substance out the window into the street" suggests that Berliners relieved themselves in chamber pots, which their landlords would empty for a price.[86] Students who did not wish to clean their rooms could hire one of the 500–600 licensed freelance servants who waited on street corners, wearing black armbands and white medallions.[87] To heat their rooms, Berliners burned wood and peat, available in "heaps" for a considerable sum. A heap of pine cost 11 to 24 taler; beech, 29 to 36 taler; peat, only 8 to 10 taler. If a student lit his own stove and heated his room only in the early morning and evening (an incentive to go to class), he could get by on a quarter-heap each of peat and wood, for about 1 taler a month.[88] Since 1829, Berlin had had gas lighting, but oil lamps and candles were still widely used. A student who saved on heat and light could afford more to eat; a midday meal at a Berlin restaurant cost 5 to 15 *Silbergroschen*; an evening meal, 5 to 10. Student rooms had no facilities for cooking. In the absence of running water, washing one's clothes consumed a good chunk of a student's allowance: 1 to 2.5 *Silbergroschen* for a shirt, 1 to 2 for underwear, and 0.33 to 1 for a pair of socks.[89] It took ingenuity to stay warm, clean, and well nourished enough to study.

As the *Academic Mentor* warned its small-town readers (or, more likely, their parents), Berlin could lead unwary young men astray. A university regulation forbade students to visit smoking houses, *especially* the one at Dorotheenstrasse 58.[90] Because of the "daily example of immorality" that certain neighborhoods offered, the *Mentor* urged students to avoid them, carefully describing their locations. "In some less frequented streets," wrote the author, "[there are] houses open to lust, whose nymphs try to poison the budding force of life."[91] There is no record that Müller or his students ever fell victim to these temptations.

The driving force of these scientists' lives was their passion to learn, and the Berlin University and the cultural community gave them a priceless opportunity. By 1833, Berlin was becoming the "German Athens" of which Prussia's leaders had dreamed.[92] The city possessed seven *Gymnasien* to prepare the most promising students for university study. The Academy of Sciences, founded in 1700, had thirty-eight regular members, and the Society of Scientific Researchers met twice a month to discuss controversial new findings.[93] The Royal Museum, opened in 1830, displayed classical sculptures and European paintings that had previously been scattered in private collections. And the opera house, built in 1741–1742, was larger than those of Rome, Paris, and London.[94]

Located across from the opera house, the university formed an integral part of this cultural milieu. Constructed by the Dutch architect Johann Boumann in 1754–1764 for Prince Henry, the brother of Frederick the Great, its main building stood proudly on Berlin's grandest street, Unter den Linden.[95]

The Royal Palace, Library, Art Museum, and academies of arts and sciences were just minutes away. Three stories high, with two side wings, the university building was separated from the street by a tall iron fence and a gate that could be locked in times of emergency. On the ground floor were administrative offices and the 300-seat lecture halls in which most professors held their classes. One floor up lay the "magnificently decorated" Aula, a large auditorium for use on festive occasions. Most of the building was occupied by museums, open both to scholars and to the general public, and by the living quarters of those who ran the university and its museums. When the university opened in the fall of 1810, its ten lecture halls took up less than a quarter of the ground floor.[96] In this allocation of space, the institution expressed the aims of its designer, Wilhelm von Humboldt: the inseparability of research and teaching and the acquisition of knowledge for its own sake.

Though controlled by the Prussian government, the university was run by its rector, the deans of its four faculties (theology, philosophy, law, and medicine), and its Academic Senate. In 1834, it employed eighty-eight full and associate professors and thirty-seven lecturers. During Müller's first semester (summer 1833), 1801 students were enrolled in the four faculties, 341 of them in medicine. In the early 1830s, theology and law attracted the most students (550–600 each), but philosophy (with 300 or fewer) had the most professors, forty-six (there were six

Figure 1.2 The main university building, Berlin, 1835. Frontispiece, Jean Eckenstein, *Der akademische Mentor für die Studirenden der Friedrich-Wilhelms-Universität in Berlin* (Berlin: Wilhelm Schüppel, 1835).

in theology, nine in law, and twenty-seven in medicine).[97] Today, many of these Berlin philosophy professors would be regarded as scientists.

At the Berlin University, the semesters were long and the teaching and administrative loads, heavy. In an institution that prided itself on combining teaching and research, the faculty barely found time to investigate. The winter semester ran from mid-October until late March; the summer one, from mid-April until mid-August.[98] To qualify as a university lecturer, scholars had to complete not just a doctoral thesis but also a second independent research project of equal complexity, the *Habilitation*. Hired as rising stars in their fields, professors could teach whatever they wished, and students could take whatever courses they wanted to prepare for exams that would admit them to prestigious careers. To take a course, students simply signed up and paid a lecture fee. Theoretically, this elective system established a competitive meritocracy in which the best lecturers attracted the most students, but in practice, the senior faculty monopolized subjects, making it hard for young, innovative lecturers to find a foothold.[99] Despite these obstacles, competition for professorships and lecturer positions was fierce, for Prussian students were as passionate about research as the university's planners.

In 1833, Müller became responsible for teaching anatomy and physiology and for running Berlin's Anatomical Institute and Anatomical Museum. Considering the amount of work this involved, it is extraordinary that he did any research at all. During each winter semester, he taught human anatomy six hours a week (daily from two to three in the afternoon) and anatomy of the sensory organs, three hours. He also ran the medical students' dissecting laboratory with his colleague Friedrich Schlemm and examined all Prussian candidates for medical degrees.[100] In the summer semester, he lectured six hours a week on physiology; four, on comparative anatomy; and three, on pathological anatomy, and offered weekly public lectures on embryology and development. Three times he served as dean of the Medical Faculty (1835–1836, 1842–1843, and 1846–1847) and twice as rector of the Berlin University (1838–1839 and 1847–1848).[101] Supervising the Anatomical Institute and Anatomical Museum required considerable bureaucratic correspondence and financial planning. Only in August and September could Müller pursue his research without interruption, escaping to the North Sea or Mediterranean coast to study marine life.

It was thus not increased research time that drew Müller to Berlin, nor was it outstanding research facilities. In 1833, these were nonexistent. Instead, it was the opportunity to build a discipline as he saw fit. He began with a few assistants in the back rooms of a dissecting hall.

As professor of anatomy, Müller lectured not in the main university building but in the Anatomical Institute (or "Theater"), ten minutes away, where he directed the gross anatomy lab. Since 1828, the Anatomical Institute had occupied a building behind the Garrison Church near the Old Stock Exchange.

Today, this area across the Spree River from the "Museum Island" lies just south of the trendy Hackescher Markt district. Müller's colleague Friedrich Schlemm co-directed the Anatomical Institute, which had its own prosector (an assistant who dissects preparations to be used in teaching), concierge, and custodian. It maintained a medical library and collections of preparations and surgical instruments, but its main purpose was to guide medical students through the dissection of over 200 cadavers a year, supplied by the poorhouses and the Charité

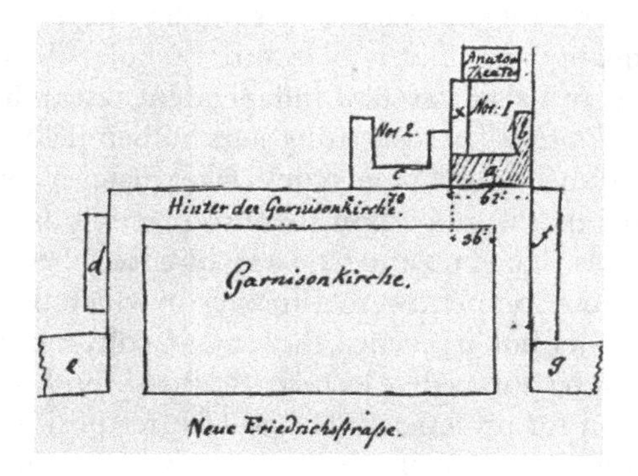

Figure 1.3 Location of the Berlin Anatomical Institute in 1832, as drawn in a letter to the Cultural Ministry discussing possible repairs. Source: GSPK, Rep. 76Va, Sekt. 2, Tit. XIX, nr. 11, vol. IV, "Bauen und Reparaturen an dem Anatomie Gebäude," 133, printed with kind permission of the Bildarchiv Preussischer Kulturbesitz.

Hospital. Müller's students also dissected the bodies of suicides, criminals, and "the so-called unfortunate."[102] Because of the necessary transportation and cleanup, as well as the time of two full professors, lecture fees were steep: 4 *Friedrichsd'or* (about 22 taler) for the lab, and 2 *Friedrichsd'or* (about 11 taler) for Müller's anatomy lectures. Müller and Schlemm shared the lab fees. Students dissected their cadavers on the ground floor, and Müller lectured upstairs.[103]

During Müller's lifetime, universities did not usually provide any designated space for professors—let alone students—to conduct experiments. Consequently, Müller and his young collaborators made some of their most crucial microscopic observations in small, dark, foul-smelling rooms adjacent to the medical students' dissecting hall. Friedrich Bidder (1810–1894), who in 1834 was visiting from Dorpat (today, Tartu, Estonia), wrote that "Müller's workroom [*Kabinet*] in the dark, cramped, musty Anatomical Theater, which was then located behind the Garrison Church and enclosed by tall buildings, consisted of two small, at best simply furnished chambers [*Stuben*]."[104] Bidder recalls that at Müller's human anatomy lectures, held every afternoon from two to three, about 200 students "occupied three sides of the lecture hall in ten rows that sloped upward as in an amphitheater, while the fourth wall consisted of a single gigantic window."[105] This description creates a vivid picture of the space in which Müller taught, but it may not be entirely accurate. Bidder wrote about Müller's Anatomical Institute at least thirty-five years after he worked there, and his portrait belies other students' references to the building's unworthiness. Du Bois-Reymond, who started working with Müller in early 1840, recollects that "the old Anatomical Institute lay on a narrow street, #1, behind the Garrison Church, and apart from remoteness—if one considers this a desirable quality in a facility for studying human anatomy—it otherwise lacked almost everything one could wish from such an institution."[106] In his memorial address for Müller, he called the Anatomical Institute "a foul-smelling hole which in Berlin represented an anatomy building."[107]

In the dissecting hall, 150–200 medical students at long tables struggled over twenty cadavers, whose parts Müller personally distributed. According to Bidder, Müller left the students largely to their own very limited devices, appearing for only half an hour to cast a look here and there. But Bidder's own response to Müller shows how generous the professor could be to a student in whom he saw talent:

> Hardly fourteen days had passed when J. Müller, whose appearance in the dissecting hall always caused a great sensation, finally came to the table where I was busy at work. After he had seen my preparation, he electrified me with the words, "*Herr Doktor*, come up to my workroom. You'll be able to work better there than here in all this confusion."[108]

Despite heavy demands on his time, Müller was willing to help a promising student.

But Bidder's account, one of the most-cited descriptions of Müller's research environment, must be read with a great deal of caution. Like the other scientists who worked with Müller, Bidder wrote about his experiences decades later. The observations just quoted come not from an 1834 journal but from memoirs prepared after 1869, by which time Bidder's memories had evolved considerably. And like all of the Müller students discussed in this book, Bidder presented himself as favorably as possible, here as a skillful dissector rescued by Müller because he stood out so dramatically from the other students. His portrait of Müller cannot be extricated from his own, self-flattering account of his scientific development.

Considering the Anatomical Institute's condition, it is not surprising that Müller and his students sought other work space—not just in the university's main building but also in their own homes. In the 1830s, most European scientists worked where they lived. When Müller moved to Berlin in 1833, he rented an apartment from the widow of the philosopher Georg Wilhelm Friedrich Hegel (1770–1831), who had died of cholera a year and a half earlier. Müller quickly arranged for his student Jakob Henle to move in downstairs. Their house, am Kupfergraben 4A, was located on the Spree about halfway between the Anatomical Institute and the university's main building, across from what is now the Pergamon Museum. When Hegel's son returned home in the fall of 1835 Henle had to move out, and he joined Theodor Schwann at a cheap boarding house on the corner of Friedrichstrasse and Mohrenstrasse. About twelve minutes by foot from the university's main building, this corner today houses fashionable shops. Müller and his students of the 1830s worked where they lived, sharing their few precious microscopes. Together, they ate, slept, traveled, dissected animals, and exchanged ideas on all subjects. Life and work occurred in the same space.

Physiology and Anatomy in Berlin, 1833–1848

In 1833, the mineralogist Weiss had warned that Müller was not an experimental physiologist at heart.[109] By 1840, his suspicions proved right. Müller performed most of his physiological experiments between 1828 and 1834. In one last innovative physiological study, *On the Compensation of Physical Forces in the Human Voice Box* (1839), he used a severed head and Nanny's piano to study the way that the human voice produces particular tones. But from the time that he established himself in Berlin, he focused increasingly on comparative anatomy. This dedication to observing animal structures involved no drastic change in his scientific outlook, however.[110] Müller viewed experimentation as a subset of observation, and he never opposed interventional experiments.

As a researcher, Müller was interested in all of life's phenomena. His studies encompass a multitude of issues as daunting as the wealth of life forms he observed. Inevitably, the historian seeking patterns in his research becomes Johannes Müller, searching for order in a dazzling array of forms. But there is a plan.[111] While Müller cared intensely for all of his projects, from blood chemistry to sea urchin development, each was a means to an end. Throughout his career, he chose topics that promised to reveal the organizing principles of life.

In his doctoral research of 1822, while watching the opposing movements of insects' legs, Müller was seeking an overarching plan of which every living being formed a part. He had begun this quest for patterns after his early training in nature philosophy, and it remained fundamental to his scientific outlook. He loved observing fish, he wrote, because "no other class of vertebrates is so rich in special, typical forms that allow such a deep, penetrating look into their organization."[112] Later in his career, he made his quest evident even in his titles, such as his 1852 paper, "On the General Plan in the Development of Echinoderms." Seven years later, Darwin's presentation of a plausible mechanism for evolution would shake naturalists' belief in a vast organization of unchanging species, but Müller died the year before *The Origin of Species* was published, and he never lost his faith in a fixed hierarchy of organisms.

In his first lecture at the Bonn University, Müller had confessed that "there is something religious in the investigation of nature."[113] His students, particularly Rudolf Virchow, noticed that Müller's science had a devotional quality. Raised a Roman Catholic, Müller had considered becoming a priest when he was a boy and had entered the Bonn University with the intention of studying theology.[114] Within days, he had resolved to study medicine instead, but he never lost his awe for life's forms. In Müller's eyes, a perceptive, intelligent person had a sacred obligation to build scientific knowledge. Like Kant and Schelling, however, Müller did not envision an external Creator who had designed nature according to divine laws.[115] He believed that nature's diverse forms had arisen as a result of the unique structure of living matter. In this respect, he can be considered a vitalist, but his belief in "life force" drove him to conduct experiments. As a physiologist, he supported any study that might show how organic matter gave rise to life functions.

Müller's enormously influential *Handbook of Human Physiology*, written and rewritten between 1833 and 1844, shows his simultaneous commitments to vitalism, philosophy, and rigorous science. Created for fellow and would-be investigators, his *Handbook* became the most respected physiology text for much of the nineteenth century.[116] The organization of this work shows how in Müller's mind, views that seem contradictory today composed a unified scientific worldview. He begins with a discussion of why organic matter differs fundamentally from inorganic, then proceeds to chemical analyses of the blood and lymph. He describes in detail the circulatory, lymphatic, respiratory, digestive, endocrine, nervous, and sensory systems in a wide variety of animals, but explains that the presence of a soul makes each organism an indivisible unit. In the second volume, which appeared in 1837–1840, Müller wrote that "everything that feels and moves itself voluntarily according to its own desires has a soul."[117] In his detailed discussions of the sensory organs, he calls the eye "an optical tool [*Werkzeug*]" and the ear "an auditory tool," referring readers to optical and acoustical studies.[118] Yet his philosophical sections cast doubt on the explanatory power of the physical laws he invokes. The same textbook that discusses the behavior of light and sound waves proposes that living organisms possess a life force for which physical laws cannot account.

Müller's *Handbook*, which began appearing shortly after he reached Berlin, can be seen as part of his effort to redesign anatomy and physiology as linked fields. A second, crucial aspect of this campaign was his foundation of a new journal in

1834, the *Archive for Anatomy, Physiology and Scientific Medicine.* This journal contained yearly reports on anatomical and physiological research throughout Europe and quickly became one of its most respected scientific periodicals.

As the key method of anatomical and physiological studies, Müller promoted microscopy. He valued the microscope above all other laboratory instruments, and his faith in microscopy unites his diverse studies.[119] According to Müller, an eye unaided by a microscope was "unarmed."[120] In 1838, he applied his student Theodor Schwann's cell theory to pathology, demonstrating in his book *On the Fine Structure of Pathological Tumors* (1838) that tumors consist of cells. More than any of his other works, Müller's tumor book showed how microscopic studies could be of use in the clinic. In his discussions of tumors, he stressed the importance of microscopy for diagnosing pathological growths. Virchow, who would develop cellular pathology as a field, believed that Müller's work gave pathologists "the strongest impetus" to use microscopes.[121]

As a method to be used in conjunction with microscopy, Müller advocated chemical analysis. He regarded it as a crucial means of identifying and classifying tissues and an essential tool of pathological anatomy. By chemical analysis, Müller meant melting or boiling tissues, extracting them with alcohol or ether, and recording their reactions to strong acids, all of which he did to tumors he found. His students would claim that he knew little about chemistry, but he valued the field and wanted to learn. In May 1842, Carl Mohr (1806–1879), a friend of Müller's publisher Jakob Hölscher, wrote that Müller was holding up the 1842 edition of his *Handbook of Human Physiology* until he heard about the organic chemist Justus Liebig's (1803–1873) latest results.[122]

Müller's desire to examine living things through a microscope determined the kinds of observations he could make and, consequently, the way that he viewed life. Because, at the time, one could scan only very fresh preparations, he was always in close contact with his research animals in their native environments. While his microscope revealed life at the cellular level, it also kept him in touch with living organisms.

Müller's microscopic and chemical studies inspired a generation of researchers, but he performed them in an anatomical context, for he believed that only by comparing animals' solutions to life's challenges could he explain what life was. Theoretically, anatomy (the study of living structures) differs from physiology (the study of life functions), but Müller's work reveals how closely related these two fields are. For Müller, identifying a structure meant being able to say what it did, which invariably involved comparisons with other animals.

As a result, observations of animals and experiments on animals functioned as interdependent methods in Müller's science. In *An Introduction to the Study of Experimental Medicine* (1865), seven years after Müller's death, the physiologist Claude Bernard would define observation and experimentation as different concepts. According to Bernard, observation was passive; experimentation, active.[123] Müller never made this distinction. In his 1826 study of vision, he asserted that "observation is itself the most important operation in physiology."[124] In the introduction to his 1830 book on genitalia, where he openly discussed his methods, he wrote that experimentation was never a good thing in and of itself. "I have always been a friend," he wrote, "of methodical, thoughtful, well-planned, or—what amounts to the same thing—philosophical treatments of a subject." But he specified that by "philosophical" he did not mean dogmatic. "I demand first, that one be tireless in observation and experience," he explained.[125] For Müller, philosophy, observation, and experiment merged to create a scientific feedback loop in which one thought by seeing and saw one's thoughts.

In Müller's 1826 study of the visual system, he argued that one could understand vision only by observing the simplest forms of eyes and examining every animal that had them. This same attitude runs through most of his physiological projects, defining the kinds of observations he made and the scientific knowledge he built. Introducing his 1830 comparative study of genitalia, he wrote that previous hypotheses about sexual organs had yielded no knowledge, and that real understanding would come only by observing "how they develop step by step" from one animal to another.[126] In his studies of marine life, the notion of "belonging" is central as he attributes embryos to this or that family according to structural details. In 1851, when he discovered tiny slugs developing in the intestinal cavities of Holothuria (sea cucumbers), he became anxious and deeply depressed, for his understanding of life's structure could not account for the appearance of one species inside another. "It is fortunate," he wrote, "that these observations . . . were not made earlier, for they could have disrupted the pace of science and been used to support confused perceptions and ideas."[127] Because knowledge mirrored the structure of life, a glitch in the animal kingdom threatened the order of science.

Like a living organism, knowledge of life had to form a unified whole. Following Kant's philosophy, Müller believed that plants and animals differed from inanimate objects in that each part of a living being served the entire organism. In the prolegomenon to his *Handbook of Human Physiology*, he wrote, "Kant says: the cause of the type of existence of each part of a living body is contained in the whole."[128] Because of this outlook, Müller was always most intrigued by the physiological systems that unified animals. The nervous system interested him so greatly because it created wholeness through communication. The brain, he wrote, "brings together all the different energies of the different parts of the nervous system into the unity of a self-aware, self-determining individual."[129] But this emphasis on unity never stopped him from performing intricate dissections and chemical analyses. He saw no contradiction, because for him, physiology and chemistry provided a means of learning how animals functioned as living units. What counted was the live individual and its relation to members of other species.

From the mid-1830s onward, the project that consumed most of Müller's energy and attention was the classification of marine life. In the late 1830s, he developed new classificatory systems for the myxinoids (hagfishes) and plagiostomes (cartilaginous fishes such as sharks and rays). In the 1840s, he continued this work with studies of the cyclostomes (lampreys) and ganoid (scaly) fishes. In his letters to Retzius, Müller referred to his project as "a natural system of the fishes" and "a system of fishes based on anatomy."[130] Müller was particularly intrigued by echinoderms, spiny animals with radial symmetry such as sea urchins and starfish. On his research trips, he used a net that he had specially designed to scoop floating echinoderm embryos from the sea surface, a technique that he called "pelagic fishery."[131] Müller focused on organisms that fell along the borders of older classificatory systems, attracted by the challenge that they presented.

Müller's way of writing about animals shows his fascination, love, and respect for the organisms he studied. In his depictions, the animals he observes come alive so that the reader can sympathize with him and the creatures he is studying. When discussing lung parasites, he takes a moment to describe the bird they are attacking: "This owl came from Lapland and lived for a winter in Stockholm. She was faint-hearted and narrow-breasted, fell ill around Christmas-time, and died later of shortness of breath."[132] While primarily interested in what is growing in her lungs, he also cares about the gasping owl herself.

Like a lover, Müller follows the forms of developing animals, seeking an enigmatic essence that attracts him. In 1891–1892, the British novelist Thomas Hardy would create a character who spent his life chasing an idealized female form, an essence he believed jumped from woman to woman. Hardy set his novel on the fossil-filled shore of southern England. His frustrated character, a sculptor, spends his life trying to capture evasive forms in stone and becomes involved with three "incarnations" of his "well-beloved" (young Avice Caro, her daughter, and finally her granddaughter) in a futile attempt to possess his ideal. Early in the novel, he is unsure whether "the migratory, elusive idealization he called his Love who, ever since his boyhood, had flitted from human shell to human shell an indefinite number of times, was going to take up her abode in the body of Avice Caro."[133] In an 1854 article, "On the Different Forms of Marine Animals," Müller wrote about the bristle worm larva *Mitraria*, "I have been pursuing her for a long time; since I first saw her in Marseilles, she has appeared to me in Trieste and lately in Messina, always in a different way [*Art*]."[134] Müller's passionate quest for meaning and order in the diverse forms of marine life anticipates the longing and frustration of Hardy's sculptor. Both are driven by the urge to capture elusive forms.

Müller's determined, systematic pursuit of individual organisms sometimes suggests the work of a detective. Although he went to the seaside each August, his encounters with particular animals depended on random chance. "A lucky accident led me to the larvae of Holothuria," he wrote in 1849.[135] Occasionally, his letters and articles read like police reports describing the movements of criminals. In 1838, seeking a shark described by Aristotle, he wrote to Nanny from southern France, "The animal I am seeking is certainly here."[136] In the same article in which he mentioned the "lucky accident," he informed readers that "I found [the Holothuria] again when I resumed my observations this summer in Nice, and I learned its final goal." He concluded his essay by admitting, "I was unable to pursue this Echinoderm any further."[137]

On first view, Müller's multitudinous research topics suggest anything but order. How he moved from one problem to another is one of the most interesting questions his—or any scientist's—work raises. Müller's writing indicates that in many cases, personal and intellectual motives merged, and feelings of personal connection raised new questions to pursue. In outlining his work, one must recognize the contributions of scientific contacts, personal urges, and sheer accidents. The temptation to impose a ready-made narrative is always great. In making sense of his science, one must pay close attention to his language. The words with which he describes it suggest its motivating force.

Müller's Favorite Words

In the more than 250 articles and books that Müller published between 1822 and 1858, he repeated a few favorite words. Four of these, which recur throughout his works, suggest his style as a scientist, thinker, and writer.

The first is *Mannigfaltigkeit* (multiplicity or diversity), a key term for the nature philosophers inspired by Kant's writings. Müller applied this word to every system he investigated, using it to describe not just the outer forms of animals but also the microscopic structure of their tissues and the kinds of sounds they could make. In 1836 he reported that in breast cancer tumors, "the fibers are interwoven in the greatest variety [*mannigfaltigsten*] of different directions, without any sense of order."[138] When discussing the human vocal apparatus in 1839, he observed that "every voice box, either freely or when pressed, can produce the same tones in the greatest variety [*mannigfaltigsten*] of different ways."[139] Müller used this word most often in his comparative anatomy papers and lectures, trying to convey the wonder of life's innumerable forms. Young Friedrich Bidder complained that in Müller's physiology lectures, "He only twice offered his listeners the opportunity to look at anything," whereas:

> in the comparative anatomy lectures, in contrast, he used an exceedingly rich variety of material. No statement was ever made that was not substantiated with an appropriate preparation, and the thoughtful interrelation of facts occurred in such a lucid, convincing way that there, for the first time, the colossal task of demonstrating the rule of a predominating idea in that multiplicity [*Mannigfaltigkeit*] of forms came alive before my eyes.[140]

Invoking a central idea realized through manifold forms, Bidder recapitulates Müller's anatomical philosophy, very likely in Müller's own words. To Müller, life's extraordinary diversity of forms presented an irresistible challenge, and he communicated his fascination to his students.

Müller's second favorite word conveys a response to this multiplicity: *Verwirrung* (confusion or disarray). Often, Müller refers to confusion in the opening paragraphs of his articles to justify the analysis that will follow. He begins his 1838 book on tumors by describing the multiplicity (*Mannigfaltigkeit*) of their forms, then claims that previous studies have led only to confusion (*Verwirrung*) because no pathologist has investigated tumor structure chemically or microscopically.[141] He opens an article on the characteristics of cartilaginous fish by stating that "the confusion which still exists in the natural history of cartilaginous fishes may be attributed to the generally imperfect descriptions which have been given of the species."[142]

Closely associated with this confusion are Müller's frequent references to *Zweifel* (doubts), which appear in his introductions and conclusions. For Müller, doubts, like confusion, were meant to be eliminated. He asserted in his 1831 paper on spinal nerve roots that "these . . . experiments . . . leave no remaining doubt about the truth of Bell's Law."[143] In physiology, doubt created the motivation to do experiments. If Müller's scientific writing reflects his real thinking, he worked to organize, differentiate, and clear things up. Throughout his career, Müller feared disorder as a kind of death, proclaiming in his 1824 lecture that without philosophy as an organizing guide, experimental physiology would "grow into a chaos of facts [*Kentnissen*] . . . in which there is no living thought."[144]

Consequently, whether he was doing physiology or anatomy, Müller sought characteristics in every animal that distinguished it from other animals while defining its relationship to them. His third favorite pair of words reflects this quest for unique qualities: *eigentümlich* (characteristic of an individual) and *Eigenschaft* (a characteristic). The first sentence of his *Handbook of Human Physiology* reads: "Physiology is the science of the characteristics [*Eigenschaften*] and appearances of living bodies . . . and of the laws according to which their functions take place."[145] Müller became interested in pathology when he began gathering tumors from doctors at the Charité Hospital for his museum collection, and the passion to classify drove his pathological studies just as it had motivated his comparative ones. When he used a microscope to study tumor tissues that until then had appeared similar, he reported that "with closer investigation, I noticed many distinguishing characteristics (*viel eigentümliches*)."[146] In the summer of 1852, while studying microscopic marine life in Trieste (an Italian city on the north Adriatic coast), he wrote to Nanny, "When you look at these objects again and again, you always find something new in them to distinguish."[147] For Müller, doing science meant observing closely to identify differences, a method that could be applied to any field.

Müller's fourth favorite word, *Rätsel* (puzzle or riddle), appears throughout his works and suggests his determination to reveal life's grand plan. Its connotations vary according to his mood, conveying either delight in life's richness or a feeling of helpless inadequacy about his inability to explain it. Frequently, he refers to an organism as a "rätselhaftes Tier" (a puzzling animal), as though animals themselves were riddles to be solved.[148] For Müller, they *were*, since each animal offered clues about a far-reaching design. Each animal was a puzzle, and each was a piece in a larger puzzle that he increasingly despaired of solving.

Müller's Museum

To comprehend the relationships among all living animals, Müller spent much of his time and money expanding the university's Anatomical Museum. The opportunity to control Prussia's finest comparative anatomical collection had been a key factor attracting him to Berlin, and the museum quickly took priority among his scientific activities. A twenty-five-year visual experiment, Müller's collection became the system through which he studied the organization of life.

As Cultural Minister von Altenstein must have noticed, however, the Anatomical Museum was not the only one at the Berlin University that collected skeletons and preserved fish. It shared the university's main building with two other museums, the Zoological Museum and the Mineralogical Cabinet. To fill his anatomical ark, Müller had to compete for financial resources with the faculty members running these other collections.[149]

As integral parts of the Berlin University, its three museums aimed to inspire future studies. They displayed all that the natural world had to offer and the ability of Prussian naturalists to collect and classify it. The mineralogical collection, organized by the natural philosophy professor Christian Samuel Weiss, occupied the first floor of the main building's left (west) wing. On the third floor of the right (east) wing and much of the main wing lay the Zoological Museum, run by the natural philosophy professor Heinrich Lichtenstein (1780–1857). Built from private collections, including that of the Royal Palace, it contained 466 whole mammals, 4,000 stuffed birds, 7,000 amphibians, 1,500 fish, and 150,000 insects. The public could view these specimens, "systematically arranged, and labeled with their scientific names," on Tuesdays and Fridays from noon to two, as long as they obtained tickets the previous afternoon. Admission was free, and tipping was forbidden.[150] Both Lichtenstein and Weiss lived in good-sized apartments on the premises, so that when the directors expanded their collections, the struggles for space got personal. In Müller's museum, as in Lichtenstein's, animals were "systematically arranged." But the purpose of Müller's—officially, to display "human and comparative anatomy and healthy and pathological conditions"—was to show patterns and relationships.[151] Like Lichtenstein, he wanted to collect as many animals as possible, but by arranging them, he hoped to learn what life was.

Berlin's Anatomical Museum predated both Müller and the university, having begun with the king's 1803 purchase of the anatomist Johann Gottlieb Walter's (1734–1818) private collection for 100,000 taler.[152] Like the Zoological Museum, the extensive, government-financed collection was a matter of Prussian pride. In his 1834 guide to Berlin, Freiherr von Zedlitz wrote:

> [The Anatomical Museum] lies on the middle floor of the left wing of the university building. . . . At the end of 1833, the whole museum contained about 10,000 preparations. The visitor's attention is called mainly to the wealth of nerve preparations, a long array of monstrous births, and about 500 animal skeletons. . . . For doctors and medical students, the museum is open from 2–3, and in the summer or by special appointment from 9–12. It is open to the public . . . in the summer from 4–6 and in the winter from 3–4 or 5, daylight permitting. . . . Tickets may be obtained on the day prior to admission. . . . Boys will be admitted only in the company of their fathers or teachers, and of the female sex, only midwives will be granted admission.[153]

Under Müller's direction, the collection would grow from a haphazard array of specimens into a thorough, zoologically oriented collection in which most known animals were represented. A prosector, a helper, and a custodian assisted Müller, the former two positions occupied by his finest students. But as the guidebook entry indicates, admission was a touchy subject. The Prussian Cultural Ministry believed it was funding a public museum, but if Müller had had his way, only those capable of appreciating his treasures would have crossed the threshold.

In the mid-1830s, visitors entered the museum at their own risk. The wooden beams of the eighteenth-century roof were rotting and could have collapsed at any time. In 1834, a large chunk of plaster fell out of the museum's ceiling, and the Cultural Ministry decided it was time for serious renovations. When the building inspectors saw the old palace's condition, they said that it was a miracle no one had been killed. Müller participated actively in the building's redesign, demanding the best possible space for his collection. The mineralogical and zoological collections were to share the third floor, and the Anatomical Museum was to occupy the second floor of the horseshoe-shaped building's west wing. This area contained a high-ceilinged room, which Müller arranged to have split into two levels, since square footage meant more to him than an elegant setting. During the renovations—which lasted for years—he convinced Weiss and Lichtenstein to keep the anatomical collection in their own apartments.[154]

Müller's letters to Retzius during this time indicate how chaotic the rebuilding must have been. In July 1843, he wrote to Retzius that the renovations would be finished in early October and that he "would finally again have the pleasure of an ordered museum." But in February 1844, he complained that the move was taking up the whole winter semester.[155] Despite his desire for order, things were getting out of hand. In the fall of 1841, Müller had sent Retzius a kangaroo skeleton that was unfortunately missing an arm. After Müller and his assistants unpacked the collection in its new quarters, he wrote "with great delight" that the kangaroo's arm had been found. In August, however, he had to admit that he had spoken too soon. His custodian, Thiele, had thought that the arm belonged to a kangaroo, but when he had inspected it closely, he had realized that it was from another animal.[156]

Because of the rate at which Müller acquired new specimens, the Anatomical Museum soon outgrew these rooms as well. Throughout the 1840s and 1850s, he and his students dissected, scanned tissues under microscopes, and performed physiological experiments in dark, cramped corners of the west wing, surrounded by bleached, bloated organisms floating in alcohol. Müller wanted a new building for the museum and the Anatomical Institute, but although the Cultural Ministry admired his work, it was more than they could afford. Despite Müller's efforts to obtain more space, the museum remained in the west wing for the rest of his life.

To understand what the museum meant to Müller, one must imagine comparative anatomy before Darwin's *The Origin of Species.* During the same years in which Müller was building his collection (1833–1858), Darwin was formulating his theory of natural selection. Müller believed that species were distinct and that there were no transitional forms between them. He did not think that animals had changed since the Creation. He was convinced, however, that one could learn the nature of life by aligning all known animals and finding the perfect place for each one. In an 1830 letter to Retzius, Müller asked his friend about the position of a snake's (Coecilia's) tear glands relative to its eyes, explaining, "This character must be very decisive for the placement of Coecilia in a system."[157] By lining up animals, one could "watch" a physiological system develop from its simplest to its most complex form. Müller did not see this development as a process that had occurred over time, but as an order established by God. He could understand it only if all of the life forms were there, every letter of the message, every piece of the puzzle.

Consequently, Müller's letters to the Cultural Ministry often read like shopping lists, requesting considerable funds to buy rare specimens. In January 1848 he asked Minister Eichhorn for 517 taler to buy a chimpanzee skeleton from Congo, a giant armadillo skeleton from Brazil, and a model Pleiosaurus from England. Müller emphasized that the museum's annual budget did not allow for the acquisition of such precious objects, and that this was a priceless opportunity, since they were seldom on the market. He mentioned other museums (Halle, London, Paris, Haarlem) that had comparable specimens, appealing to Eichhorn's Prussian pride.[158] In such letters, one can hear Müller's tone of urgency. He *had* to have one of each animal.

When Eichhorn couldn't or wouldn't underwrite his purchases, Müller bought animals with his own money. He paid his assistants out of his 2500 taler salary, and much of the remainder seems to have gone for rare anatomical specimens.[159] In 1835, Müller received almost 700 taler to build the museum collection, but his list of purchases—including a "magnificent orangutan" for 300 taler—suggests that he probably bought some specimens with his own income.[160] When Müller received a bonus in 1842 for declining an offer from Munich, he spent the money on his collection and raises for his assistants, Reichert and Peters.[161] In time and money, he gave everything he had to create a complete representation of animal life.

Müller's compulsion to collect all known living forms recalls Noah's divine mission to bring two of each animal into his ark. In Genesis, God commands Noah:

> Of every living thing, of all flesh, you shall bring two of every kind into the ark, to keep them alive with you; they shall be male and female. Of the birds according to their kinds, and of the animals according to their kinds, of every creeping thing of the ground according to its kind, two of every kind shall come in to you, to keep them alive.[162]

Following orders, Noah collected live animals with the intention of breeding them. In contrast, Müller's specimens were dead, living only in the memories of the people who viewed his collection. For Müller, assembling the remains of all known animals and aligning them for close observation was an essential task, since it promised to reveal life's plan. Like Noah, Müller assembled a representation of life in order to preserve it.

As a compulsive collector, Müller also resembled his contemporary Honoré de Balzac (1799–1850). Both writers admired Cuvier; both worked in frenetic bouts, fueled by strong coffee; and both wrote copiously, trying to characterize entire systems of "animals." Like Balzac's *Comédie humaine*, Müller's articles offer a literary world with recurring "characters" who are described from varying points of view. In his 1842 introduction to this family of novels, Balzac explained:

> I saw that from this perspective [of Cuvier's and Geoffroy Saint-Hilaire's debates about how animals' forms were related to their environments], society resembled nature. Doesn't society make out of man, according to the environments in which his action occurs, as many different men as there are varieties in zoology? . . . In all times, then, there have existed, and there will exist, social species just as there are zoological species. If Buffon has produced a magnificent work, trying to represent all of zoology within one book, isn't there a work of this type to be written for society?[163]

In this statement of purpose, Balzac represents himself more as a recorder than a creator, more as a historian than a teller of tales. If these are the criteria for his *Comédie humaine*, then his fiction can fulfill its goal only if it is complete. Like Müller's museum, Balzac's representation will succeed only if it contains one "animal" of each type.

As museum director, Müller acted more like an archivist than a teacher. If he had had his way, no one would have entered but skilled comparative anatomists, and his will to protect the collection far exceeded his desire to show it. When young Emil du Bois-Reymond began approaching Müller about doing research, he complained to his friend Eduard Hallmann, "The Zootomical Museum is without exception inaccessible to anyone."[164] His artistic talent and eagerness to experiment soon won Müller's confidence, however. The next June, when he had become Müller's research assistant, he wrote gleefully to Hallmann about Müller's misery when a group of Very Important Persons invaded the museum:

> Undoubtedly, Müller was the unhappiest person there. For in the pathological museum, in his workroom, the diplomatic corps—women and children included—was supposed to take a seat. Müller found himself treated like rabble, and Krause said to him, in the museum, in my presence, "these are all people of the highest rank who know how to handle themselves!" Namely to his objections when the diplomatic corps wanted to settle its weight on his anatomical jars. . . . Now Müller was faced with Rita and Christina and so forth, weeping and wailing before his cabinets full of the loveliest carcinome, steatome, hydrocephalus, and hemicephalie. But he could only fall back on reflex movements like a decapitated *Col. Natix* and cry incessantly, "O Spinoza! Is this a punishment or a test?"—To no avail; he had to get out, and over the cabinets full of the most disgusting science there appeared, as if by magic . . . a delicate tent of white muslin with laurel garlands. In front of Thiele's

> elephant stood a second Fabricius, the bust of Old Fritz, separated from the monster by a pink silk niche with some flowers. But Müller was so spiteful, he wanted to know nothing of all this splendor.[165]

Such insults were rare, but Müller's reactions in this literary depiction suggest his overall attitude. Once young du Bois-Reymond had proved himself, Müller gave him the keys to the museum, but he wanted the diplomats out.

After Müller's death in 1858, his successor, the anatomist Karl Bogislaw Reichert (1811–1883), guarded the collection just as jealously. When Reichert died in 1873, the Cultural Ministry planned to unify the Anatomical, Zoological, and Mineralogical museums, moving them out of the main university building into their own quarters. Then, in the 1880s, what Müller had most feared occurred. The collection was divided, some specimens joining the Natural Sciences Museum in 1889, the rest going to the new Anatomical Institute that Müller had hoped to realize, built on the grounds of the Veterinary School.[166] His collection never revealed the great plan of life.

Living to Work

What motivated Müller to strive at the pace he did is one of the most fascinating questions his story raises. Writing to Nanny in 1846 from Helgoland, a remote North Sea island, he confessed, "I can't wait for the sunrise, which will bring [me] new material." Three years later, he wrote from Nice, "I can't spare a single minute that might keep me from examining a glass full of living things before nightfall."[167] In late 1848 Nanny reported that Müller would insist on traveling without interruption, living only on coffee, and would work on his feet for seven hours without anything to eat or drink.[168] In Helgoland, Müller and his students would gather specimens from daybreak until 3:30 in the afternoon, then pause for a brief meal and do microscopic work in the evenings. The students could barely keep up with him and were unanimous about his monstrous energy. Müller never feared having more work than he could handle and, especially in his late twenties, preferred working on many different projects at once. What he feared was running out of work, a possibility that haunted him increasingly with age.

In his studies of marine life, Müller restricted himself to certain methods, using strategies that required more effort and provided less information than he might have obtained through laboratory experiments. Instead of breeding marine organisms and systematically watching their embryos develop, Müller preferred pelagic fishery, skimming random netfuls of plankton out of the waves, then checking to see which pieces of the developmental puzzle he had scooped up. Müller was well aware of this method's limitations, admitting in an 1852 article that it required "great endurance and patience, for some days and even weeks bring almost nothing, or nothing that one is looking for, and then there are days when the material in the fine net is so rich, that the day is not long enough to work through it all."[169] His letters express even less confidence than his scientific assessment. In 1853, he wrote to Nanny, "Naturally, my way of fishing will slowly go out like a light when it is fully exhausted." By 1855, he confessed that he was "ever more certain that no more discoveries will bloom out of my pelagic fishery."[170] While he longed to solve the riddle of animal development, he clung to his net out of a stronger compulsion. As long as he could keep fishing, there was work to do.

Considering all the hours he put in, Müller did not make much money. In the 1820s he had lived in poverty, and throughout his life, he feared becoming poor again. Since neither he nor Nanny came from a wealthy family, money was always a pressing issue. When he was earning 800 taler a year as a full professor in Bonn, he received an offer from Freiburg with a 1,500 florin (about 1,500 taler) salary. Fearful of losing a respected researcher, the curator of the Bonn University asked the Prussian Cultural Ministry to give Müller a raise. "Müller comes from a family of very small means," he wrote, "and he owes his brothers and sisters. . . . One cannot live more frugally or modestly than this family does."[171] In Berlin, Müller spent much of his salary on the Anatomical Museum, and the result was that his museum purchases and his research trips—which were not funded by the university—left little money for household expenses. In 1853, after Müller turned down another lucrative offer from Munich, Nanny wrote to their son in exasperation, "He blows away 1,500 taler as if it were a feather, but it's often hard to get five *Silbergroschen* out of him."[172] If the Müllers lacked money, they did so by Johannes's choice. Life and work drew upon the same account, and he feared that the funds would run dry.

Family responsibilities added to Müller's financial challenges. As the oldest of five children, he had become the head of his family at age eighteen, when his father had died. For the next ten years, he had received help from his family, who made considerable sacrifices to finance his studies, but he had also become their leader. By the late 1820s, he was the authority figure to whom they turned to settle their differences.[173] Once Müller prospered in science, his younger brothers lived off of him, Philipp starting an unsuccessful business and Georg emigrating to America—probably fleeing debts—leaving his wife and son behind. Both brothers may have had problems with alcohol.[174] The need to care for his siblings' families as well as his own put Müller in a position of great pressure.[175]

Not surprisingly, Müller couldn't sleep. As he aged, his insomnia grew more acute, but he experienced sleepless nights even as a young man. In his 1826 study of fantasy images, he wrote that he performed self-experiments partly because "sleepless nights would become shorter when I could wander amidst the forms created by my own eyes as though I were awake."[176] In September 1828, he wrote to Nanny that he had been unable to sleep for several nights in a row, and referred to "the loneliness of night."[177] In the late 1820s, Müller's self-experiments and insomnia became a vicious cycle; when he felt himself succumbing to an urge to sleep, he used coffee to stay awake and study his fantasy images. After his move to Berlin, his self-experiments ceased, but his insomnia didn't. In September 1846, he wrote to Nanny from Helgoland, "I'm not sleeping much better here than I was in Berlin; my bed can only be compared to an oyster crate, and I am truly happy when the sun rising behind the dunes reaches my bed and my face."[178] In 1849, he wrote to his daughter from Nice:

> Our nights [are] long . . . I usually get up a couple of times and walk around, look at the moon, listen to the sound of the sea beneath our window, gaze at the light reflected from the sea into the leftover sea water in the glass from yesterday's excursion, drink some water, and look at the clock.[179]

It has been claimed that the restless, anxious Müller wandered the Berlin streets at night, but while this romantic image makes for a good story, it is unsupported.[180] Whether Müller stayed awake to experiment, or experimented because he was awake, he could find no peace. Possibly he saw sleep as a waste of time, but the tone of his letters suggests he had little choice in the matter.

Müller's descriptions of his sleepless nights betray heart-rending loneliness, but unlike the stereotypical "lonely scientist," he never saw science as something to be done alone.[181] He felt lonely because he wanted to share his scientific ideas with people who understood them. In his 1824 article on fetal respiration, he wrote that he had performed the experiments "in the company of several of my friends."[182] Sometimes he took his family along on research trips; he wrote to Retzius in 1842 that Nanny and the children were learning Italian for their sojourn in Italy.[183] By 1850, he regarded Max as a student-partner, proud that his twenty-year-old son could do dissections and use a microscope competently.[184] On his seaside research trips, however, his family members were not his preferred companions. In 1852, he told Retzius, "If I can't find anyone else, my wife will accompany me."[185]

More often, Müller made a family of his collaborators, since his need for intimacy was best assuaged by those who shared his scientific interests. In a moving letter to Karl Ernst von Baer in January 1828, he confessed, "If you knew how lonely I am living here [in Bonn] and how little I can share my joy in nature with anyone else, then you can well imagine how much I gain from a far, far-off friend."[186] He wrote to Retzius, "I often think with heartfelt joy of the days that we shared in Berlin and the hours in which I first saw and got to know you, most honored friend."[187] As he aged, Müller's need for companionship grew stronger. In 1852 he wrote to Retzius, "I haven't had any desire to travel alone for a long time now, especially to the sea. It's hard when you're alone to get through the idle hours; one doesn't always have something to do."[188] Müller's workaholism, anxiety, insomnia, and loneliness acted synergistically, so that as he aged and his best-loved students left him, he grew increasingly depressed.

According to some historians, Müller was a cyclic manic-depressive. Certainly his furious activity, recurring breakdowns, and insomnia suggest this kind of mental illness. But *was* Müller mentally ill? It is wrong to associate a troubled nineteenth-century scientist with a disease defined in the twentieth century, recasting his personality in a modern mold. Doctors of his own time interpreted his collapses quite differently, and there were always strong external reasons for his depressions. There is no evidence whatever that his disease, if he had one, was "an inherited part of his nature."[189]

Müller's first breakdown, in 1827, was most likely a collapse from overwork, and his second one, at the end of his physiological period in 1840, is also attributable to external causes. In 1839, his close friend Karl Windischmann died of tuberculosis. Johannes, Nanny, and their two children left their home of six years for a new apartment at Cantianstrasse 5.[190] In the spring of 1839 and the fall of 1840, Müller's two best-loved students left him to assume their own academic posts: Theodor Schwann, to replace Windischmann in Louvain, Belgium; and Jakob Henle, to begin work in Zurich, Switzerland. In May of 1840, Cultural Minister von Altenstein died, and was replaced by Johann Eichhorn (1779–1856), a reactionary who lacked Altenstein's interest in science.[191] It is hard to imagine anyone who would not feel uprooted under these circumstances.

After 1840, Müller performed almost no physiological experiments, inexplicably leaving the field when he was king. Some historians have accepted du Bois-Reymond's claim that Müller broke down in 1840 when he realized that he could no longer be first in his field.[192] When one considers his attachment to his students, however, his realization that they had moved on appears more significant. After 1840, Johannes Müller begins referring to himself in his letters to Nanny as *Vatermännchen* (little father-man).

Müller and the Revolution of 1848

Müller's crisis in 1848 is even easier to explain in practical terms.

In the late 1840s, after three years of poor harvests and economic recessions, peasants and industrial workers were starving all over the German-speaking territories.[193] Social tensions had been building in Prussia ever since the hesitant King Friedrich Wilhelm IV had assumed power in 1840. In late February 1848, Parisian workers took to the streets to challenge the rule of the Citizen King Louis Philippe, and the Prussians decided to follow their example. The driving force of the 1848 Berlin uprising was not proletarians, however, but middle-class liberals who demanded freedom of expression, an end to the military and police state, and a unified Germany with a constitution and national legislature. With the growth of industry, the time had come in which a divided Germany was bad for business.[194]

On 18 March, the king announced that he would lift censorship, convene the Prussian *Landtag* (Assembly), and support a general reorganization and national constitution. An enormous crowd gathered outside the Royal Palace to thank the king, but when troops attempted to disperse the people, a few soldiers fired their weapons and fighting broke out. In the battles that night, over 180 people died. On 19 March, the king agreed to withdraw his troops from Berlin, and a citizens' militia took their place. The Hohenzollern dynasty and its aristocratic officer corps yielded to the liberals and radicals almost without a fight—which made their return to power that much easier.[195]

Starting in May 1848, the National Assembly convened in Frankfurt with representatives from most German-speaking states. Their goal was to draft a constitution for a unified German nation. But from the beginning, a split developed between the liberal majority (drawn mostly from the wealthier, educated middle class) and the radicals (the lower middle class and workers). The radicals wanted a democratic republic, whereas the liberals favored a constitutional monarchy. Although the liberals welcomed the workers' support in breaking the aristocrats' power, they meant to protect their property rights and feared rule by the masses. The radicals believed the revolution would bring lasting social improvements only if every individual could vote.[196] While they argued, the king and army regained control. In November, Friedrich Wilhelm IV named the reactionary Count Brandenburg as prime minister, and he quickly recalled the army to Berlin. The king then dismissed the Prussian Assembly he had promised, and produced a constitution that made his rule supreme.[197] In late March 1849, when the Frankfurt Assembly completed its deliberations and voted to make Friedrich Wilhelm IV the German kaiser, he declined the imperial crown, ending the liberals' hopes for a constitutional monarchy. Neither he nor the rulers of the other most powerful German states recognized the constitution drafted in Frankfurt, and by July 1849, the last public protests had been crushed.[198]

As rector of Berlin University in 1847–1848, Müller had to maintain order and mediate between the king and the radical students and lecturers, who were led by his former pupils Robert Remak and Rudolf Virchow.[199] Liberal and radical academics played a leading role in the revolution of 1848, so that Virchow was only one of many who deprived Müller of sleep.[200] It would have been an excruciating job even for the most adept politician, which Müller was not. The anatomist found himself caught between the government that had made his career possible and his former students, who were determined to overthrow it.

Müller's students and subsequent historians have unanimously called him a conservative, but it is important to evaluate this label in the context of his life.[201] As an almost self-made man who had risen with government help, Müller viewed the Prussian government as a generous, reasonable bringer of order, and could not understand the students' resistance to it. Recalling his own poverty, Müller feared that social chaos would deprive him of everything he had attained, both his financial security and his precious anatomical ark. He valued science above all, and his own experience had shown the king and Cultural Ministry to be supporters of research and academic freedom.

As rector, Müller had the right to cancel and reinitiate classes, and his prime goal was to keep the university open. He could also permit or deny students the right to assemble and assign them spaces for their noisy meetings. When street fighting broke out, he warned them to stay out of it, but 200 gathered before the closed Aula and asked his permission to arm themselves. While he almost certainly denied it, many refused to obey. People of all classes and ages helped to build the barricades, but well-dressed young men led the fights.[202]

In academic robes, Müller and the deans of faculties walked down the street to the Royal Palace, where they asked the king to withdraw his troops from the city as the students demanded. Müller spoke only good of the students, saying that they were his friends and that they would prove themselves worthy in the hard days ahead. He then had to face much of the student body, assembled in the Aula, and tell them how the meeting had gone. During his report, wild cries erupted that a student had been killed by soldiers at the Oranienburg Gate. Again, the students demanded arms. In the funeral procession for the 183 victims—including some students—staged as a reconciliation between people and king, seventy-eight-year-old Alexander von Humboldt marched beside Müller, probably for moral support.[203]

In his dealings with the radical and mainstream faculty, Müller did not communicate well. On 27 March, he called a full faculty meeting without advertising the agenda, then asked his colleagues to send the king a proposal to convene the Prussian Assembly. The step would have been a conservative one, defining the past week's fighting as a Prussian problem and denying the liberal and radical desires for German unification and a national assembly. People demanded to speak, but he allowed no discussion, only a vote. Of the 107 lecturers present, ninety-eight voted in favor and seven against, including Rudolf Virchow. Du Bois-Reymond abstained. When the students got wind of these proceedings the next day, there was such an uproar that Müller abandoned the plan.[204]

In the spring of 1848, most of the radical students and faculty wanted to reform their institution as well as their country. But as a senior professor, Müller had no desire to change the university's governing structure. He feared that rule by the academic masses would endanger the progress of science. After a 30 May meeting to discuss university reform, Müller wrote to the cultural minister that lecturers and associate professors should not vote at faculty meetings, form organizations, or draft resolutions. If they ever received voting rights, he noted, they would outnumber full professors 100 to 59. They would then work to promote their own interests, selecting young scholars for open positions, and science would become the "handmaid" of partisan politics. The junior faculty, he declared, were demanding the rights of full professors without having earned them through research. Far more skillful in dealing with his superiors than with his students and colleagues, Müller timed his letter to coincide with a changing of the guard at the Cultural Ministry. It went straight to the hands of the new archconservative minister, Adalbert von Ladenberg (1798–1855).[205]

At the Cultural Ministry, things had become almost as chaotic as they were at the university. On 19 March, the day after the worst fighting, Minister Eichhorn had resigned and had been replaced by aristocratic Count Schwerin (1804–1872). Schwerin's supervision of the university, however, lasted less than four months. On 8 July, the more qualified Ladenberg took over, determined to enforce the will of army, aristocracy, and king. When planning a ceremony in late July, Müller ceded to the students' wish that two black, red, and gold banners (the colors of German unity) be hung from the balcony. Ladenberg got word of the plan and summoned Müller to his office, scolding him for his laxity and demanding that a black-and-white Prussian flag be hung between the banners. Müller obeyed, but when he appeared on the balcony and delivered a speech on the necessity of free academic institutions for the new philosophical-empirical direction in science, students hurled stones at the Prussian flag.

After the ceremony, Ladenberg again commanded Müller to appear, and this time threatened that if he could not control the students, the government would intervene, taking the most drastic measures.[206]

Müller held on until the end of his term, struggling to appease students, minister, and king. The day his rectorship ended, he left with his family for Koblenz, where he suffered a full mental collapse. For several months he remained almost sleepless, took no comfort in his family, and could do no work.[207] He found relief only when he traveled to Ostend, Belgium, with his former student Theodor Schwann and slowly resumed his studies of marine life. This time his recovery took longer. It was not until April 1849 that Humboldt could write, "Müller's return [to Berlin] makes me very happy."[208]

But Müller never fully recovered from the horrors of 1848. In the 1850s, he suffered from more chronic, less acute depressions, beginning with his discovery of slugs in sea cucumbers. The possibility that one animal could develop inside another threatened his whole concept of life's organization. Struggling to account for this find, he wrote to Retzius, "I still feel as though my poor head has been smashed to pieces by all the pain this matter has given me."[209] While he was puzzling over this quandary in June 1852, his mother died. Nanny wrote to his sister, "Müller is crying."[210] Soon afterward, his son Max passed his doctoral exam and decided to become a physician rather than a researcher, as his father had hoped. These sorrows combined with a growing fear that his pelagic fishery would bring him no further discoveries. In August 1854 Müller confessed to Retzius, "Because I've been exploiting this fine-netted fishing so intensely for so many years, I'm soon going to have to abandon it altogether, since I now almost always encounter the same old friends among the sea animals."[211]

In 1855, when he survived a nightmarish shipwreck but his student Schmidt did not, Müller fell into a depression from which he never emerged. Feeling responsible for the boy's death, he went personally to comfort Schmidt's mother. When the young zoology student Albert Gunther saw Müller in the summer of 1855, the shattered scientist told him, "I cannot speak with you. My mind is gone and with the dead."[212] Ironically, Müller himself was saved by an act of pelagic fishery when rescuers scooped him out of the North Sea, clinging to a piece of wreckage. His letters describing the incident attest to his power as a writer.[213]

Making People See

Müller acquired the fame that he did partly because he arranged words as well as he arranged animals. His clear, forceful writing brought him success throughout his career, not just in his self-promotional letters to the cultural minister but also in his textbooks and articles. Müller's way of doing science defies the stereotype of scientists as people who work better with images than with words. Like many successful scientists and fiction writers, he dealt equally well with both.[214]

When describing an unknown phenomenon or coining a new term, scientists think as creatively as poets, and like poets, they incorporate the languages of many fields. In the late 1840s, seeking a name for the hordes of tiny microorganisms he was observing in seawater, Müller consulted the philologist Jakob Grimm (1785–1863), who had compiled a dictionary of the German language in addition to his well-known collection of German folktales. Together, the scientist and the linguist devised the term *pelagischer Auftrieb* (oceanic upwelling) for these microorganisms. They have since become known as plankton.[215]

Müller thought carefully not just about the words that he used but also about the abilities of words and images to represent animal forms. "When I call these [sea cucumber embryos] slipper-shaped or horn-shaped," he wrote in 1852, "the expressions themselves give a general, accurate image of their external form."[216] Müller used all of his senses to do physiology, especially when he considered how the human senses worked. In the 1840 edition of his *Handbook of Human Physiology*, he wrote that "an idea [*Vorstellung*] is related to a sensation [*Empfindung*] much as a sign is related to a thing, but a sign that arises only for a specific purpose and whose like [*Art*] is then independent of the sensation."[217] His sensitivity to language helped him to understand how the nervous system worked.

In his scientific writing, Müller refers to literary works. Not surprisingly, he often quotes Goethe. Most nineteenth-century German scientists did, and Müller admired Goethe's comparative anatomical studies. In October 1828, Müller met Goethe, and the two discussed Müller's studies of fantasy images.[218] What is significant is which Goethe Müller quotes. In his 1826 study of fantasy images, he invokes the character Ottilie from *Elective Affinities* (1809). In this novel, Goethe depicts a chemical substitution reaction among four strong personalities. Watchful, self-destructive Ottilie is the magnetic lover who pulls impulsive Eduard away from his sensible wife, Charlotte. When discussing his fantasy images in 1826, Müller incorporates a full paragraph of *Elective Affinities*, describing Ottilie's realistic vision of Eduard while she is falling asleep. "How happy I was," writes Müller, "when I rediscovered that in *Elective Affinities*, one of the people [*Menschen*] whose senses are most powerful also knew how to give verisimilitude to an artistically rich image through intense self-observation." Later in the same essay, Müller quotes Ottilie's journal, in which she records epigrams and insights about human behavior. "One may think any way one likes, but one always thinks by seeing," she proclaims.[219] Müller uses Goethe's novel as though it were evidence, quoting Ottilie not just as if she were a real person, but an authority. He seems drawn to Ottilie—and to Goethe—because both, like him, are master observers. These quotations in his early scientific writing serve as far more than well-worded formulations or examples. For Müller, literature is a museum full of specimens offering evidence for his ideas.[220]

Like a literary writer, Müller does his best to describe conditions so that his readers can re-create them with their own senses. Whether recording conditions in vivo or in vitro, Müller's writing is intensely visual, allowing the reader to see what he sees in detail. In his early study of fetal respiration, Müller observes that a prematurely born rabbit "crept and threw itself anxiously around over the dead body [of the previous fetal rabbit], tried to pull itself to its feet, opened its mouth, and pushed its head ceaselessly against the glass wall of the tank."[221] This observation is not data, nor is it an essential result. Müller includes it so that readers can see the experiment for themselves, drawing them right into his work space.

In a science built on close observation, the ability to communicate one's perceptions plays a central role. Müller's wonderful descriptions of the animals he has observed appeal straight to his readers' visual imaginations. In his writing, a tumor becomes a honeycomb, and a fungal growth in the lungs resembles "the rind of a Swiss or Dutch cheese."[222] A deadly fungus "turns silkworms into mummies."[223] Müller composes his scientific articles not just to report but to re-create and inspire. Like an explorer's diary, his writing offers worlds that only he has seen to the waiting eyes of his readers.

Müller's comparisons become most ingenious when he describes the delicate, quivering embryos he is observing under his lens. Several of these larvae, he notes, are shaped like small ships. Others look like pyramids or easels. In his studies of marine life in 1848–1852, he refers several times to "the rococo larvae of Helsingör," writing that "when regarded superficially, the *Auricularia* look like a coat of arms with rococo ornamentation along the border."[224] Considering the violence of 1848 and his own subsequent collapse, it is not surprising that Müller saw rococo forms in the microorganisms he studied, recalling the aesthetics of a more stable era.[225] In July 1848, when he felt personally responsible for the chaotic Berlin University, he wrote that a starfish seemed to be supporting its larvae on its arms "as one imagines the celestial spheres resting on King Atlas's shoulders."[226]

In his scientific works, Müller appeals not just to the readers' vision but to their other four senses as well. In an 1836 study of the digestive system, he uses these words to describe a small cube of egg white that has been exposed to digestive fluid for twelve hours: "The surface of the small piece of egg white was translucent and gruel-like and crumbled when touched by a finger; the core was cheese-like and had retained its color but was easily crushed."[227] Six years later, examining a sick codfish, Müller reports that its swim bladder contains "a large quantity of yellowish-white, gooey, stringy, sticky material."[228] When introducing the lymph in his *Handbook of Human Physiology*, he describes its smell and taste as well as its appearance.[229] In his eagerness to convey the positions of slugs in sea cucumbers' intestinal cavities, he writes, "One can get the sense of this relationship best by inserting one's finger deep into one's mouth and closing one's lips around it."[230] His scientific communications are structured more like virtual reality programs than simple presentations of data.

As a writer, Müller pursued many of the same goals as realist novelists, offering readers accurate, detailed descriptions of the world he depicted, yet aware that he was creating this world through his descriptions.[231] He began his article on the mysterious appearance of slugs in sea cucumbers much as Balzac might have: "In Trieste, in the Bay of Muggia, there live a great number of Holothuria of the species *Synapta*."[232] Although describing a real rather than a fictional sea, Müller knows that he is telling a story.

As a professional observer, Müller felt duty-bound to communicate all that he experienced, but his letters to his friends and family engage a reader's imagi-

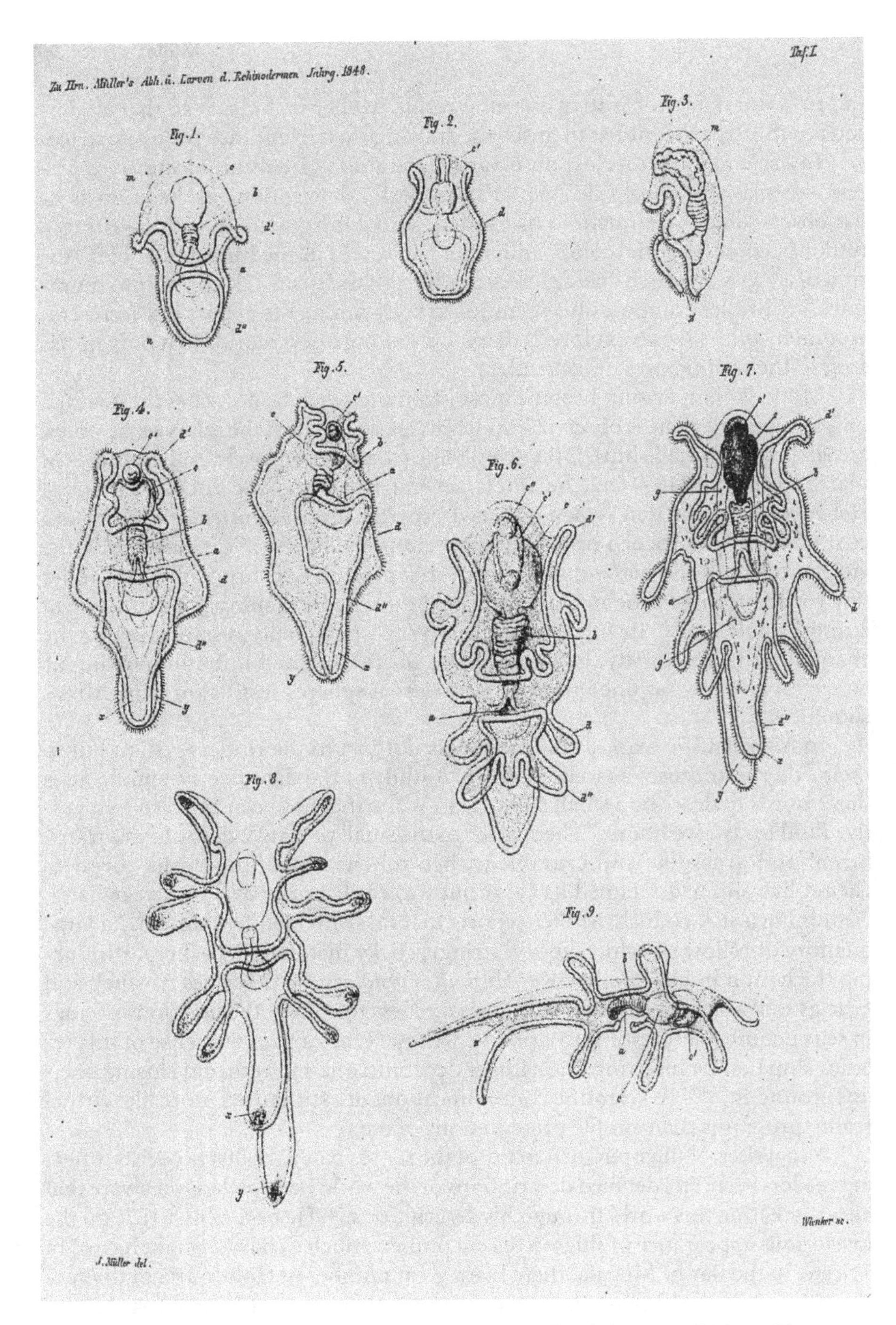

Figure 1.4 Müller's rococo larvae, *Bipinnaria* of Helsingör and Marseille. Source: Johannes Müller, "Über die Larven und die Metamorphose der Echinodermen," *Physikalische Abhandlungen der Königlichen Akademie der Wissenschaften zu Berlin* (1848), plate I.

nation just as his scientific publications do. His most powerful description is the one that he claimed not to want to share: that of his 1855 shipwreck:

> I want to spare you the description of this catastrophe. What was most horrible was that when the fire of the machine hit the water, there was an explosion, and in the same instant, the ship fell precipitously into the depths, the broken bow leading the way, and the whole, heaped-up, howling mass of humanity on deck followed it into the churning water, carrying me along with it. I was pushed back up to the surface; I swam after small, floating pieces of wreckage and looked around for the best one, groping my way from plank to plank until I finally got hold of a floating stairway, on which I quickly lay down, then just as quickly fell off again. I held on hard and thought that I was holding onto this piece of wood for you.[233]

In this, the worst experience of Müller's life, he felt obligated *not* to write, yet he wrote anyway. While he wanted to spare Nanny, his need to record and work through his observations was stronger. Müller wrote as he lived and worked, out of a compulsion to organize and communicate.

Like Müller, his students wrote well, and their skill at creating narratives helped their scientific careers. Their eyewitness accounts of Müller sound so convincing that for 100–150 years, many historians have accepted them without question. As I present these descriptions in subsequent chapters, I will ask readers to keep Müller's voice in their ears. One must remember *why* it was crucial to study plankton for seven hours at a stretch, and to obtain a giant armadillo for the museum. Müller worked unceasingly because he loved nature, a quality he shared with his students. But for the most part, they rejected his notion that comparing animals would reveal the way that life worked. Since they could obtain jobs only with his recommendation, they often stifled their objections, recording them long after the fact. Separating their thoughts from Müller's is a delicate interpretive task, a work of literary as well as scientific analysis.

2

Cells and Selves

The Training of Jakob Henle and Theodor Schwann

An assistant who presides so long in the class and laboratory ends up as the most inconvenient fixture precisely because he is the most convenient.
—Emil du Bois-Reymond to Carl Ludwig, 17 July 1868[1]

In the previous chapter, I described Müller's scientific aims, methods, and values and the conditions under which he and his students worked. In this and the next five chapters, I will present the students' viewpoints, emphasizing each pupil's relationship with Müller, Müller's influence on his science, and his motivation for depicting Müller in a particular way.

For thirty years, Müller attracted some of Europe's brightest young scientists, inspiring them to investigate the forms and phenomena of life. But how did he do it, and how did they respond to his ideas? How did this troubled investigator motivate scientists whose work would move in such different directions?

First and foremost, Müller aroused students' intellectual curiosity. Most wanted to explore life at the cellular level, and as an expert microscopist and physiological experimenter, Müller seemed to be the best possible guide. In 1832, Curator Rehfues of Bonn wrote:

> [Müller] constantly surprises his listeners with the novelty of his thoughts and investigations; he binds them indissolubly to the scientific world, whose entire depths he opens up before them, so that they hang on each one of his words and looks.[2]

For less sensitive, more pragmatic students, a second factor may have predominated: Müller's reputation as the greatest anatomist and physiologist in the German territories and the academic power that he commanded. For twenty-five years, Müller's recommendation settled job searches, and winning his respect could be the key to academic success. A third factor, Müller's magnetic personality, probably exerted a lesser attraction than his exciting research and reputation, but as Rehfues's description shows, his personal style captivated listeners who might not otherwise have been drawn to comparative anatomy. Working with Müller meant more than microscopy or fishing; it meant an intimate relationship with a person obsessed with life's forms.

The German word *Doktorvater* conveys the relationship between Müller and his students far better than the English "adviser." In almost every sense, Müller became the father of his young assistants, assuming responsibility for their lives as well as their research. Sometimes he viewed these helpers as anonymous pawns, as can be seen from his letters to Retzius. Soon after moving to Berlin, he reported that he had one "young man" studying the adrenal gland and "another young man" studying the glands along the intestinal tract. In 1839, he told Retzius, "I've sent a young man to collect cartilaginous fish on the Mediterranean coast," and in 1854 he asked Retzius if he knew "a young man" who could collect fossils.[3] But Müller's most gifted students became his closest friends, apart from Anders Retzius and Nanny.

Between Müller and his young assistants, however, lay the tension that underlies every relationship between teacher and student, making their dynamics complex. Müller sought soul-mates who would help him discover life's plan, but the students sought independence as well as knowledge. Because of the way he handled his "young men," it is difficult to say where his thoughts ended and theirs began.[4] Müller's science was a science of closeness, creating intimate knowledge of individual organisms and relying on equally intimate knowledge of his young helpers.

Perhaps because of Müller's closeness to his students, they represented him as a foil when they tried to define themselves. If we can believe his pupils, the innovations for which Müller is best known—studying bodily phenomena with the techniques of physics and chemistry, and spreading the use of microscopes—were really introduced by the young assistants to whom he was so closely attached. When historians downplay Müller's influence on his students' work, they faithfully reproduce his students' own stories. Only by comparing these self-asserting primary accounts can we understand the experience of working with Müller. This chapter examines the comments of Jakob Henle and Theodor Schwann, both of whom met Müller while he was teaching in Bonn, rejoined him in Berlin, and worked with him until 1839.

Müller's Favorite Student

Born in 1809, lively, talented Jakob Henle was the pride of his family, the oldest child and only son of a rabbi's daughter and a Jewish businessman.[5] In 1820, the Henles moved from Mainz to Koblenz, the city where Müller was raised. A year later, when Henle was twelve, his parents converted from Judaism to Christianity, a step they had long been planning. They left it to their son and oldest daughter to decide whether they wanted to be Catholic or Protestant, insisting only that the children justify their decisions in writing. "Without hesitation," Henle chose to be a Protestant, an interesting move in the predominantly Catholic Rhineland.[6] Henle later wrote to his younger sister, Marie, that he could not reproach their father for converting, since "this way we had a happy, carefree youth and got an education and entered into relations that would otherwise have been impossible for us."[7]

In the early 1820s, many educated German Jews saw conversion as a route to social advancement. For instance, Karl Marx, from the western German city of Trier, was baptized in 1824, at the age of six.[8] The long struggle against Napoleon had created feelings of solidarity among German-speaking people, who gladly embraced their own literature, music, and art. To some degree, the Napoleonic Wars had built support for Protestant Prussia, the only German land outside of Austria that could rival the French militarily. This last trend may explain young Henle's choice of a new religion. His personal letters show an identification with Christianity, never any regret for his lost Judaism. In 1830, he sent his family an extensive Christmas list, including "a nice-looking set of surgical instruments," and he told them how much he was looking forward to visiting them "for Christmas."[9]

Henle possessed daunting gifts for languages and music, and he and Müller developed their relationship partly because of their fine ears. Henle spoke German, English, French, Italian, and Danish. He played the violin, viola, and cello, and had a beautiful singing voice.[10] A frequent guest of the Mendelssohn family, he became a good friend of Felix Mendelssohn-Bartholdy (1809–1847), who shared his love of music and Jewish background.[11] Fifteen-year-old Henle first met Johannes Müller not through science but music, for the young student used to sing duets with Nanny Zeiller at her soirées in Koblenz.[12] Müller had met Nanny while visiting his family in 1821, and until they married in 1827, he often traveled to Koblenz from Bonn to enjoy her company and hear her lovely voice.

Between songs, Müller turned Henle's attention to the natural sciences. As a high school student, Henle had considered studying literature, even becoming a Protestant minister, but he had not thought of anatomy until he met Müller in 1824.[13] When he entered the Bonn University in October 1827, Henle began visiting the newlywed Müllers regularly, maintaining their musical relationship. Scientific friendship with this lighthearted young student may have helped Müller recover from the depression he suffered in the summer of 1827.

From the time that Henle began studying anatomy, he had found himself fascinated by the purposefulness (*Zweckmässigkeit*) of the human body, the suitability of its form to its functions. In 1827 he wrote to his parents:

> What to other people seems dry in my subject, for instance, the basic, necessary, almost painstaking study of all the bones and muscles, fills me with astonishment and delighted wonder at the extraordinary functionality [*Zweckmässigkeit*], which can be followed right down to the smallest parts. I know of no more beautiful nourishment for fantasy than mentally composing the beautiful structure of the human body out of all the individual bones and muscles.[14]

Henle shared Müller's passion for putting things in order, and he was attracted by his teacher's vision of comparative anatomy. The aspect of Müller's science that drew him was the quest for patterns stemming from nature philosophy. In 1829 Henle wrote home:

> The thought of a comparative anatomy, which in Müller's system explains the different forms and characteristics, seized me quite powerfully. It is certainly the highest aim of comparative anatomy to be a special, reasoning anatomy which demonstrates the progress of all forms, from their simplest development, throughout the array of animals, explaining each fixed step in any animal, and at the same time leaving no curiosity unnoticed.[15]

Henle studied medicine in Bonn from 1827 to 1830 and from 1831 to 1832, spending the spring semester of 1830 and winter semester of 1830–1831 in Heidelberg. Because Müller conducted experiments in his own home, Henle participated in his work as well as his social life.

As a research partner, Henle made a delightful companion. Almost everyone who ever worked with him commented on his lively sense of humor. In letters to his family, Henle laughed as hard at himself as at anyone around him, and he never took his science too seriously. Writing home from Heidelberg, he depicts himself before a large stack of books, smoking a pipe and wearing a ragged nightshirt, uncertain whether it is Wednesday or Saturday.[16] He refers to his studying, which Müller saw as a holy calling, as "working like an ox."[17]

One of Henle's letters home from Bonn reveals as much about his living situation as about his personality:

> Money! Money! Money! I have practically nothing left, and I owe my friend Mathieu 10 taler. Money! I've had to pay 46 taler for lecture fees alone, a whole lot for books and twenty for household expenses. Money! Otherwise I'm doing fine, but Money! Money! Money! [The letter continues in this vein for two long paragraphs, interspersed with cries of 'Money!"] . . . Next time I'll tell you about my classes. Money! Money! Money! Cloth! Bearskin! Calendars! Biot's physics! Ham! Sugar! Money! Money! Money! Money! Money! Ham! Money! Cloth! Money! Bearskin! Money! Calendars! Money! Physics! Money! Sugar! Money! Your ever-loving Jakob.[18]

As might be expected, Henle's father did not respond well to such letters. After one conflict during Henle's stay in Heidelberg, the student offered the senior businessman a detailed eight-week budget including 14 taler for rent, four for laundry, four for borrowing musical instruments, and two for boot polishing.[19] As a student, Müller had suffered the same need for money, but he had lacked a father who could fulfill it. As collaborators, the young professor and student understood one another well, but their class origins gave them different social expectations.

Like Müller, Henle participated actively in the Bonn University *Burschenschaften*, which in the late 1820s still attracted idealistic students.[20] In the spring of 1828, he became interested in these patriotic fraternities and remained involved with them for two years, embracing their nationalistic ideals. Unlike his teacher, Henle also accepted their tests of manhood and honor, fighting duels in December 1828, February 1830, and March 1830.[21] Why such a bright, vivacious young scientist would repeatedly risk his life is unclear. At the time, duels among *Burschenschaft* members were common. Facial wounds made political statements, and students inflicted them intentionally, stuffing the wounds with horsehair to make the scars bigger.[22] Although good with a knife at the dissecting table, Henle did considerably less well in the open air. After winning one duel, he jabbed his sword triumphantly into the ground to celebrate the victory—and ran it through his own foot. In the contest of March 1830, he received a deep wound to his right cheek that branded him for life. The recovery forced him to spend two weeks in bed, and his family was horrified.[23]

Figure 2.1 Jakob Henle, with his head turned to hide the scar on his right cheek. Source: Friedrich Merkel, *Jacob Henle: Ein Deutsches Gelehrtenleben* (Brunswick: Friedrich Vieweg, 1891), frontispiece.

It is not known who advised Henle to leave Bonn for Heidelberg in the spring of 1830, about a month after he had recovered from his facial wound. Heidelberg offered extensive opportunities for clinical training, but going there meant interrupting his research with Müller. Possibly his worried family effected the transfer, but it is more likely that Müller himself sent him packing. From the beginning, Henle had heeded Müller's advice on most academic matters, including his choice of career, and the young professor may have sent his favorite student away to save his life.

By this time, Henle had internalized Müller's love of natural forms and needed no urging to study. He wrote to his family that he rose between four and five and worked until breakfast. In July he asked his sisters for 10 gulden to buy "a small case of instruments for dissecting insects and other cute little animals."[24] In Heidelberg, Henle served on numerous wards, where he sought cadavers to dissect. Among those he autopsied were a friend's maid who had just died of brain fever and a baby he had "cured to death," then bought from the mother for a taler.[25] Henle criticized his professors for focusing too much on practical matters without thinking through their theoretical systems, an attitude very similar to Müller's.[26] Like his teacher, he believed that the best science combined philosophical theorizing with painstaking observations. Henle stayed in Heidelberg for a year, returning to Bonn in May 1831. From this time on, he turned away from his *Burschenschaft* friends and devoted himself fully to Müller and his anatomical research.

Scientific Intimacy

Despite the differences in their backgrounds, Henle and Müller harbored similar scientific ambitions. Their determination to advance academically first drew them together, then pulled them apart. Henle's letters suggest that the fun-loving student may have been even more career conscious than his professor. When Henle selected a thesis topic, his and Müller's urges acted synergistically, revealing a lot about the way their academic goals affected their scientific choices. On 3 June 1831, Henle told his parents that he would write his dissertation on the bones in birds' heads because "up to now the subject has been dealt with very little and can make a good name for me, if I'm willing to work through it systematically." A week later, he wrote that he had abandoned the topic because "Müller has suggested another matter that is of more general medical interest and has been worked on even less." By 30 June, he informed his family that he had changed his subject again, since he had just made an anatomical discovery [of the pupil membrane] "that, if I'm the first to present it to the medical world, will certainly make a bit of a name for me."[27] For his doctoral research, Henle wanted a subject that would bring scientific glory as well as intellectual satisfaction, and Müller helped him to find one.

During June 1831, while Henle was selecting a thesis project, he wrote home:

> The most beautiful evenings are the ones I spend with the Müllers. Every day, Müller is friendlier and more sincere with me. . . . On the whole, the

> tone between us is much less forced. Müller lets me take part in all of his projects and ideas, and Nanny often tells me how happy it makes Müller that he again has someone who can take a truly sincere interest in him and his work. Now I believe all these assurances I once received with such mistrust.[28]

This initial "mistrust" of Müller never emerges in Henle's extant letters; one encounters it only here in retrospect, after it has supposedly been eliminated. Apparently, Henle had at first doubted Müller's willingness to share all of his scientific thoughts, and may have suspected that the senior scientist—Müller was eight years older—wanted to exploit him. As Nanny's comments indicate, Müller experienced the friendship with the budding scientist as a joyous return to life after his crippling depression of 1827. Witty, perceptive Henle became the ideal companion, one who understood his work and could share every aspect of his existence.

From the beginning, Henle seems to have known that sticking with Müller would bring academic success. In the same letter in which he announced his final choice of research topic, Henle listed his reasons for studying anatomy. The first was "the friendship of Müller, which can be such an advantage in getting started." After describing Müller's invitations to use his instruments and books, to work "practically under [his] own eyes," and to investigate a problem suitable for submitting to a journal, Henle wrote: "So I'll let him introduce me to the world, and hang my name on his, so that he can drag me forward." Still, his tone remains playful. Once he has completed his dissertation and taken his medical exam, he assures his parents, "I'll look for a nice little position, try to advance as quickly as possible, and marry a young, beautiful, clever, rich woman who speaks French, plays the piano, and knows how to handle horses."[29]

Müller and Henle were right in recognizing the importance of the pupil membrane, and Henle did well to shuffle topics until he found one of real significance. When Henle defended his thesis on ocular anatomy in August 1831, one examiner said it was the best dissertation he had ever seen.[30] Once published, the study attracted considerable attention, and prominent anatomists disputed the results.[31]

When Henle passed his doctoral exam, Müller offered him two gifts: a trip to Paris and the privilege of calling him *Du*, the German "you" for family members and close friends. At the time, a *Du* relationship between an adviser and ex-student was unusual. Müller did not start saying *Du* to his friend Anders Retzius—a colleague of his own age and rank—until he had known him for six years.[32] He addressed his former student Emil du Bois-Reymond as *Sie*, the formal "you," even when the physiologist had worked as his assistant for almost ten years.[33]

When Henle learned of the Paris trip, he wrote to his sister Marie, "I felt as though my soul had been struck by lightning."[34] For two months, he and his mentor traveled together on a sort of scientific honeymoon. Nanny did not accompany them. In Paris, they shared a suite in a hotel near the Botanical Garden. At the nearby Natural History Museum, French scientists opened their anatomical collections and workrooms to the Berlin visitors. After meeting Müller, Cuvier instructed his assistant Laudrillard, "Give this gentleman everything he wishes."[35] The closeness of Müller and Henle's relationship at this time comes through in Henle's amusing description of Alexander von Humboldt's visit. Just as Müller's patrons had presented him to the great explorer, Müller introduced his young student, Henle:

> You should have seen the way we equipped our poor room to receive our honored guests. First we put everything into the most beautiful order. But then we found that here, a weak spot on a chair; there, a rip in a tablecloth, had to be covered by a book, and so we finally decided on a system of Charming Disorder, vowing that we would rather look sloppy than poor. . . . The Minister was seated on a night- or reclining throne covered with flowery red velvet. The other gentlemen got the same, only stuffed with feathers, so that they settled to a height corresponding to their eminence. Müller and I set up camp on two cane chairs with seven legs between them. We were both dressed all in black, including our linen, since our perfidious *femme blanchisseuse* [laundress]had left us in the lurch.[36]

Henle's witty description shows the intimate conditions under which the teacher and student lived, and the fun they must have had dealing with social challenges in the foreign capital. In social situations like this, the jovial Henle served Müller as a catalyst, putting his introverted mentor at ease. At times, however, the task was more than Henle could handle. During the Paris trip, he wrote to his friend Mathieu (who by then had married Henle's sister Marie), "For me, Müller is a little too square for hanging around in bars [*Müller ist mir für die Kneiperei doch etwas zu solid*], and I avoid going with him to visit groups of people."[37] Müller seemed happiest when they interacted one-on-one.

After the research trip, the two young scientists continued to collaborate as intensely as ever. In November 1831, Henle wrote, "I'm working a great deal with Müller. Every day he's either here, or I'm there."[38]

In April 1832, when Henle's doctorate was officially awarded, the medical faculty made sure to celebrate the popular student's success properly. The ceremonial procedure suggests the nuptial quality of the relationships involved. In awarding Henle his doctorate, the dean placed a ring on the young man's finger. Henle wrote that he appeared "in Müller's short pants and my own legs, which reaped general admiration." At the lively masquerade that followed, "professors, doctors, students, and artists soon ended in each other's arms." Müller told Henle it was the best time he had ever had. Describing his last days in Bonn, before he went on to Berlin, Henle wrote that "with Müller . . . I grew ever more cordial, and in the end our fondness finally became something like a pledge of eternal friendship [*unsere Zärtlichkeit artete endlich gar in einem Schmollis aus*]."[39]

Müller expressed his feelings for Henle in direct, passionate terms. After receiving his doctorate, Henle left to prepare for his Prussian state medical exam in Berlin, and in his first letter to the absent student, Müller wrote:

> It is now becoming serious, my dear heart's own Henle [*mein lieber Herzens-Henle*], that you should go. I miss your company a great deal, for it has grown so dear to me, but if I just think that you, in leaving us, are making great strides and collecting new treasures, then I am satisfied and am happy to hear from you, often and copiously, from afar. And I don't want to be kept waiting long for an answer. Just be sure to hold us dear in your thoughts. So please move on [to Berlin] accompanied by a thousand good wishes, by our love, and followed by our warm, caring, but not too anxious sympathy, and come back to us like a good friend with your old devotion.[40]

When Müller says "we," he means himself and Nanny; Henle and Müller were never lovers. But their relationship extended far beyond shared assumptions about theories, methods, and research topics. As an emotional bond, it fulfilled Müller's longing for scientific intimacy; the extroverted Henle, however, does not appear to have felt the same need. The connection between them was intellectual, social, sentimental, and, to both men, extremely useful.

Three Rhineland Scientists in Berlin

In early 1831, while Henle was in Heidelberg, Müller had found another young companion. On a walk in the countryside, he had encountered Theodor Schwann, a twenty-year-old medical student who had been attending his physiology, pathology, and comparative anatomy lectures. Like Henle, Schwann came from a far wealthier background than Müller; his father ran a successful publishing house. The fourth of thirteen children in a devout Catholic family, Schwann had studied at a Jesuit school in Cologne before entering the Bonn University in October 1829.[41] In one of his physiology lectures, Müller had impressed Schwann with his discussion of Bell's Law, the notion that spinal nerve roots initially directed toward the back carry sensory information, whereas those initially going down toward the belly carry motor signals. Schwann suggested some experiments that might test this hypothesis, and Müller responded that he was performing exactly those experiments on frogs. As Schwann recalls, "He invited me to come to him as often as I wanted to do experiments."[42] That spring and summer, Schwann worked intensely with Müller, studying the organization of sensory and motor nerves in frogs. What excited Schwann was not Müller's comparative anatomy but his experimental physiology, so that from the beginning, he and Henle were drawn to different aspects of Müller's science.

But like Henle, Schwann soon interrupted his research with Müller to study at another university. In the fall of 1831, he transferred to the University of Würzburg so that he could work with Johannes Schönlein (1793–1864), a respected clinician who advocated the use of new diagnostic tools such as the stethoscope.[43] The introspective, deeply religious Schwann was not involved in any fraternities, so the reasons for his move seem to have been purely academic. During his year in Heidelberg,

Henle had also wanted to study with Schönlein in Würzburg, but Müller had advised against it. Henle had protested to his parents, "You must certainly see that the reasons he gives against [studying in] Würzburg are really no reasons at all, but almost personal prejudices."[44] It is interesting that Müller let Schwann go to Würzburg in spite of these "prejudices" when he had advised Henle to stay away. Müller respected Schönlein's achievements, for eight years later, he would lead a campaign to bring the distinguished clinician to Berlin. In 1830, the University of Würzburg was considered politically active, and Schönlein himself was a known liberal.[45] Müller may have feared that Henle would return to the *Burschenschaften*, but more important, he missed Henle's company. He and Schwann never developed the same kind of emotional tie.

Henle moved to Berlin in May 1832, almost a year before Müller. In November, when the anatomist Rudolphi died and Müller campaigned to win his vacant chair, Henle sent him any information that might give him an edge. When Müller arrived in April 1833, having secured the position, Henle moved from his room on Dorotheenstrasse into the house of the recently deceased philosopher Hegel with the Müllers. On the second floor, Henle lived as a lodger with Hegel's widow; on the third floor lived Nanny and Johannes.[46] Considering how extensively the two scientists collaborated, in a time before telephones it made sense to live in such proximity. But the choice reflects their attachment as well as their need to communicate. Müller wanted Henle as close as possible.

Soon after his arrival, Müller made Henle his helper at the Anatomical Museum. The museum job paid only 10 taler per month, at least three of which must have gone for rent. In the fall of 1834, when Müller's prosector D'Alton took a job in Halle, Müller made Henle prosector, which brought him a salary of 480 taler per year. In addition, he appointed Henle editor of the *Archive for Anatomy and Physiology*, so that Henle became responsible for overseeing all the articles published. This time-consuming but educational work paid an additional 200 taler. The position teaching anatomy at the Art Academy, which paid 400 taler and for which Henle had also hoped, went to Müller's colleague Froriep, whose father, a respected physician, had lobbied the cultural minister.[47] Despite Müller's help, Henle complained to his family about the restrictive, competitive university structure, which pitted highly gifted young scientists against each other in political battles for scarce jobs.[48] No matter how talented a scientist was, he could not become a lecturer until he had completed his *Habilitation*, a minimum of three years after receiving his doctorate.[49] Since there could be no more than one lecturer or professor for any given subject, bright young researchers could not support themselves through teaching until those just ahead of them had moved on.

In October 1834, when Müller made Henle prosector, he appointed Schwann museum helper.[50] Still attracted by Müller's physiology, Schwann had left Würzburg for Berlin as soon as Müller had arrived. As his two most trusted assistants, the new Berlin physiologist had chosen two *Rhineländer*, two young men with whom he had already worked closely.

As scientists from the Rhine region, Müller, Henle, and Schwann were all foreigners in Berlin. According to German stereotypes then and now, *Rheinländer* are jovial and lively; Prussians, severe, time-obsessed, and militaristic. While these cultural images distort reality—Müller's personality shows the limits of the Rhineland stereotype—there can be no doubt that common feelings of uprootedness united the three scientists. In Henle's first reference to Schwann in his letters home, he mentions that the younger student is "also a *Rheinländer*."[51]

Unlike Müller, Henle fulfilled the image of the fun-loving *Rheinlander*. Despite his limited income, he played in a string quartet, went to the theater, and attended parties and balls. At a "brilliant costume ball at Professor Schlesinger's" Henle appeared in drag:

> Straight from the hands of the court hairdresser, I sallied forth in a blond wig with feathers and flowers in my hair and an assembly of curls half a foot in diameter over each sideburn. In addition to this, I was made up with a little bit of red and a whole lot of white, with a striking beauty mark on my forehead. I had a pretty low-cut *schali*-dress from banker Magnus's wife, who is my height and fortunately, at present, is expecting. Around my neck and throat, *à la Lippelehnchen*, a coral necklace heightened the dazzling brownness of my complexion. Under the dress I wore a hoop-petticoat, and under the petticoat, some sofa cushions. I had white silk stockings, lace-up shoes with ribbons, a buckle, a brooch, a cambric handkerchief, a fan, and what was most wonderful, no mask. It took a while for people to recognize me.[52]

Henle was writing for his family, especially for his sisters, whom he knew would be interested in the details of his toilette. Clearly, though, he delighted in the splendor of his costume. He thoroughly enjoyed appearing in Frau Magnus's maternity dress.

While the trio of Rhineland scientists shared a powerful interest in cells, no three personalities could have been more different. Friedrich Bidder, who worked with Henle for two semesters in a small room at the Anatomical Institute, wrote:

> Whereas in the room occupied by Müller and Schwann, silent, unbroken calm usually reigned, interrupted at most by deep sighs, in the room where Henle and I worked, it often got quite loud, partly because of our manipulations of the cadavers with saws, hammers, and chisels, but more often because of the delightful humor and sparkling wit with which Henle accompanied this work. Not even Müller could keep from breaking into hearty laughter once Henle really got going.[53]

Bidder's eyewitness account of Henle and Müller seems to speak for itself. As long as Henle was in the group, Müller had a good time. But Bidder's description cannot be trusted. In this passage, he portrays himself as the lively companion of Henle, but in 1840 Henle called Bidder "a pretty nice, poor, sick person who needs a cure in Ems . . . [and] is allowed to speak only a little and very softly."[54] Considering how much Bidder's self-representation differs from Henle's depiction of him, Bidder's portraits of Henle and Müller may be equally misleading. Bidder's and Henle's references to Schwann, on the other hand, coincide exactly. In a letter of October 1835, Henle called Müller's sighing young assistant "a good, quiet, hardworking person whom I hope to get to know better by living with him."[55] Like Müller, he saw no boundary between living and working space, expecting to share both with his collaborators.

Homemade Science

In October 1835, Henle got the chance to know Schwann better. Forced to leave his room downstairs from Müller, maybe because Frau Hegel's son had returned home, he moved into "a less than third-rate guesthouse" at 66 Friedrichstrasse. Its owner, Herr Hilgendorf, had divided the floors over his pub into ten little apartments that he rented to less than affluent customers. In jest, Henle called the place "the Hotel Hilgendorf." When Henle's brother-in-law Mathieu came to Berlin to take a Prussian civil service exam, he moved into an adjacent Hilgendorf room, and when he left, Schwann replaced him as Henle's neighbor. "Neighbor," though, is not quite the right word, for their living situation more closely resembled that of a dormitory than that of a hotel or boardinghouse. Whenever a room opened up, Müller's assistants arranged for an old friend from Bonn to move in, so that a colony of mutually supportive, high-spirited Rhineland students soon developed.[56]

Though the assistants had their own rooms, they spent so much time in each other's living quarters that in 1882, in his memorial address for Schwann, Henle retained a detailed picture of Schwann at the Hotel Hilgendorf:

> I see him there before me, a young man of less than middle height, with a beardless face and an almost child-like, always cheerful expression; plain, dark-blond hair that somehow always stood up; and a fur-trimmed night-shirt. [He lived] in a narrow, rather dark rear room on the third floor of a less than second-rate restored building (the corner of Friedrich- and Mohrenstrasse), which he often failed to leave for days at a time. [He was] surrounded by only a few books but by innumerable glass bottles, flasks, test tubes, and homemade primitive apparatus.[57]

Henle's own room was just as much a work space. One visitor described it as follows:

> Jars and containers of preparations and books of every shape and form; live crabs, leeches, and frogs lie around in orderly disorder, lending a unique charm to the pleasant still-life around us. A muscle-man, a bust of Goethe, a porcelain nymph, and a full-grown skeleton receive us cordially, proclaiming: "only one man can live here."[58]

In these rooms, Müller's assistants studied the forms of cells, determining how animals moved, breathed, and digested. Together, they walked twenty minutes twice a day between "home" and "work"—but since they experimented at home and ate at the institute, these concepts were hard to distinguish.

Interestingly, Henle often called his work space at the institute his "office" (*Büro* or *Amt*). He wrote to his family that he worked at his "office [*Büro*]" until four in the afternoon, eating his midday meal in his "room [*Zimmer*]" at 4:30, and that he spent forty minutes a day traipsing back and forth between his home and his "office [*Amt*]."[59] Since Müller never used this word for their work space, Henle's use of the term may reflect the difference between his and his adviser's class backgrounds. Henle's letters to his father were directed to a successful businessman with distinct concepts of "office" and "home."

Henle's recollection of the Berlin Anatomical Institute closely resembles du Bois-Reymond's in his memorial address for Müller. Describing it in 1882, Henle may have merged his own memory of it with the physiologist's phrases of 1858. In the 1830s, recalls Henle, the three Rhineland scientists worked in "that euphemistically labeled Anatomical Institute behind the Garrison Church."[60] Whether Henle, Schwann, and Müller worked at the institute or the museum depended on which lectures Müller was giving and what specimens were available. Henle remembers days on which the three worked at the institute from early morning until late afternoon, stopping around noon to enjoy a second breakfast with wine in "the director's room [*Zimmer*]."[61] In the winter semester, Bidder recalls working in Müller's *Kabinet* from eight until one, then returning at two to hear Müller's anatomy lectures. In the summer, Bidder and the others dissected in "the spacious halls" of the Anatomical Museum.[62] Henle's and Bidder's use of the word *Zimmer* for both the assistant's own living quarters and Müller's workroom suggests how the different spaces played the same roles. Anatomy and physiology flourished in spaces for teaching, exhibiting, and performing the experimenters' own life functions. If these young scientists became intimate friends, it was partly because their working conditions demanded it.

The Favorite Student in Jail

Working hard, developing his own ideas about how anatomy should be studied, Henle began to believe that he deserved his own professorship. Then, to his horror, he was offered a job as anatomy professor in Dorpat. In 1834, this Russian city east of the Prussian border struck the Berlin scientist as Siberia. Its university, founded in 1802 by Czar Alexander I, employed many gifted German scientists who later moved on to Prussian schools. As early as March 1834, Henle knew that the Dorpat administration would ask Müller to recommend a candidate and that he would probably be offered the job. The position paid 1500 taler per year, but Henle feared that he would be unable to do research in this primitive setting.[63]

Bidder's memories of the Dorpat Anatomical Institute show that Henle's concerns were probably right. In Dorpat, there was "no usable microscope" and no money to buy one. In the absence of a dissecting room, the anatomy professor worked in the same space in which the museum collection—a few skeletons—was displayed. Out of his own salary, the scientist was expected to buy the alcohol for preserving new specimens, which was expensive because of a tax for transporting it through the city gates. This tax could be lowered if the alcohol were rendered undrinkable—by the introduction of a body part, for instance—but the mysterious disappearance of alcohol in transit seems to have been an ongoing problem.[64] The letter offering Henle the job indicates his teaching load by describing the duties of his predecessor, Martin Heinrich Rathke: three to five hours per week of physiology lectures, five of pathology, an unspecified number in zoology, and the direction of the Anatomical and Pathological museums.[65]

In a spirited move, Henle used the offer to improve his position in Berlin, asking Cultural Minister von Altenstein for a 100 taler raise. The minister asked him why he didn't just take the job, and he replied that it would prevent him from doing research. Altenstein then asked, "What does Professor Müller think?" and Henle answered that Müller "would hate to see me go [*dass er mich ungern scheiden sehe*]." The minister said that he would see what he could do.[66] But before Altenstein could respond, Henle lost the unwanted job. On 2 July 1835, he was dragged from his bed and thrown in jail.

The Prussian police arrested Henle because of his *Burschenschaft* activity four to six years earlier. Officially, *Burschenschaften* had been forbidden since 1819, but they had persisted until 1833, when an attack on the main guard post in Frankfurt had led to a government crackdown.[67] Their combination of nationalistic and liberal values threatened the Prussian administration, and in a sweep during 1833–39, the government targeted former members whom they regarded as potential troublemakers. When the University of Dorpat learned of Henle's arrest, they withdrew their offer "because the government regards foreign *Burschenschaften* as dangerous."[68] Henle confessed to his parents "how unbounded was my joy at this sad event."[69]

But the charges against Henle were no joke. From prison, he wrote that he had known for six months that he had been under investigation; he had been told when he had applied to become a lecturer.[70] Still, he was shocked when three policemen awakened him from a deep sleep and led him away. Henle remained in jail from 2 July to 28 July 1835, but even this did not crush his sparkling personality. To reassure his parents, he wrote that he was living no worse behind bars than he did normally, and that it was better to be restricted by dead bolts than by gout. He called his time in prison "the Babylonian captivity."[71] Reportedly, a guard at the Hausvogtei prison told Henle's brother-in-law Mathieu, "That was a good guy [*Das war eine gute Seele*]. You can't imagine how we miss him here. If only we could have him back!"[72]

They almost did. After Henle was released, the investigation continued, and in April 1836, he was suspended from his job as prosector and forbidden to lecture at the university. Then, on 5 January 1837, he was sentenced to six years in jail.

From the time of Henle's arrest, Müller had fought for his release, using his influence with Minister von Altenstein. Henle's biographer claims that Müller was too naïve to handle such a sensitive political situation and that he thought he was doing enough simply by saying that Henle was indispensable as his prosector.[73] But, indirectly, Müller did save Henle. Müller's admirer and protector Alexander von Humboldt—to whom he had introduced Henle on the Paris trip—intervened, and on 2 March 1837 obtained a pardon directly from the Prussian king. Later that year Henle was able to resume his job as prosector, and in 1838, he began holding lectures. When Henle was released from the Hausvogtei on 28 July 1835, the "naive" Müller told him that his four weeks in prison had done him more good than if he had written a thick book, for he now had the attention of people "in all circles."[74]

But why did the Prussian government target Henle in the first place? Müller himself had been just as active in the Bonn University *Burschenschaften*, and at least 600 people were under investigation.[75] By 1835, however, Müller was becoming internationally known as a physiologist, and his arrest would have embarrassed the Prussian regime. The young lecturer Henle, on the other hand, was relatively powerless. And with the prominent dueling scar on his cheek, he looked like a *Burschenschaft* member. His connection to Müller saved his career and maybe his life, increasing his debt to his mentor.

While the government dissected Müller's dissector, science continued. In August 1837, perhaps to celebrate his assistant's liberation, Müller invited Henle on another research trip, this time to Leiden and London to study museum collections of cartilaginous fishes. Once again, they performed all of their daily activities together, even writing their letters home at the same time.[76] While they were in Dover, Henle expressed a great desire to see Edinburgh. Müller wrote to Nanny, "He was only too eager to see the Scots with their naked legs. As he was taking his clothes off, I suggested that he take a good look at himself and make do with that instead of the Scots."[77] Henle and Müller never went to Edinburgh, but they shared a lasting fascination with animal forms.

Henle never lost his attraction to Müller's comparative anatomy. Like his teacher, he took a great interest in the vocal system, perhaps because of his love of singing, and he performed his own research on the structure of the larynx. His *Comparative Anatomical Description of the Larynx* (1839) incorporates what he most loved about Müller's comparative anatomy. Systematically, he studied "the developmental steps" of this organ along "the step-ladder of organisms," not expecting any "continuous progress," but rather "growth in all directions, as if from a common central point."[78] This analysis of laryngeal development from reptiles to mammals was pure research, not a means to any higher scientific or philosophical end. In the late 1830s, Henle approved of the direction that Müller's science was taking, with its emphasis on observation, collecting, and comparative analysis.

Microscopy in the 1830s

Even though Henle and Müller's focus, by this time, was primarily anatomical, and Schwann's, physiological, the three agreed that microscopy was crucial for learning what tissues looked like and how bodies worked. From 1834 to 1839, Müller and his young assistants scanned every sort of animal tissue, recording the diverse structures of living cells. When Henle published General Anatomy in 1841,

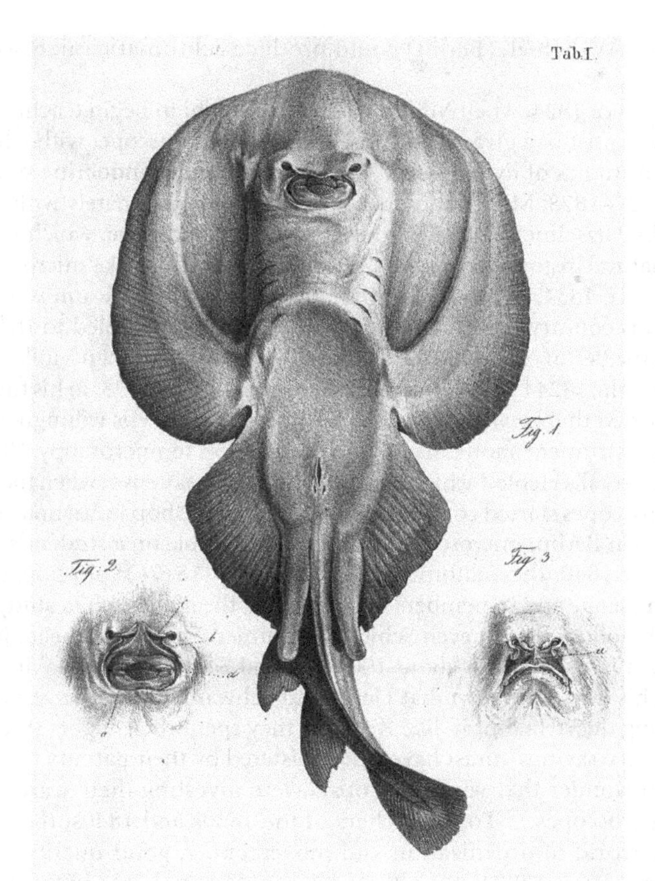

Figure 2.2 Like Müller, Henle took an interest in marine life and had a gift for conveying animals' personalities. Depicted is *Narcine*, a new species of electric ray. Source: Dr. F. G. J. Henle, *Ueber Narcine, eine neue Gattung electrischer Rochen nebst einer Synopsis der electrischen Rochen* (Berlin: G. Eichler, 1834), plate I.

its pages of practical advice on how to use microscopes won great appreciation and respect. This knowledge and the discoveries he made with it were the fruits of his fifteen years of training with Müller.

In the mid-1830s, Müller and his assistants benefited from a new type of microscope that eliminated chromatic aberration, the appearance of rainbow patterns along the edges of structures that prevented scientists from obtaining sharp images. The Fraunhofer firm in Munich had offered achromatic instruments since 1811, but the best, most widely imitated new model was the Paris optician Charles Chevalier's microscope of 1824.[79] By the early 1830s, the firms of Georg Oberhauser (Paris), Simon Plössl (Vienna), and Philipp Heinrich

Pistor and F. W. Schiek (Berlin) could produce achromatic microscopes relatively cheaply.[80]

In the fall of 1824, when Müller had left Rudolphi to begin teaching in Bonn, the Berlin scientist had given Müller a Fraunhofer microscope, with which he began his own studies of living tissue. While observing the endocrine system in the winter of 1827–1828, Müller had used this instrument alternately with a finer one produced by Utzschneider and Fraunhofer. This microscope, which belonged to the Bonn natural history department, had allowed him to make micrometric measurements.[81] In 1834, Bidder wrote that the microscope Schwann was using, left behind by his countryman C. E. von Liphardt when he traveled to Italy, was the only one at the Berlin Anatomical Institute.[82] If this was true, then Müller must have been keeping the 1824 Fraunhofer instrument at home. By 1838, in his tumor study, Müller reported that he was using a Schiek microscope.[83] His willingness to invest in a Schiek instrument shows his ongoing dedication to microscopy. "Those were happy days," recalls Henle, "which today's generation may envy, when the first good, handy microscopes started coming out of Plössl's workshop in Vienna and Pistor's and Schiek's in Berlin, microscopes that were affordable on a student's savings."[84] In November 1840, the anatomist Albert Kölliker (1817–1905) bought his own Schiek microscope and remembers sitting up half the night with it, studying earthworms and mollusks.[85] But even Schiek's instruments were not cheap. In 1834, an achromatic microscope cost about 100 taler, and Schwann was making 10 taler a month.[86] This does not mean that Henle and Schwann did not have microscopes in their living quarters, but if, like Kölliker, they spent their nights studying cells, their "students' savings" must have been bolstered by their parents.

It is no wonder that young scientists were investing their scarce money in Schiek's microscopes. [87] To researchers of the 1830s and 1840s, they revealed a fascinating world of organization and movement. A good-quality instrument could magnify images 500 times, although it could be used only on thick, unstained sections.[88]

In his highly acclaimed anatomy textbook, Henle wrote that "general anatomy today is mainly microscopic anatomy," and he included an informative section on microscopy, optical illusions, and the preparation of specimens.[89] In this practical section, he explained why a compound microscope that combined the magnifying power of several weak lenses offered a sharper image of a larger field than a simple microscope with one powerful, highly curved lens. He pointed out that "the most essential point in microscopic work is the use of light," a fact that might escape beginners. Direct sunlight, he warned, could create no end of illusions because of the way it was reflected from uneven surfaces. The resulting stripes could be mistaken for cell membranes. But Henle also attacked those anatomists who rejected microscopy because of the illusions it could create, commenting that most errors were not caused by bad optics but by bad judgment.[90] The microscopy that he had learned with Müller led him to his scientific achievements.

In his own microscopic investigations, Henle studied the body's membranes, loosening the tissue with water, scraping it with a scalpel—or even a fingernail—and examining the freed material. In 1837, once his pardon was secure, he completed his *Habilitation* on the absorptive surface of the intestine. Henle's 1838 article "On the Extent of the Epithelium in the Human Body" reads like a travelogue, offering descriptions of almost all the body's surfaces. Henle takes the reader on a journey, describing cells in the tear glands, nostrils, tongue, salivary glands, tonsils, esophagus, stomach, penis, vagina, mammary glands, and brain. The trip is not rigorously directed except for a general movement from top to bottom and back again. Along the way, Henle tells readers how to scrape a moist membrane with a scalpel and examine the freed material under a microscope, so that they will see "cells laid out next to each other as in the tiniest mosaic."[91] Although the study is purely descriptive, it works toward an important conclusion. Based on his observations, declares Henle, the skin and moist membranes consist of cells. They are not secretions of tissues inside the body.

Henle's study of pus and mucus, which also appeared in 1838, continued these observations of epithelial tissues. In this article, he stressed how important microscopic analysis had become for the study of disease, since it allowed investigators to characterize pathological products such as pus. He distinguished three arrangements of cells along the body's surfaces: those resembling brick walls, clusters of cylinders, and mops of wiggling cilia. By observing vast numbers of different tissues, Henle attempted to determine why the body starts producing pus when wounded, or excess mucus during a cold. Like other pathologists of the 1830s, he speculated that the leakage of blood into the tissues caused pathological growth, but his extensive observations also showed him that cells with nuclei were not unique to epithelial tissue and could be found all over the body.[92]

Henle is best known today for microscopic investigations he conducted during his last year with Müller, published as *Studies in Pathology* (1840). In one of this book's four essays, "On Miasmata and Contagia," Henle asserted that some diseases are caused by microscopic living organisms. This work encouraged the scientists who developed germ theory in the 1870s.[93] Henle never substantiated his claim with systematic experiments, and like most nineteenth-century physicians, he believed that many diseases were caused by local concentrations of bad soil, water, and air. His work stood out, however, for its creative imagery and argumentation. If a thorn pricks someone's finger, he wrote, causing inflammation and pus formation, and then pricks the finger of a second person, causing the same pathological response, this does not mean that the thorn is transferring the disease. It is transferring the *cause* of the disease.[94] In this research, Müller's combination of theorizing and observing served Henle well, but in the late 1830s, their friendship dissolved. During the same years, Schwann would also distance himself from Müller, but for very different reasons than Henle.

Relentless Logic

Quiet, serious Theodor Schwann used the microscopes of the mid-1830s better than just about anyone, but he and Henle were attracted to different Müllers. Both welcomed their teacher's call for minute observation, but whereas Henle delighted in functional structures, Schwann preferred systematic experiments. The older pupil had been drawn by Müller's comparative anatomy; the younger one, by his nerve root studies. Henle recalls that Schwann had an "inborn drive" to experiment and a special gift for handling technical apparatus. His brother, a goldsmith, advised him in the construction of physiological and electrical equipment. During his years in Berlin (1833–1839), Schwann prepared all of his own experimental setups. According to Henle, no one who ever saw Schwann at work and watched "his sober gaze [*die Nüchternheit seiner Beobachtung*]" could doubt the reliability of his findings.[95] Müller influenced Schwann's science by teaching him to use a microscope and urging him to conduct experiments, but once the student began experimenting, his views diverged increasingly from his mentor's.

In his thesis project, carried out in 1833–1834 under Müller's direction, Schwann proved that developing chick embryos needed oxygen, specifying to the hour when atmospheric air was necessary and how deprivation affected the developing animal. To pump the air out of the incubation chamber, he designed and built his own apparatus.[96] In Schwann's 1834 article on the subject, one detects a scientific style radically different from Henle's and quite distinct even from Müller's.[97] Schwann opens by asking the questions he wants to answer: (1) Can development occur if respiration is hindered? (2) If so, to what point can it occur? He then lists each experiment he conducted, for instance, incubating sixteen eggs in hydrogen and six in atmospheric air as a control. He describes in detail the appearance of failed incubations and concludes that the critical period occurs between the twenty-fourth and thirtieth hour, since eggs deprived of oxygen could be rescued after twenty-four hours, but not those deprived of it for thirty. In his first scientific communication, Schwann impresses the reader with the accessibility of his writing and the clear progression of his logic. Developing his own voice and style, he already sounds different from Müller.

After passing the Prussian state medical exams in the summer of 1834, Schwann dedicated himself entirely to research with Müller.[98] Rather than practicing medicine, he lived off of some money he had inherited from his family. Schwann began recording his experiments in a laboratory notebook whose first entry is dated April 1835. In October and November of that year—just after moving to the "Hotel Hilgendorf"—he conducted studies of muscle tension, since his observations of the striations in muscle tissue had made him curious about the way muscles worked. To learn how muscle force varied with the degree of contraction, he kept the stimulus constant and measured the length of contractions under increasing loads. Again, he built his own apparatus, in which frog legs hoisted weights when they twitched.[99]

In his studies of protein digestion, which he conducted partly with Müller, Schwann again took a quantitative approach. Systematically, he diluted stomach acid to determine whether the acid alone was sufficient to digest protein, or whether another, unknown factor was involved. He also attempted to measure how much acid or unknown substance was needed to digest a given amount of protein.[100]

The titles and opening sentences of Müller's and Schwann's digestion papers, which appeared back to back in the 1836 volume of the *Archive*, suggest the growing differences in their scientific styles. The first, by "Prof. Dr. J. Müller and Dr. Schwann," is titled "Experiments on the Artificial Digestion of Coagulated Protein." The second, by "Dr. Th. Schwann, Assistant at the Anatomical Museum in Berlin," is called "On the Nature [*Wesen*] of the Digestive Process."[101] The first article begins, "Through the experiments carried out by one of us on the action of dilute acid on muscle tissue and coagulated protein, it has become highly improbable that the acids in digestive juice described by Prout, Tiedemann, Gmelin, and Dunglison, acetic and hydrochloric acid, can have the functions in the rapid breakdown of food ascribed to them by many prominent physiologists."[102] Schwann's article commences, "Through Eberle's brilliant discovery we have become acquainted with a factor in epithelial tissue treated with dilute acid which carries out digestion just as well in an artificial way as in a natural one."[103] We cannot know to what degree Müller, who may have suggested the experiments; Schwann, who performed them; and Henle, who edited the journal, contributed to the language of these articles. From the bylines, one may presume that Müller had more say in the first; Schwann, in the second. The title and opening of the first strike the reader with their reluctance to announce a result and attribute credit. The title mentions only experiments, not a discovery, and the first sentence, which continues for half a page, states only that the results contradict the findings of four other scientists. In contrast, Schwann's article promises to explain the digestive process and gives exaggerated credit to another investigator. In either case, the style might be attributed to standard contemporary formulas of scientific writing, but these two articles appeared in the same journal volume and were edited by the same person. Together, they suggest that Müller and Schwann saw their research in different ways.

The differences between these protein digestion articles do not stop with the opening sentences. The joint essay is narrated in the third person ("Prof. Müller") and the first person plural ("we"), and mentions Schwann nowhere except the byline. It even employs long, clumsy phrases ("one of us, who has continued these investigations in the next article") rather than naming the young scientist.[104] It is divided into four sections to help the reader distinguish the different types of experiments conducted, and it offers detailed visual descriptions of the ways that acids corroded cubes of protein. What it doesn't do is convey the significance of the studies, something that Schwann's essay accomplishes far better.

Throughout his article, Schwann refers to himself as "I." He begins by explaining the shortcomings of Eberle's stomach acid study, which never revealed the substance responsible for digestion. He then carefully describes how he extracted digestive juice from an ox's third and fourth stomachs so that readers can repeat his experiments. Unlike Müller, Schwann represents his experiments as responses to questions. After showing that hydrochloric acid alone cannot dissolve protein, he outlines five possible roles of acid in digestion: (1) It may act as a simple solvent, making protein vulnerable to another substance; (2) It may combine chemically with another substance, so that the resulting compound digests protein; (3) It may dissolve the products of protein undergoing digestion; (4) It may decompose in order to combine with accumulating digestive products;

or (5) It may predispose digestive products to decomposition without itself decomposing. Schwann then describes the experiments he performed to eliminate the four incorrect possibilities. In a reader-friendly progression of logic, he proves that only the fifth alternative can be true, then begins describing experiments to determine how the unknown digestive factor works. Before and after each new round, Schwann asks the questions the experiments were designed to answer and gives the solution that each revealed. He speculates that digestion is analogous to fermentation but warns readers about the limits of this analogy. Whether or not his thoughts and experiments actually followed such a smooth progression, Schwann structured his article so that it would represent an idealized investigator's reasoning. The essay is a masterpiece of scientific logic, and Müller never wrote anything like it.

Intrigued by the chemical transformations occurring in plants and animals, Schwann began experiments in early 1836 to determine whether fermentation and putrefaction were caused by live organisms. By constructing a special apparatus, he showed that a cube of meat in a glass flask would not decompose if sealed off from atmospheric air or exposed to air that had been heated to a high temperature. He concluded that a heat-sensitive factor in the air, possibly the germ of a living organism, caused fermentation.[105] Before Schwann's article appeared, the French physician Charles Cagniard-Latour (1777–1859), who had been working independently, presented the idea to the French Academy of Sciences. But whereas Cagniard-Latour had based his claim on microscopic observations, Schwann relied on his technical expertise to design setups that controlled air flow.[106] His skill as a technician, combined with his "sober gaze," led him to fundamental discoveries about the way the human body works at the cellular level.

While conducting these early studies of respiration, digestion, and muscle contraction, Schwann discussed his work with Müller, and Müller must be given credit for having inspired Schwann to do experiments at all. But Schwann's gift for building innovative technological setups and his step-by-step logical style were his own contributions. When describing how he got his most famous scientific idea, he acknowledged the role of another scientist, but it was not Johannes Müller.

Seeing Cells

Whether achromatic microscopes caused scientists to see animal cells in the 1830s is a matter of debate.[107] With their thick walls, plant cells were relatively easy to make out, but to detect cells in animal tissue, microscopists had to be looking for them. Significantly, Schwann came to his cell studies in late 1837 after investigating muscle movement, digestion, and fermentation. His proposal that all living organisms consist of cells (the cell theory) was part of his greater scientific aim. He wanted to show that living phenomena had physical, material causes and were not the manifestations of some mysterious life force.[108] This idea contradicted Müller's most fundamental scientific views.

Credited today with formulating the cell theory, Schwann tells his own story of how it emerged. In a speech given at an honorary dinner in 1878, Schwann reconstructed his moment of enlightenment forty-one years after the fact:

> One day when I was having dinner with Schleiden [in October 1837], that illustrious botanist indicated to me the important role that the nucleus plays in the development of plant cells. Suddenly, I remembered having seen a similar structure [*un organe pareil*] in cells of the chorda dorsalis, and at that very instant I grasped the extreme importance the discovery would have if I succeeded in showing that, in the cells of the chorda dorsalis, the nucleus plays the same role that it plays in the development of plant cells. . . . This fact, if solidly established through observation, would imply the negation of a vital force common to animals and would make it necessary to admit the individual life of the elementary parts of other tissues and a common means of formation through cells. This recognition of a principle, later verified by observation, constitutes the discovery I had the good fortune to make. . . . I invited Schleiden to accompany me to the Anatomical Theater, where I showed him the nuclei in the chorda dorsalis cells. He saw [*reconnut*] a perfect resemblance to the nuclei of plants.[109]

In Schwann's account, the moment of discovery is a moment of connection as he reinterprets his recollected observation in the light of another find. Schwann saw animal cells not while looking through a microscope, but while talking to another scientist.

His inspirer, Matthias Jakob Schleiden (1804–1881), had come to science by an indirect route. The son of a wealthy physician, Schleiden had first studied law, but while practicing in Heidelberg, he had become so depressed that he had attempted suicide. In 1833, Schleiden had started studying science, and in 1837 was collaborating with Müller in Berlin, exploring the development of plants.[110] The idea that both plants and animals were made of cells emerged from the interaction between Schleiden and Schwann.

Schwann acknowledged, however, that Müller and Henle had played crucial roles in his discovery, since he had observed the chorda dorsalis together with Müller, and Henle's membrane studies had shown him that epithelial cells could grow even when not in direct contact with blood vessels, proving that they were not secreted by the blood.[111] Yet in the influential book he wrote describing his findings, Schwann defined his view in opposition to Müller's comparative anatomy.

The notion of animal cells excited Schwann because it suggested that all living things conformed to physical laws. The title of his work, *Microscopical Researches into the Accordance in the Structure and Growth of Animals and Plants* (1839), indicates the way he wanted other scientists to interpret his findings. Schwann argued that like crystals—which are inorganic and cannot be driven by any life force—all plant and animal cells follow "one common principle of development."[112] He compared the emergence of new cells in plants, in the chorda dorsalis, and in cartilage, showing that even in these very different tissues, the nucleus and cell wall or membrane behaved similarly. According to Schwann, all cells grew out of a formless, extracellular substance that he called the *cytoblastema.* The analogy to crystal growth served as a leitmotif throughout his work, although he warned readers against its limitations. He employed it to convey his most essential point: the common cellular, crystal-like structure of all living things suggested that physical and chemical laws, not some indefinable life force, controlled life functions:

> The elementary particles of organisms, then, no longer lie side by side unconnectedly, like productions which are merely capable of classification in natural history, according to similarity of form; they are united by a common bond, the similarity of their formative principle, and they may be compared together and physiologically arranged in accordance with the various modifications under which that principle is exhibited.[113]

In this description, he deliberately invoked the rows of animals in Müller's museum, carefully arranged according to their forms. But unlike Müller, he believed that the animals had acquired their positions in this arrangement because of the physical laws of matter, not because of qualities unique to organic matter which were realizing nature's intention and design. In the final section of his work, Schwann argued powerfully against the notion of life force, asserting that "it conduces much more to the object of science to strive, at least to adopt the physical explanation."[114] By 1839, Müller's preference of form to function was frustrating Schwann, but there were stronger reasons to leave Berlin.

A Religious Crisis

From both academic and economic standpoints, Schwann's position in Berlin was worse than Henle's. It is little wonder that he wanted to leave. On 6 December 1838, when he received an offer from the Belgian University of Louvain, Schwann wrote to Cultural Minister von Altenstein. He explained that his position in Berlin was precarious since he could not complete his *Habilitation* and lecture at the university, where Henle was the designated lecturer in the only subjects he was qualified to teach. Even though he could not advance academically, Schwann explained, he had stayed in Berlin so as not to lose "the favorable opportunity to further [his scientific] training, which presented itself here more than any place else."[115] He had thus resolved to build his reputation by publishing scientific articles, which he listed for Altenstein's benefit. For five years, he had lived on the museum helper's's salary of 120 taler, so that he had had to draw on his paternal inheritance for the other three fourths of his expenses. This inheritance was now exhausted, and his parents had nine other surviving children, only one of whom (his older brother Peter, a theologian in Braunsberg) could support himself economically. Knowing Altenstein's dedication to science, Schwann pointed out that since qualifying as a physician, he had devoted himself to research rather than practicing medicine. Schwann asked the minister for an associate professorship of physiology in Bonn, stating that it was "more honorable to dedicate one's energy to the fatherland" and that Bonn provided a better research environment than Louvain.[116] Müller, he wrote, supported the idea. While Schwann's letter is respectful, directed to a specific, powerful reader, his frustration with his position is clear. To work with Müller and Henle, the young scientist was sacrificing his independence, and with his scientific views diverging from his teacher's, the daily interactions with Müller no longer merited this sacrifice.

When Altenstein denied Schwann the Bonn professorship, the young microscopist took the job in Louvain, leaving Müller in April 1839.[117] According to Henle, Schwann did it for religious reasons. What Henle called "the Catholic University of Louvain" had reopened in 1817; the nation of Belgium was much younger, having been recognized only in 1830. Both had emerged with difficulty out of power struggles between Catholics and Protestants.[118] "A devout Catholic," in Henle's description, Schwann was "at first disturbed" by "the consequences arising from the development of cell theory."[119] Schwann's own writings tell a different story, but religion had a much greater effect on Schwann's scientific choices than it ever did on Henle's or Müller's.

In 1838, the same year in which Schwann performed the research for his cell book, he experienced a religious crisis.[120] At the time, Prussia was beginning a campaign—which would last for decades—to suppress the Catholic religion in all of its territories. In 1837, Prussian authorities arrested the archbishop of Cologne, a terrible insult to the people of the Rhineland.[121] Müller, a Catholic *Rheinlander*, and Henle, a *Rheinlander* who was Protestant by choice, seem not to have been disturbed by the incident, but it upset the more pious Schwann. In his 1838 letter to Altenstein, Schwann mentioned that Bonn was close to his paternal city of Neuss. Louvain, too, was closer to home than Berlin. But more significantly, both were Catholic cities, whereas Berlin was the heart of Prussian Protestantism. In Louvain, Schwann would have to teach in French, but in terms of religion, Belgium was less foreign than Berlin.

While it is tempting to believe Henle, the relation between Schwann's science and religion is more complex than he indicates. In an 1837 letter to his brother Peter, Schwann wrote that "the question of the soul's location . . . will be pretty much banned from physiology" if the soul acts on the brain just as mechanical, chemical, and electrical stimuli do. He complains that "Müller, on the other hand, is trying to show that the psychical principle extends throughout the entire body."[122] In his *Microscopical Researches*, Schwann reasoned that "matter with the powers inherent in it owes its existence to a rational Being," arguing that there is no need for life force when God himself has given matter the power to organize itself and perform life functions.[123] Müller's notion of life force mixed theological and scientific questions in a way that muddled both. If we can believe Schwann, the moment in which he realized that cells composed all living things liberated him to practice science while maintaining his religious beliefs.

But can we believe Schwann? The sharp drop-off of his research after 1839 suggests that Henle's view has some merit. Claims that Schwann "abandoned

Figure 2.3 Theodore Schwann. Source: Théodore Schwann, *Lettres de Théodore Schwann*, ed. Marcel Florkin (Liège: Société Royale des Sciences de Liège, 1961), 65, printed with kind permission of the Société Royale des Sciences de Liège.

rationalism and became a mystic" after leaving Müller's group misrepresent his thinking.[124] In 1844, he conducted a study on the role of gall in digestion as rigorously logical as his 1836 experiments, outstanding not just for its organization but for its sensitive descriptions of the dogs on which he experimented.[125] Although Schwann performed relatively little research in Belgium, he taught anatomy and physiology for the rest of his life, never abandoning his commitment to rigorous science. But there can be no doubt that like Müller, the scientists who developed the cell theory were "deeply troubled." In 1838, the year of Schwann's crisis, Schleiden again attempted suicide because of his "total lack of sound religious philosophy."[126]

The Students Rebel

If influence made Henle and Schwann anxious, Müller certainly exerted a great deal of it. He convinced Henle to study anatomy and medicine rather than literature or music, so that Henle owed Müller the fact that he was a scientist at all. According to Henle, Müller also gave Schwann "the first stimulus" to do microscopic studies.[127] Schwann recalls that "dealing with Müller gave you the most extraordinary encouragement to follow up every new thought you ever expressed to him with investigations and experiments."[128] Schwann's research benefited not just from his own insights but from Müller's amazing memory and ability to connect ideas. When he joined Müller in Berlin, Schwann remembers, his teacher reminded him of an idea about respiration that Schwann had ventured during a walk in Bonn four years earlier. Schwann himself had long since forgotten it.[129] Yet while Müller encouraged his students, he never exerted his influence by stifling their thoughts. He left Schwann to develop cell theory on his own, he told Henle, because "he wanted everyone to have his say."[130] He made his favorite students scientists by respecting their own scientific thinking.

Within the possibilities offered by the Prussian university system, Müller did everything he could to support Henle and Schwann. Henle came to Berlin in 1832 with a letter of introduction from Müller, so that he was welcomed by the senior anatomist Rudolphi and most of the medical faculty.[131] Once established in Berlin, Müller presented his students to prominent foreign visitors. Both Henle and Bidder recall being invited to Müller's house to join "a pretty respectable gathering of old and young intellectuals, beautiful and brilliant women, Germans, foreigners, Russians, Poles, Dutchmen and Frenchmen."[132] Most significantly, Müller promoted his assistants' work in his letters to the cultural minister. On 8 January 1836, he sent Altenstein an issue of the 1835 *Archive for Anatomy and Physiology*, pointing out "an important observation" made by Henle and "several characteristics [of sperm] discovered by Henle." On 3 June 1836 he offered another issue of the *Archive*, pointing out work by "our valiant [*wacker*] Dr. Schwann, assistant at the Anatomical Museum" and praising Schwann as a "talented young man."[133] From such references, Müller gained little personal advantage, except maybe a reputation for generosity. By calling attention to his students' work, he can only have been trying to increase their chances of winning jobs.

Still, there is some question about Müller's willingness to give Henle and Schwann credit. Schwann recalled in 1858 that "the credit [*Eigenthum*] for the discoveries we made was mutually respected in the highest degree, and Müller's sense of fairness always deserved the greatest esteem."[134] Henle was not so sure. In his letters home, his sense of humor exposes the tensions in his relationship with Müller, as in one account of his editorship of the *Archive*: "On the title page it says, 'edited by J. Müller in conjunction with several other scholars.' At present, these several other scholars are represented by me."[135] Henle was paid for his work, but he wanted to see his name in print.

Müller used Schwann's name more liberally. Soon after Schwann conducted his studies of animal cells, Müller began praising the work in his own publications. In his 1838 tumor study, he cites both Schleiden and Schwann for their research on cells and refers to "the principle discovered by Schwann governing the reproduction and growth of animals."[136] While working with Schwann and Henle in 1833–1839, Müller produced a new volume or updated edition of the *Handbook of Human Physiology* almost each year, incorporating their findings. In the second volume of the *Handbook* (1840), Müller wrote that Schwann's discovery of animal cells "belong[ed] to some of the most important progress ever made in physiology."[137] In April 1840, du Bois-Reymond wrote to his friend

Eduard Hallmann that Müller was calling Schwann's *Microcopical Researches* "the foundation of any future general physiology."[138]

Conceivably, Schwann's achievements threatened Müller. In December 1839, when du Bois-Reymond was just getting to know his teacher, he told Hallmann:

> I've really come to respect Schwann since I read something in Müller's physiology book. It's funny how Müller first mentions him: "the assistant, Herr Schwann"; then it's "Schwann observed," etc., and finally the "little" Schwann, as you call him, grows way up over the big fat Müller's head and on up into the heavens, and the aforementioned does almost nothing but cite Dr. Schwann's observations.[139]

Du Bois-Reymond's letters to Hallmann give a biased representation of Müller and are composed from a distinct perspective. Hallmann, the reader, had worked as Müller's assistant but had left on bad terms with him. Du Bois-Reymond, the writer, was just getting to know the prominent physiologist, heedful of Hallmann's warnings and eager for Müller's approval. Telling Hallmann of the giant's weakness and the "little" scientist's rise to power comforted them both. Still, du Bois-Reymond's picture of the growing student and shrinking mentor may accurately represent Müller's feelings.

On some occasions, Müller declined to give Schwann credit. Schwann later praised Müller's fairness because the senior scientist acknowledged any experiments that Henle or Schwann had thought up themselves, but Müller's demanding teaching schedule kept him from performing all the experiments of his own design. When his assistants carried out experiments that he had assigned them, Müller used only his own name, with the result that some additions to the *Handbook* during 1837–1840 describe Schwann's results without acknowledging his assistance.[140] Henle had hoped to "hang his name" on Müller's, but in the process, Müller hung his own name on Henle's and Schwann's.

By the late 1830s, Henle was growing increasingly frustrated with Müller's personal failings. Here he differed from Schwann, whose objections were confined to scientific theories and academic restrictions. A major source of the tension between Müller and Henle was their different attitudes toward teaching microscopy. In contrast to Müller, Henle believed that all medical students should have a chance to look through a microscope and that the instrument should be used regularly in teaching, not just for research.[141] In an advisory report in 1883, more than forty years after he had worked with Müller, Henle wrote, "In 1837, as a Berlin lecturer, I was the first who displayed compound microscopes and familiarized students with objects in their natural form."[142] Müller, on the other hand, reserved microscopy for those gifted students who first proved their devotion and skill at the dissecting table.[143] Certainly Henle's students appreciated his efforts to teach microscopy. When Eduard Hallmann listed the medical courses that du Bois-Reymond should take in Berlin, he wrote, "Henle's general anatomy class is very much to be recommended, because it offers many microscopic demonstrations."[144] Albert Kölliker, who studied with Henle and Müller in Berlin in 1840, supports Hallmann's claim. In his comparative anatomy and physiology lectures, recalls Kölliker, Müller gave students "the broad view," showing the connections between different living things. Henle, in contrast, provided a microscopic view, which Kölliker greatly appreciated.[145] Sixty years later, in his memoirs, he claims:

> I [can] still see that long, narrow hallway [*Vorplatz*] next to his lecture hall in the university building where Henle, for lack of any other space, and with barely five or six microscopes, showed us the simplest things, but things so powerful in their novelty: epithelial cells, flakes of epidermis, ciliated cells, blood corpuscles, pus cells, semen, then preparations extracted from muscles, tendons, nerves, sections of cartilage, slices of bone, and so on.[146]

The neuroanatomist Wilhelm Waldeyer (1836–1921), who spread the idea that the nervous system consists of individual cells, believed that with the possible exception of Jan Purkinje's microscopy courses in Breslau, Henle's were "the first of their kind."[147] Although Müller gave his handpicked assistants every opportunity to use microscopes, he did not see microscopy as an essential part of every medical student's education.

Possibly the students' enthusiasm for Henle's classes aroused Müller's jealousy. In February 1839, Henle wrote to his family:

> More and more [Müller's] egotism betrays itself, his fear of the powerful and what always seems to accompany this, his lack of consideration for the little people. Even though I'm still very attached to him, his actions and thoughts displease me. Because I can't conceal this from him, I've lost his trust. The fact that he no longer particularly desires my advice in the steps he takes publicly—on the contrary, he tries to hide them from me—doesn't hurt me, but it makes the relationship more anxious, tense, and uncomfortable.[148]

The six-year struggle for academic power in Berlin destroyed the closeness built up in Bonn.

Perhaps because Henle was himself so frustrated with Johannes Müller, the observant student saw Nanny's unhappiness with her husband. Henle's letters shed new light on a relationship that Müller's biographers depict as near-perfect. Because of Henle's musical activities, he knew Nanny well, and his comments indicate that like his mentor, she began to annoy him. In 1835 he wrote to his family, "We have long agreed that Müller's wife has not found with him the true, comfortable [*gemütliche*] satisfaction she is seeking. But she is happy in the thought that she possesses the foremost anatomist in Germany and might not exchange it for the most tender feeling [*Gemüt*]."[149] The next year, Henle reported that "our sweet little [*niedliches*] life here has suffered a small disruption due to a break between Müller and Froriep about lectures [*in betreff einer auditorischen Angelegenheit*], a break in which the wives are involved."[150] Müller's academic and social lives were so intricately interwoven that his relationship with Nanny touched most aspects of his work. Henle's snide remarks about Nanny reflect his irritation with his own academic predicament.

By Henle's thirtieth birthday in July 1839, the anatomist felt that he was too old to be a prosector and longed for independence from Müller. But in the absence of any job offers, he had to stay on until the professor seemed to "crush" him. In April 1840 he wrote to his friend A. Schöll:

> I had now been shut out of the Anatomical Institute once already and had to fear being shut out permanently, the more the success of my lectures gave me a claim to it. I had only one lecture left, for which physiology was a prerequisite and which in turn prepared students for clinical lectures. If Müller and Schönlein wanted to, they could crush me between the two of them. . . . Müller's longing for an offer for me and his joy when it finally came were too unconcealed for me to stay any longer in the position I had been holding. To have done so would clearly have insulted him.[151]

At the time, Henle had just received a call to work as professor of anatomy and physiology in Zurich. He gladly accepted the offer, which arrived just in time.

Du Bois-Reymond, who delightedly informed Hallmann of any changes in Müller's personality, wrote in December 1840 that "since Henle left, Müller has become another man."[152] But three months later, du Bois-Reymond reported trouble between Müller and Karl Reichert, the young anatomist who had replaced Henle as prosector:

> At the Anatomical Institute, where we had passed the first quarter with Reichert as though in heaven, [Müller] thought he could introduce the old Henle-like tone [*den alten Henle'schen Ton*], with which he failed bitterly. . . . He came to believe that Reichert had no idea how to handle a knife and wanted to instruct him continually.[153]

While one must read du Bois-Reymond's letters with caution, they depict a situation so tense that even if he vastly exaggerated the relationship among Müller, Henle, and the younger assistants, it could not have been good. Apparently Müller, who may have missed his old intimacy with Henle, took out his anger on Henle's replacement. Later that year, Müller drew closer to his old friend Anders Retzius, writing, "You have grown even more dear to me than you already were. I have no friend whom I have ever loved as much as you."[154] Ten years earlier, these phrases would have been directed at Henle.

Personalities evolve with time, and a letter from Gustav Magnus to Henle suggests that the Henle-Müller friendship broke down as much from Müller's withdrawal as from the student's fight for independence. Shortly after Müller's death in May 1858, Magnus wrote:

> If you mourn the loss twice over because your relationship with him was never cleared up, you're wrong! From what I know of the relationship and Müller's attitude toward you, no change entered into your relations with each other that did not also enter into his relations with everyone else. The Müller of the time of his arrival and that of the last years are

> two very different people. You'll recall that even during the time you were here, he was changing from year to year, and it kept on like that until 1848. . . . From that time on, . . . he withdrew more and more from any kind of social dealings.[155]

Müller's depression and anxiety may have undermined his trust in his favorite student.

In Zurich, soon after leaving Müller, Henle published his respected textbook, *General Anatomy*, which presented anatomy from a microscopic perspective. Du Bois-Reymond wrote to Hermann Helmholtz in 1853, "Every one of us has Henle's A*natomy*. . . in his hands."[156] With the physician Karl Pfeufer (1806–1869), Henle founded the *Journal for Rational Medicine*, "rational" meaning a combination of theoretical and practical approaches. Like Müller, Henle believed that good science must involve some philosophy as well as direct observation.

In 1844, Henle moved from Zurich to a more prestigious professorship of anatomy and physiology in Heidelberg, where he focused on pathological studies. Stressing the resemblance of pathological symptoms to normal bodily functions, he argued that illnesses were not things in themselves but altered versions of normal physiological processes. In his *Handbook of Rational Pathology* (1846–1853) Henle called disease "the expression of a typical force [*Kraft*] under unusual conditions."[157] This approach to medicine sounds remarkably similar to that of Rudolf Virchow, whose relationships with Müller and Henle will be discussed in chapter 5. But while Virchow stressed the physical environment as a cause of disease, Henle looked to the nervous system, through which all stimuli seemed to affect the body. As pathologists, the two became strong opponents.

From a socioeconomic standpoint, Henle did not marry as "well" as most Müller students, perhaps because of his musicality and belief in romantic love. In Zurich he developed a relationship with Elise Egloff, the governess of a friend's family. After one of his sisters coached her so that she could fulfill the social role of a professor's wife, he married her in 1846. They had a son and a daughter, but sadly, the love match ended after only two years when Egloff died of tuberculosis. A year later Henle remarried, this time selecting Marie Richter, a military officer's daughter, with whom he had five more children. They raised their family in Göttingen, where Henle became professor of anatomy in 1852 and remained for the rest of his career.[158]

When Müller died in 1858 and his position was divided, Henle was offered his chair of anatomy, the most prestigious one in Germany. The former student turned it down. He dreaded the heavy teaching load and administrative burden of a senior professor in the Prussian capital, and he liked his position in Göttingen.[159] Perhaps his harrowing year and a half under investigation had permanently colored his view of Berlin. [160] Three decades after his happy collaboration with Müller in Bonn, he had become one of Germany's foremost anatomists, but he did not want to be identified as Müller's follower.

Of all of Müller's students, Henle most closely retained his teacher's scientific values. He loved studying the forms of animals and never openly attacked the notion of life force. His determination to offer all students microscopes continued and expanded his mentor's vision, even if it upset Müller in the late 1830s. Their close relationship broke down not because of scientific differences, but because of a destructive struggle for control. This was not the case with Theodor Schwann.

Rebelling in Retrospect

Like Henle, Schwann began questioning Müller's judgment while still working for him, and his opposition intensified with time. In the early 1840s, the young physiologist du Bois-Reymond looked to Schwann as a role model, a student close to Müller who rejected vitalism and advocated rigorous physiological experimentation. In 1841 du Bois-Reymond wrote to Hallmann: "To me, Schwann is a sort of demigod. His way of doing experiments [*seine Art Versuche anzustellen*] is unparalleled. The theoretical part of his book [*Microscopical Researches*] . . . is immeasurably great. What a style! Such calm and clarity will never recur."[161] Schwann's withdrawal from active research disappointed du Bois-Reymond terribly, and by 1849, his letters to Carl Ludwig and Hermann von Helmholtz contain snide comments about Schwann's latest work. Du Bois-Reymond reported to Ludwig, "I made Schwann's acquaintance in Neuss. I can tell you confidentially that in a while we can expect from him a logical but crackbrained opus; a theory of organic creatures. Once more a star that falls, falls, falls, and disappears."[162] He called Schwann "the sleeping lion."[163]

Still, du Bois-Reymond turned to Schwann for material when preparing Müller's memorial address. Despite his disappointment, he hoped that Schwann would substantiate his thesis that Müller had promoted physiological experimentation but had never really practiced it because of his vitalism and obsessive love of animal forms. Since du Bois-Reymond delivered the address on 8 July 1858, and Schwann did not write to him until 22 December, du Bois-Reymond was seeking comments for the written record, not the public speech. He printed Schwann's entire letter in the notes to his published address, explaining to readers that he had asked the former student "to put together his memories of Joh. Müller" and that Schwann had "expressed his view on an issue not unimportant for the history of physiology, namely to what degree he should be regarded as a pupil or independent contemporary [*selbständiger Zeitgenossener*] of Müller."[164] Du Bois-Reymond's introduction to the letter suggests the prompting the invitation must have involved. Twenty years after Schwann and Müller's collaboration, the experimentalist urged his former hero to declare independence.

This time, Schwann did not let du Bois-Reymond down. Halfway through his long reply, he directly addressed the main question:

> If you ask me whether I see myself as a pupil or an independent contemporary [*Coetan*] of Joh. Müller, that depends on what you understand by pupil. I became acquainted with physiology through his lectures and his

> personal conversations with me. . . . On the other hand, even when I was a student in Bonn, my thought went in a very different direction from his.[165]

Schwann explains that Müller believed organisms contained a life force that made them fundamentally different from inorganic matter. Müller did conduct important physiological experiments, but for the wrong reason: to understand this non-existent life force. In this way, Schwann remarks, "He established the right methods in German physiology while affirming the principles of nature philosophy."[166]

As Schwann proceeds, his urge to distinguish himself from Müller grows more apparent. "The physical approach," he claims, "which I beat into physiology and which means pursuing real explanations for life functions . . . was something Joh. Müller never had." Schwann's own study of muscle contraction, he claims, was "the first time an apparent life function was shown to be subject to mathematical laws that could be expressed in numbers." In Schwann's estimate, Müller had little or nothing to do with his discovery of animal cells. "That the formation of cells must constitute a general principle of development, and that there could be one principle of development, was something with which Joh. Müller did not at all agree."[167] To make matters clear, Schwann ended his letter by writing that "Joh. Müller . . . never claimed to have taken part in my cell studies, but confined himself to applying them directly to tumors."[168]

Schwann believed that Müller's simultaneous embrace of life force and cell theory made no sense, so that ultimately, his mentor's science did not work. If all living things grew and reproduced according to an observable mechanism, how could one believe in a mysterious life force? In a lecture of 1878, Schwann stated that he had never believed in vital force, seeking final causes "not in the creature, but in the creator."[169] Like du Bois-Reymond, he presented himself, not Müller, as the person who had overcome fuzzy notions of life force through rigorous physiological experiments. If anything, he had accomplished this in spite of his mentor.

Despite their differences, Müller retained the greatest respect for Schwann, and perhaps also a lingering fondness. After his breakdown in 1848, he turned to Schwann, who took him to the Belgian coastal city of Ostend, where he could explore the local sea life and recover his will to live.[170] During this difficult time, Müller did not visit Henle.

Schwann's and Henle's Müller

In Schwann's and Henle's complex relationships with Müller, social factors played a significant role. While their Rhineland culture united the three scientists, class differences created variations in their attitudes. As a shoemaker's son, Müller did not belong to the working class, but his family never enjoyed the income or prestige of a businessman's or publisher's household. If Müller had not become a scientist, he would have been a saddle maker. Henle and Schwann would have been upper-middle-class professionals. Müller exerted his greatest influence over their lives by leading them to careers in scientific research. At the time, so few academic jobs were available that they might otherwise have ruled out this unpromising choice. Since Prussia's administration of the Rhineland had allowed Müller to become a scientist, he viewed the Prussian government and his patrons, Schulze, Rehfues, and Altenstein, as providers of opportunity. As members of a more privileged class, Henle and Schwann felt less gratitude. For their hard work, they expected professorships and the social prestige that went with them. To endure poverty for the privilege of working with Müller was one thing at twenty but quite another at thirty. They believed that they deserved their own jobs.

In Henle's case, this social and academic pressure combined with emotional factors to undermine what had been an intimate friendship. Müller loved his young student and wanted to live and travel with him as well as work with him. Each used the other: Henle relying on Müller to rise in science; Müller relying on Henle for labor that might have been better acknowledged. As an anatomist, Henle never openly refuted Müller's beliefs in life force and a great plan of life, and he did not seem to find these views particularly repugnant. His relationship with Müller degenerated from a close emotional bond to active dislike because he felt that Müller's determination to control him was crippling his growth. Seeing Henle's desire to escape, Müller may have felt betrayed. Never again would he be on such intimate terms with a student.

Müller shaped Henle's science by inspiring his love for comparative anatomy and microscopy. Of all of the students considered in this study, Henle differed the least from Müller in his theoretical assumptions and his broad, observational approach to the life sciences. This is not surprising, considering that he was only eight years younger. Unlike du Bois-Reymond, who will be considered in the next chapter, Henle left no detailed portrait of Müller for posterity, even though for a time he was closer to Müller than du Bois-Reymond ever was. Henle's comments on Müller in his personal letters are expressions of anger he dared not show. To win his own professorship, he had to have Müller's recommendation, and until he got one, he had to survive from day to day.

Müller may not have loved Schwann as much as he did Henle, but he did value him as a scientific soul mate. Schwann got to know Müller by expressing interest in his research, and he quickly won his professor's confidence. Müller influenced Schwann by urging him to perform experiments, not knowing—perhaps not caring—that the results would lead the student away from his teacher's views. During the 1830s, Schwann developed as an innovative experimentalist and skilled microscopist, rejecting Müller's core beliefs and challenging his mentor in print. But unlike Henle, Schwann seems to have harbored no anger toward Müller, and he didn't blame Müller for thwarting his career. He left Berlin for religious and economic reasons, waiting for Müller's death—and du Bois-Reymond's invitation—to express his full opinion of Müller's science.

Although Henle and Schwann differed in their evaluations of Müller's science and mentorship, they shared experiences with him that his later students did not. Both knew him in Bonn, while he was a young professor excitedly investigating many physiological problems at once and less burdened by administrative tasks. Before he entered the spotlight as Prussia's foremost anatomist and physiologist, he was freer to socialize with students and could interact with them almost as equals. As Emil du Bois-Reymond stood trembling in 1839, waiting to meet the great Johannes Müller, he could never have envisioned such a relationship. In the year of Schwann's departure, an ambitious new student awaited Müller who would work with him for nineteen years and shape our understanding of him to this day.

3

Emil du Bois-Reymond as a Scientific and Literary Creator

Müller has kept me occupied at the museum, carrying out what is in his opinion the highest activity of the human intellect, namely, classifying fossil vermin.
—Emil du Bois-Reymond to Carl Ludwig, 7 August 1849

Writing as Action

Emil du Bois-Reymond, who worked with Müller from 1840 until 1858, wrote the official story of his mentor's life. When Müller died in April 1858, the Academy of Sciences asked him to deliver a memorial address, and he spent over a year revising his talk into a substantial biography. In this account, du Bois-Reymond depicted Müller's work so as to promote his own scientific views.[1] Like Schwann, he believed that physical laws could explain life functions, which should be investigated through rigorous quantitative experiments. Consequently, du Bois-Reymond portrayed Müller as a scientist who inspired experimental physiology but failed to practice it himself (at least, after 1840), losing himself in useless speculation about how life force realized itself in a grand plan. For du Bois-Reymond, science meant action: creative, innovative adaptations of technology to investigate physiological phenomena.

In March 1841, Emil du Bois-Reymond wrote to his friend Eduard Hallmann that he had just read *Hamlet*, "indisputably the profoundest work that has ever been written . . . this story [is] nothing more than the most wonderful realization of your views on the relationship between reflection and action. To me, Hamlet is essentially the person who is destroyed by reflection."[2] It is not surprising that du Bois-Reymond, a twenty-two-year-old Berlin medical student, would be thinking about *Hamlet* in the early 1840s. At the best Prussian high schools, students read and translated passages from Shakespeare's plays. Hermann von Helmholtz translated Shakespearean passages at the Potsdam *Gymnasium*, and in an 1839 letter to his parents criticized a "frighteningly bad performance of *Hamlet*" he had seen at a Berlin theater.[3]

When du Bois-Reymond described *Hamlet* as a comparison of reflection to action, he was probably thinking about Hamlet's "to be or not to be" speech early in act III:

> Thus conscience does make cowards of us all,
> And thus the native hue of resolution
> Is sicklied o'er, with the pale cast of thought,
> And enterprises of great pith and moment,
> With this regard their currents turn awry,
> And lose the name of action.[4]

To du Bois-Reymond, action meant experimentation, and from the time that he met Müller in late 1839, he styled his scientific identity in opposition to Müller's contemplative, non-interventional comparative anatomy.[5] Two months after he praised *Hamlet*, about the time that he began using his first galvanometer, du Bois-Reymond wrote to Hallmann, "Action, action [*Tun, tun*], that's the solution for me now!"[6] Very likely, he formulated this sentence with Shakespeare in mind.[7]

As a lover of irony, du Bois-Reymond knew he was quoting a play that challenged the opposition between language and action. In *Hamlet*, action is represented largely by murder, and the characters who advocate it (Hamlet's usurping uncle Claudius, for example) are not the most sympathetic. More important, the play points to the power of speech, sometimes equating actions with words. Gertrude's plea to Hamlet, "these words like daggers enter in mine ears," recalls his father's murder, in which Claudius pours poison into the ear of the sleeping king.[8] As the play goes on, Hamlet increasingly emphasizes the force of language, wondering, as he dies, "what a wounded name (things standing thus unknown) shall live behind me."[9] Shakespeare's play excited du Bois-Reymond not just with its apparent call to action, but also with its illustration of the power of storytelling. Action meant taking control through words as well as experiments.

In the mid-nineteenth century, du Bois-Reymond helped to create the history of science as a field.[10] In his scientific books as well as his popular lectures, he spoke actively about history. In the introduction to *Animal Electricity*, a work that consumed his entire life as a scientist, he wrote, "The historical arguments have multiple purposes. For the one part, they are justified in and of themselves as historical studies of [scientific] literature."[11] In that book du Bois-Reymond described his misconceptions and wasted efforts, as well as his successes, because he wanted readers to experience his science for themselves. In an 1872 lecture on the history of science, he declared, "The best way to explain [*mittheilen*] a science is to tell its history . . . one should teach a science and its history at the same time."[12] He believed that science was the highest form of culture, and that by narrating the history of science, he was describing the history of humankind.

A careful writer, du Bois-Reymond was intensely aware of his power to create a cultural memory of science. More than any other of Müller's students, he heeded Madame de Merteuil's warning that one must always write for someone else. In the spring of 1851, du Bois-Reymond admonished Helmholtz for having written an incomprehensible report on the velocity of nerve impulses:

> I must tell you, by the way, that I'm not at all satisfied with your description [*Darstellung*]. I've read through your paper and abstract a few times, and have been unable to grasp what you really did or how you did it. . . . Please don't be offended, but you must absolutely take more care to step back from your own point of view [*Standpunkt des Wissens*] and put yourself in that of people who don't know yet what your work is about or what you're trying to argue.[13]

Unlike Helmholtz's paper, du Bois-Reymond's memorial address for Müller was the work of a highly skilled writer who for two decades had consciously constructed each letter, essay, and lecture for a particular reader, always with a special aim in mind. Like his laboratory setups, du Bois-Reymond's writing was *zweckmässig*: highly functional, and appropriate for a particular purpose.[14]

Hope

Inside the front cover of a laboratory notebook dated 1 February 1850–March 1856, du Bois-Reymond jotted down a poem by Goethe:

> Schaff, das Tagwerk meiner Hände,
> Hohes Glück, dass ich's vollende!
> Lass, o lass mich nicht ermatten!
> Nein, es sind nicht leere Träume:
> Jetzt nur Stangen, diese Bäume
> Geben einst noch Frucht und Schatten.[15]

Goethe's early poem, titled "Hope," has the ring of a Protestant hymn. With its naïve purity, it presents an interesting contrast to the dense scribblings and detailed drawings within. If du Bois-Reymond copied the poem shortly before performing the experiments recorded inside, he was finding consolation in Goethe's youthful work while he was acutely aware that his own youth had passed. During 1850–1856, du Bois-Reymond's research had lost momentum, and with Müller in relatively good health, he saw no prospect of a professorship in Berlin. Meanwhile, his peers Hermann von Helmholtz, Carl Ludwig, and Ernst Brücke had assumed respectable academic jobs. With its celebration of hard work and long-term goals, du Bois-Reymond's transcription of "Hope" suggests that he sought inspiration by looking back at his middle-class Huguenot roots.[16]

Du Bois-Reymond grew up in a French-speaking household.[17] His mother, a minister's daughter, came from a cultured family in the Berlin Huguenot community. Descended from Protestants who had left France in 1685, when Louis XIV withdrew the Edict of Nantes, the Huguenots quickly joined Berlin's educated middle class. His mother's grandfather, Daniel Chodowiecki (1726–1801), had become the most popular illustrator and engraver in eighteenth-century Prussia, and directed the Academy of Arts from 1797 to 1801. In du Bois-Reymond's public lectures, especially after German unification in 1871, he presented himself as a patriotic Prussian, but he could never dissociate himself from his French-language background. On the morning the Franco-Prussian War was declared, he reportedly excused himself for his French name.[18] When he traveled to Paris in 1850, he bragged, "As I speak French like a patriarch, I am naturally at a great advantage."[19] Perhaps his French was a little too good. According to one source, du Bois-Reymond spoke German with a French accent all his life.[20] Despite his high social position, he remained something of an outsider in Berlin.

The second of four surviving children, du Bois-Reymond was the eldest son in a prominent, cultured Berlin family. His younger brother, Paul, became a mathematician; his older sister, Julie, married a physician; and his younger sister, Félicie, married a geologist. A third, very bright and promising son died of scarlet fever at the age of six.[21]

Like Müller, du Bois-Reymond's father, Félix Henri, was a self-made man from a region nominally but not culturally Prussian.[22] He became a powerful member of Berlin society because of his intelligence and self-confidence, but also because of his ability to win patrons.[23] Born in 1782 in the Swiss canton of Neuchâtel (then Neuenburg, a Prussian province), he escaped from an apprenticeship to a watchmaker and arrived in Berlin in 1804, unable to speak a word of German. Félix dreamed of becoming a scholar, and with help from a family who had employed him as a tutor, he gained admission to the Pépinière, where Virchow and Helmholtz later studied medicine. When Félix was dismissed for voicing a student complaint, a Swiss professor rescued him by offering him a job teaching French to military cadets. Félix continued studying medicine, then linguistics, while tutoring the children of Berlin intellectuals, but he interrupted his studies to join the war against Napoleon. Again helped by former professors and employers, he obtained a government post, and in 1816 became court counselor for all Neuchâtel matters.

A gifted, ambitious man who had suffered poverty in his youth, the authoritarian Félix du Bois-Reymond ran a frugal household in which intellectual achievements were highly prized. Like Müller, who was nineteen years his junior, he viewed Prussia as a land of opportunity. He also shared the scientist's admiration for Immanuel Kant's philosophy, which dictated that the mind's internal structure determines the kind of knowledge people can obtain. Throughout his life, du Bois-Reymond struggled with his father, who once called him "that boy who spits in my face and walks on my belly."[24] In 1840, while living with his parents and studying medicine with his father's support, du Bois-Reymond wrote to Hallmann, "It's just about impossible for me to get along with my father, a decided Kantian and besides that, the world's cleverest dialectician."[25] Félix encouraged his son's scientific studies but maddened him with his insistence that the mind imposed its order on the world.

Despite the difference in age, du Bois-Reymond's *Doktorvater* Müller probably reminded him of his ambitious, Kantian father. In 1849, Helmholtz playfully closed a letter to du Bois-Reymond by offering "my highest respects to your parents, Müller and Magnus."[26] He did not say which one was the mother. At the time, the physicist Gustav Magnus (1802–1870) owned Berlin's finest collection of laboratory instruments. In the absence of university work space, he experimented with them in his own home, am Kupfergraben 7, just three doors down from Müller's first Berlin house. Magnus came from one of Berlin's most successful business families, one member of which had provided the dress Henle wore to the costume ball. Magnus had planned to design industrial technology, but instead had become a professor of physics at the Berlin University, using his personal fortune to create a first-rate physical laboratory.[27] Magnus handpicked the students who could work with his instruments; among them were du Bois-Reymond and Helmholtz, who met in Magnus's laboratory in December 1845.[28] As Emil's mentors, Müller and Magnus functioned as his scientific parents. Du Bois-Reymond's complex relationship with Müller reflects his many conflicts with his father.

Because of his collaborations with Berlin's instrument makers as well as his cosmopolitan tastes, du Bois-Reymond spent nearly his entire career in Berlin. By the 1840s, Berlin rivaled Vienna as the cultural center of the German-speaking world, and except for occasional vacations and research trips, du Bois-Reymond refused to leave it. He studied at the French-language *Gymnasium* and the Berlin University, at the time two of the finest schools in the German territories.[29] When he enrolled in the university's philosophy department in the spring of 1837, he could not decide what sort of career he wanted, and took courses in an exploratory rather than a goal-oriented way.[30] He had a great talent for drawing and considered becoming an artist like his great-grandfather Chodowiecki. In the spring of 1838, du Bois-Reymond decided to transfer to the Bonn University; at the time, many students went to several universities while pursuing a degree.[31] Disappointed after a year, he returned to Berlin in the fall of 1839 and—with his father's strong approval—began studying medicine. It was during this year that he met Müller.

Du Bois-Reymond differed from Müller in having a father who supported his scientific studies intellectually as well as financially. Félix hoped that his son would become the scholar he had wanted to be, and he took active measures to achieve this goal. Müller's father had dreamed that his son would someday run a leather shop. Despite his clashes with his father, Emil consulted Félix throughout his formation as a scientist. It is clear from his letters to Hallmann, Ludwig, and Helmholtz that he often talked with his father and that Félix's advice mattered. When du Bois-Reymond transferred to the University of Bonn in 1838, he arrived with two letters of introduction procured by his father. When he returned to Berlin, Félix advised him to take Müller's physiology class.[32] Du Bois-Reymond's attempts to buy a good microscope also indicate the vast difference between Müller's and his student's social levels. In August 1840, du Bois-Reymond complained to Hallmann that his father thought microscopy was just a passing fad, but by December, Félix and his son had "worked out" a way for Emil to buy a Pistor microscope. Emil was to use his own money, a well-invested baptismal gift from his grandfather that, with interest, had grown to 300 or 400 taler.[33]

The du Bois-Reymonds' social network continued to serve them even after Félix was forcibly retired in 1848. In the negotiations following the political uprisings, Neuchâtel left Prussia, so that the position of court counselor was no longer available. For Emil, his father's layoff had immediate practical consequences. "Due to an unlawful stroke of the pen by one of our former ministries," he wrote to Ludwig, "my father will no longer be in a position to give me an allowance."[34] To keep Félix's son experimenting, Alexander von Humboldt took a specially bound copy of *Animal Electricity* to the Prussian king. In April 1849, Humboldt wrote to Emil that he had obtained 500 taler, "a gift from the king himself . . . partly due to the importance of your work, but also partly (and this must gladden you) because he would like to do something nice for your father, that worthy man who was forced out of service because of the Neuchâtel situation."[35] That same month Humboldt also wrote to Cultural Minister von Ladenburg, requesting help for Emil "because his father, with whom my brother [Wilhelm von Humboldt] often spoke about the philosophy of language, has suddenly lost half his income, so that the situation of young du Bois is very sad indeed."[36] It is questionable how glad Emil could have been to learn that his research support was a kind of paternal severance pay, but he took the money. As frustrating as Félix was, Emil's relationship with him made his research possible.

Félix du Bois-Reymond once said, "I was born for research."[37] Throughout his life, he remained interested in science, language, and political economy. Once his diplomatic career ended, he concentrated on scholarly work, and in 1862 published a well-received book on linguistics. When it appeared, Helmholtz wrote to du Bois-Reymond, mentioning "your father's book, which considering his advanced age surprised me more than a little." The sarcastic du Bois-Reymond called it "my father's senile busy work [*altersschwache Machwerk*]."[38] In his 1848 introduction to *Animal Electricity*, du Bois-Reymond used his father exactly as he would later use Müller, casting him as a superseded supporter who had missed crucial opportunities:

> In my father's youth, his passionate inclination for the study of nature raised him from the lower circles of rural life. . . . The storms of a revolutionary time, which now threaten to break out again, . . . caused his plans to fail and forced him to work in foreign parts. But the more gloomily he looks back, in the evening of his life, on the vicissitudes that restricted his own research, the more intensely he wants me to partake of that good fortune, from whose privation he once so painfully suffered.[39]

Through this narrative, du Bois-Reymond takes control of his family history. In 1848, Félix was sixty-six and Müller was forty-seven. Both would live for another decade or more, despite Emil's hopeful reference to "the evening of his [Felix's] life." In this portrait, the experimenting son creates a father who failed as a scientist and delights in his son's success, a description that—at best—is fictional.[40]

While du Bois-Reymond portrayed his father as a scientist manqué, Félix provided a materially comfortable, highly cultured home environment. Consequently, his son's attitude in most of his writing is not so much hope as expectation. From the time that he began studying science, du Bois-Reymond *expected* to make important discoveries, win a job as a professor instead of practicing medicine, and be rewarded for his achievements in every sense. Since science did not pay off until one obtained an academic job, du Bois-Reymond had to live off of his father during his first decade with Müller. In 1841 he wrote to Hallmann:

> In comparison to what I can pry loose from my father, my experiments and books cost me an enormous amount, so that I've already come to the point where I'm going to have to take out a serious loan. I can only hope that my bourgeois relations will let me do it *insciis parentibus*.[41]

To live like a middle-class man—which he never doubted that he deserved—du Bois-Reymond needed support from his parents. Félix, who liked the idea of a scientist son, granted an allowance that was never big enough.

As a member of the Berlin *Bildungsbürgertum* (educated upper middle class), du Bois-Reymond regarded himself as superior to most other Germans and was proud of his manners as well as his first-rate education.[42] He bragged to Hallmann about his ability to charm women: "After one soirée at Müller's, where things were as stiff as usual, Reichert asked Müller's wife [*die Müllerin*] how she had liked his assistant [du Bois-Reymond], to which Müller replied, 'Thank God, that's the first one yet who's paid any attention to my ladies.'"[43] Well aware of his arrogance,

Figure 3.1 Emil du Bois-Reymond. Source: Emil du Bois-Reymond and Carl Ludwig, *Zwei grosse Naturforscher des 19. Jahrhunderts: Ein Briefwechsel zwischen Emil du Bois-Reymond und Carl Ludwig*, ed. Estelle du Bois-Reymond (Leipzig: Johann Ambrosius Barth, 1927), frontispiece.

the attentive assistant wrote to Hallmann, "Modesty, you must know, is a bovine virtue."[44] In the revolutionary year of 1848, the impoverished young experimenter called his landlords "the philistines."[45] When writing to his friends, du Bois-Reymond never concealed his contempt for some of the artisans who supplied scientists with instruments, such as the optician Doerffel, "a mere haggler with lorgnettes for Guards officers and Asiatic tourists, rich Poles, Russians and similar folks."[46] He greatly respected the mechanics who helped him design his instruments, however. His letters contain no derogatory references to Müller's class origins, but his descriptions of his adviser must be read in the context of these other remarks.

In his insults of uncultured or misguided scientists (a category that included most of his peers), du Bois-Reymond was particularly creative. He complained to Hallmann about Müller's dissecting hall full of medical students, where "the dead company was far less repulsive than the living."[47] In early 1840, when du Bois-Reymond obtained the much-coveted keys to the Anatomical Museum, he did not get them directly from Müller. He had to go through the custodian Thiele, "that impenetrable, gravelly, spiny creature of skin and bone in our museum."[48] Never impressed by scientific prominence, du Bois-Reymond despised anyone insufficiently interested in physiological experiments. He called Johann Christian Jüngken (1793–1875), director of surgery at the Charité Hospital, a "steer" (*Rindvieh*) and claimed that the embryologist Martin Rathke was "as dumb as an ox."[49] The renowned chemist Carl Gustav Mitscherlich (1805–1871) was a "stain on Prussian science."[50] For the privileged du Bois-Reymond, failure to appreciate electrophysiology meant more than a weak understanding of science; it implied a lack of culture.

As he aged, du Bois-Reymond became increasingly conservative, but even in his youth he was no revolutionary.[51] As a young scientist, he might best be described as an elitist liberal.[52] During the uprisings of 1848, while Virchow fought at the barricades and Müller defended the museum, du Bois-Reymond stayed home, sick in bed. About a month later, he wrote to Ludwig:

> Even though I had not been behind the barricades, one was thrilled through and through with the happy awareness that one felt the courage in defiance of all the guards' bayonets, to defend the achievements one had not helped fight for. . . . I had been unwell on the days before [March] 18th, and it was partly for this reason and partly on principle that I . . . did not participate in the movements on those days. They seemed to us, and they were, nothing better than riots fomented by a nasty clique of Jewish litterateurs. . . . [I] am, for the moment, keeping as aloof from political participation as I can without seeming to lack principles.[53]

Du Bois-Reymond did not specify what these principles were, but apparently they demanded that he support the popular rebellion against the Prussian state while refraining from any actual fighting. If he was torn in two directions, illness provided the ideal solution. In early 1849 he wrote to Hallmann, "At first, I was quite intoxicated by the wine of the new time, but unfortunately the nasty reality of things soon returned me to reason, and I had the satisfaction of being one of the first in my circle of friends to be denounced as a gray-haired reactionary."[54] The activist Virchow, who spent the fall of 1848 campaigning for democracy and medical reform, never liked du Bois-Reymond. Even after they had been colleagues for decades, he addressed him as "Sie" in short, icy letters.

In du Bois-Reymond's understanding of the 1848 uprisings, Shakespeare played a pivotal role. He wrote to Carl Ludwig that in March 1848, when Berlin began "governing itself," "the whole depth of the German national character was to show itself in a scene that no Shakespeare could have made more tragic."[55] In the same letter in which he called *Hamlet* "the profoundest work ever written," he confided to Hallmann, "The [Prussian] king is a Hamlet," that is, all talk and no action.[56]

As a scientist, du Bois-Reymond sought a topic that would keep him from becoming a Hamlet, a question that would compel him to act. Well aware of his own tendency to make grand plans, and then to change them, he sometimes doubted that science was the best profession for him. In August 1840 he wrote to Hallmann, "You insist again in your last letter that I have a special gift for the investigation of organic nature, while I've reached the sad conclusion that I'm not an investigator of nature in any respect."[57] But by May 1842 he had changed his mind, reflecting, "I can't understand why—a year ago, I think—I doubted that I could become an investigator of nature [*Naturforscher*]."[58] By *Naturforscher*, du Bois-Reymond meant a scientist, a person who investigated nature by performing experiments. Upholding electrophysiology as an ideal, he gained the confidence to do science only when he began exploring animals and instruments with his own hands. The physical aspect of experimentation reassured him that he was no Hamlet.

Zapping the Horse

In early 1841, Müller told du Bois-Reymond about the Italian physiologist Carlo Matteucci's (1811–1868) new evidence for electricity inherent in nerve and muscle tissue. Matteucci's work had revived debates of the 1790s about whether animals contained their own, intrinsic electricity. In the late eighteenth century, the Italian physician Luigi Galvani (1737–1798) had produced evidence that frog nerves and muscles were electrically active, but the Italian physicist Alessandro Volta (1745–1827) had denied this, arguing that the electricity Galvani had observed had been produced by the metal electrodes used to detect it. Alexander von Humboldt had defended Galvani, showing that one dissected frog leg could make another contract in the absence of metals, but Volta's invention of the battery had won him so much respect that for several decades, his view had prevailed.[59] Matteucci's work suggested that Galvani and Humboldt might have been right after all, and Müller urged du Bois-Reymond to repeat and extend the Italian physiologist's experiments. In the opening sentences of his greatest work, *Animal Electricity* (1848–1849), du Bois-Reymond re-created this moment, revealing Müller's catalytic role in the project. By his own account, the young physiologist began studying animal electricity at Müller's "request."[60] He dedicated the first volume of *Animal Electricity* to Müller "with respect and gratitude," making clear who had given the project its initial impulse. As he was well aware, Müller shaped his scientific career. Du Bois-Reymond spent the next fifty-five years studying animal electricity.[61]

Electrophysiology consumed du Bois-Reymond's life because it provided a way to demonstrate his strongest scientific conviction. He believed that no forces operated in living organisms that were not also present in inorganic materials, and he studied the electricity of nerves and muscles to disprove the existence of life force. One may wonder why Müller, who believed living matter had unique properties, assigned his student this project. By doing so, he permanently molded du Bois-Reymond's science, giving him his lifelong research topic and his preferred experimental animal, the frog. Probably Müller wanted to know whether electricity played a role in nerve impulse transmission and muscle movement, and his technically oriented student seemed the best person to find out. Perhaps Müller didn't care about du Bois-Reymond's theoretical orientation, since the student never mentions any face-to-face debates. Like his student, he may simply have wanted to know how nerves worked, and didn't worry about the philosophical implications of the answer, as long as it was right.

Du Bois-Reymond discovered that every muscle in the frog produced electrical activity detectable with a magnetized needle. While his model of nerve impulse conduction proved wrong at the particle level, he demonstrated that nerves and muscles do conduct detectable currents that vary with their activity and soundness. In a private letter to the physiologist Henry Bence Jones, in which he could throw aside the caution demanded by scientific publications, he proposed an electrochemical mechanism that sounds a great deal like the one accepted today. Nearly a century before chemically charged ions were known to pass through channels in neuronal membranes, he wrote, "I fancy that in the muscles and in the electrical organs, centers of chemical action, or groups of electropositive and negative atoms are disposed in a certain order . . . and that, in the moment of contraction or of shock, these groups assume some other position."[62] His electrophysiological studies of nerves and muscles laid the groundwork for our current understanding of nerve impulse transmission.

Perhaps even more important, du Bois-Reymond developed the methods and instruments to measure the minuscule currents in live animals.[63] He did his most valuable research during his first decade of experimentation (1840–1850), while complaining piteously to Hallmann, Ludwig, and Helmholtz about his poverty and lack of recognition.[64] The first volume of *Animal Electricity* was published in September 1848; the second emerged in parts in 1849, 1860, and 1884, and was never completed as originally planned. Together, the work amounted to over 2000 pages.[65] His electrophysiological studies consumed so much time and energy because they were works of the most exact science.

As a physiological experimenter in the 1840s, du Bois-Reymond despised unbridled theorizing. In his demand for rigorous analysis and hands-on research, he challenged Müller's 1824 vision of the ideal physiologist. Like Müller, he had originally admired nature philosophy, but he had later rejected it much more fully. Du Bois-Reymond came to know nature philosophy through the Berlin philosopher Henrich Steffens (1773–1845), who may have inspired his transfer to Bonn.[66] In his 1877 lecture "Physiological Instruction Then and Now," du Bois-Reymond ridiculed the analogical trend in nature philosophy, recalling Oken's and Steffens's tendency to align animals with human organs. Letting it speak for itself, he read a sentence of his lecture notes from Steffens's 1837–1838 anthropology course: "Every organ of the human body corresponds to a certain animal, *is* an animal. For example, the wet, slippery tongue, which can move in all directions, is an octopus, a cuttlefish."[67] Sliding smoothly from analogy to identity, the statement recalls the recklessness with which some nature philosophers imposed order on the animal kingdom. Depicting nature philosophy as a morass from which the physiologists of the 1840s heroically escaped, du Bois-Reymond avoids mentioning these philosophers' own experiments. As in his memorial address for Müller, the deprecation of nature philosophy allowed him to create a progressive narrative.

Of the many claims of nature philosophy, the notion of life force offended du Bois-Reymond the most. In his view, when theoreticians explained animal movements by means of a force unique to living things, they not only kept scientists from experimenting; they kept them from thinking. In the introduction to *Animal Electricity*, du Bois-Reymond declared, "Force in this sense is nothing but a hidden, monstrous growth of the irresistible tendency toward personification which has been impressed upon us, a rhetorical artifice of our brain."[68] He argued that the presumed division between organic and inorganic forces was completely arbitrary, but he also represented life force as an unwanted growth or specter, suggesting the emotion underlying his convictions. In *Animal Electricity*, he called life force a "hated weed," a "non-thing [*Unding*]."[69] Du Bois-Reymond introduced the book that would become his life's work as a war against this enemy, promising that "through my investigations, . . . life force will again be driven from one of its trenches; indeed, from one that is not the least stubborn."[70]

To convince his readers that muscles functioned electrically, du Bois-Reymond used a lot more than logic. Along with experimental data, rhetoric played an essential role. The first volume of *Animal Electricity* appeared the same year as *The Communist Manifesto*, and at times du Bois-Reymond's writing sounds a lot like Karl Marx's and Friedrich Engels's.[71] Born the same year as Marx, du Bois-Reymond was no revolutionary. He called the workers' movement a "psychosis of the working classes."[72] Probably he never read *The Communist Manifesto*, but as an educated middle-class German, he used the same rhetorical strategies as Marx and Engels to turn readers against established ways of thinking. The *Manifesto* opens, "A specter is haunting Europe—the specter of communism."[73] Marx and Engels use the specter metaphor to ridicule the middle-class's fears of communism, a movement they cannot comprehend, and thus associate with everything evil. Du Bois-Reymond applies the word to life force in a very similar way. Attacking the poorly defined concept, he declares, "This specter must finally be banished."[74] Using the same kind of irony, du Bois-Reymond and Marx expose long-standing contradictions: the physiologist, those of nature philosophy; the political activist, those of bourgeois thinking. Both alternate long, sarcastic sentences with short, telegraphic ones. Du Bois-Reymond's list of everything life force can do, for instance, resembles Marx's enumerations of communism's purported horrors:

> This all-purpose servant-girl, by the way, possesses highly diverse knowledge and skills. For she organizes, assimilates, secretes, reproduces; she directs development; she absorbs and distinguishes healing and poisonous, useful and useless substances; she heals wounds and creates crises; she is the final cause of animal movement; she helps the so-called soul, at least in the process of thinking, etc., etc.[75]

Both thirty-year-old writers reveal the absurdity of their opponents' claims simply by laying them out all at once. They argue that structures so riddled with contradictions cannot stand, and that, as part of an inevitable progressive development, each must be replaced by something more solid. "Physiology will fulfill its destiny [*Schicksal*]," concludes du Bois-Reymond.[76] Like Marx, he presents his work as the first step in a revolution that will produce knowledge based on material reality.

Marx claimed that in his theory of history, he had turned the philosopher Hegel on his head, starting not with an abstract ideal but with material conditions and economic struggles. In developing the language of his new experimental physiology, du Bois-Reymond made a similar move, seizing the terms of nature philosophy and endowing them with new meaning. The literary critic Harold Bloom has called this strategy "tessera," a process in which "a poet antithetically 'completes' his precursor, by so reading the parent-poem as to retain its terms but to mean them in another sense."[77] Bloom identifies this seizing and transformation of language while analyzing poets' anxieties about their predecessors, but his insight applies equally well to du Bois-Reymond's dismantling of nature philosophy.

In the physiologist's essays and letters, one often encounters the term *Zweckmässigkeit*, meaning purposefulness or—as he used it—functionality. One could even call it one of his favorite words, since he relied on it to express his scientific and aesthetic ideals.[78] Significantly, though, it was not his term. Kant and the nature philosophers he inspired had used it to describe the unique way in which each part of an organism served the living whole. Jakob Henle had invoked it when telling his parents why he was so captivated by Müller's comparative anatomy. Consciously and deliberately, du Bois-Reymond applied it to machines, denying any fundamental difference between organic and inorganic matter. In the introduction to *Animal Electricity*, he wrote disgustedly, "As far as the *Zweckmässigkeit* of organic nature is concerned, I will not even allow myself to dignify the childishness of this idea with a response."[79] When he republished this introduction in 1869 as the polemical essay "On Life Force," he reflected, "I knew no better way to challenge the grounds for life force based on the *Zweckmässigkeit* of organic nature than by [giving] evidence for the *Zweckmässigkeit* of inorganic nature."[80] Unlike Helmholtz, who banned the word from his scientific writing, du Bois-Reymond kept it alive as a reminder of how science had changed.[81]

The disturbing frontispiece of *Animal Electricity*'s second volume (1849) depicts du Bois-Reymond's violent break with the philosophical past. In an extraordinary drawing that du Bois-Reymond made himself, a swimming horse writhes in agony, having just been shocked by an electric eel. The image illustrates a claim from Müller's *Handbook of Human Physiology* that had haunted du Bois-Reymond for years: "Electric eels can . . . even fight and weaken horses, as A. v. Humboldt described so beautifully in his *Views of Nature*."[82] Like the other superb illustrations of *Animal Electricity*, this image conveys du Bois-Reymond's tactics as well as his essential message. In the introduction to the first volume, he had described the entrenched opposition to experimentation, then written, "How could I hope to confront these oppressive influences better than by trying to make the reader a participant in the mental work, an inner eyewitness to the experiments?"[83] A gifted draftsman, du Bois-Reymond believed that he could convince readers with visual images. In the frontispiece, the tortured horse—whose every muscle is accentuated—suggests life force under attack, suffering from a jolt of animal electricity. Experimental physiology is zapping nature philosophy.[84]

Throughout his career as an experimentalist, du Bois-Reymond had problems with horses. He performed many electrophysiological experiments while living over a stable on Karlstrasse, and his awareness of his downstairs neigh-

Figure 3.2 The frontispiece to volume 2 of *Animal Electricity*, showing a horse being attacked by an electric eel. Source: Emil du Bois-Reymond, *Untersuchungen über thierische Elektricität* (Berlin: Reimer, 1849).

bors comes across in his writing.[85] In the introduction to *Animal Electricity*, he declares that "matter is not like a cart to which forces can be arbitrarily hitched or unhitched like horses."[86] In this metaphor, he uses horses to represent the life force of nature philosophers, which purportedly animated living things, then mysteriously disappeared when they died. The vibrations of horse-drawn carriages disrupted his experiments in the Anatomical Museum, and even in the 1880s, in his specially constructed Physiological Institute. For du Bois-Reymond, the horse became the perfect symbol of life force, for each was a confounded nuisance.

The Frog Kennel

Du Bois-Reymond's relationship with Müller determined more than his choice of research topic and the physiological system with which he worked. To a large degree, it also decided the spaces in which he could carry out his studies. Until he married in 1853, du Bois-Reymond performed most of his experiments in his own apartment. He recalled that in the 1840s:

> If a young person wanted to conduct physiological experiments himself, he had to do so mostly in his own room [*Stube*], where the frogs and the rabbits got him into trouble with his neighbors (we didn't dare try anything with dogs), and where many investigations were simply impossible, or could be performed only by struggling with the utmost repulsiveness.[87]

From the time he began medical school until 1845, du Bois-Reymond stayed at his parents' house, Potsdamerstrasse 36A, then quite far from the city center. When he believed he had completed his experiments on animal electricity, his father gave him his own apartment at Karlstrasse 21, alongside the Veterinary School. In 1850, he moved to a better apartment in the adjacent building (Karlstrasse 20).[88] But of course, no work space was ever good enough. When scarlet fever confined him to his room in 1840, he complained to Hallmann that it had "one window, two doors" and was "isolated at the end of a long, cold hallway."[89] Helmholtz remembered that du Bois-Reymond's apartment at Karlstrasse 21 contained "two little rooms, each with one window." One of them had "a small bookshelf and a sofa to sleep on"; the other, du Bois-Reymond's dissecting table, galvanometer, and other instruments. The workroom overlooked the Veterinary School courtyard.[90]

Like Müller's other students, du Bois-Reymond also worked at the Anatomical Institute and Anatomical Museum, so that he was forced to carry instruments and animals back and forth between the two. In 1841, he bemoaned:

> The endless going and coming, the endless schlepping of instruments and preparations between home [*von mir zu Haus*] and [the Anatomical Museum] and the other way around . . . . If Thiele isn't there [*zu Haus*],

> you've walked across the city for half an hour in the heat with a sack of frogs and ice for nothing.[91]

Du Bois-Reymond uses "at home" [*zu Haus*] for both his room in his parents' house and the Anatomical Museum. While the second sentence refers to the custodian Thiele, the phrasing suggests a merging of his living and working space. In a letter to Ludwig, he called his apartment "the dog kennel."[92]

In reality, Karlstrasse 21 was a frog kennel. In his studies of spinal nerve roots, Müller had shown physiologists the value of frogs as experimental animals. Hardy and easy to breed, the frog offered a simple, accessible model for exploring the relation between muscles and nerves: the gastrocnemius muscle (the powerful thigh muscle that allows frogs to jump) and the highly visible sciatic nerve. Du Bois-Reymond's studies of animal electricity therefore depended on a steady supply of healthy frogs. By the fall of 1841, 100 frogs shared his room in his parents' house.[93] In his building on Karlstrasse, he became known as the "frog doctor."[94]

In April 1848, with Berlin in political and social uproar, du Bois-Reymond complained to Ludwig about the "lack of frogs."[95] They were easier to catch and maintain than other experimental animals, but it could be a challenge to keep them alive, especially in winter. In *Animal Electricity*, du Bois-Reymond wrote that during four successive winters in the 1840s, he had lost 10–50 percent of his frogs in epidemics.[96] Dejected, he told Ludwig in 1871 that because of his servant's absence, all of his frogs had "frozen to death."[97] Nervously aware of his reliance on the animals, du Bois-Reymond joked with Ludwig about them even when he was at his most desperate. In 1875 the two arranged for an emergency shipment, describing their cargo as if it were sensitive contraband. Du Bois-Reymond began: "For the first time since I started sending frogs' souls to Hades thirty-five years ago, I am deadly short of sufficiently large specimens. . . . Do you have really big frogs?" The next day, Ludwig answered from Leipzig, "We have the frogs that you long for. . . . I shall hand half a dozen of the frogs to the head conductor of the train that departs on Tuesday at 11 o'clock, telling him to take them to the university directly on arriving in Berlin, and that he will get a tip of 1 taler if the animals arrive in good condition. The frogs arrived "well, though a little stiff with cold."[98]

When he could get the right animals, du Bois-Reymond produced impressive demonstrations of muscles' electrical activity, attracting curious physicists and physiologists who wanted to see the evidence with their own eyes. Among the scientists who visited the frog kennel were Humboldt, Müller, Magnus, and Helmholtz. When du Bois-Reymond discovered that a muscle in a prolonged series of contractions deflected a galvanometer needle, he called the experiment "so striking, simple, certain, and steady in its success that as long as my apparatus is in good condition and I have frogs on hand, I volunteer to repeat it in any place, at any time, as many times as anyone would like, without it once failing."[99] His challenge echoed Müller's 1831 nerve root study, which had demanded that a good physiological experiment yield "the same certain and unambiguous phenomena in any place, at any time, under the same conditions."[100] In the summer of 1846, du Bois-Reymond began seeing deflections caused by long, powerful contractions of human arm muscles. Not all of his friends who flexed their biceps could make the galvanometer needle move, but seventy-nine-year-old Alexander von Humboldt succeeded in May 1849, with Helmholtz and Müller watching.[101] Du Bois-Reymond sought real as well as virtual witnesses for his experiments, and the Berlin scientists enjoyed viewing animal electricity at work. Müller's and Humboldt's willingness to visit a student's apartment to see experiments shows that interest in electrophysiology was not restricted to scientists of a new generation. Like Humboldt, Müller supported du Bois-Reymond's work and was eager to see his results.

Humboldt, whose mother was a Huguenot, was quickly attracted to du Bois-Reymond and did everything possible to promote his career. Whereas Müller had been introduced to Humboldt as a Berlin outsider, Humboldt was part of du Bois-Reymond's cultural network from the beginning. When writing to Humboldt, du Bois-Reymond played on the senior scientist's personal experience with electrophysiology, framing his work as a continuation of Humboldt's experiments fifty years earlier. Writing to Humboldt in French, he declared:

> Physicists and physiologists will see this dream realized of an electricity operating the movements and perhaps transmitting the sensations in animal bodies, a dream that will soon have lasted a century, and of whose prestige, if I am not mistaken, you yourself, Monsieur le Baron, have never ceased to be aware.[102]

Humboldt also emphasized the continuity, using his own reputation to market the younger physiologist's work. Describing du Bois-Reymond's experiments to Cultural Minister von Ladenberg, Humboldt wrote:

> He is studying a matter, the deep natural secret of muscle movement, with which I, too, was occupied in the earlier half of my life. He has succeeded in making the discovery after which I sought in vain: he has physically demonstrated how (without any stimulating influence of metals or other inorganic substances) a person can move a magnetic needle at a distance through force of will and muscle tension.[103]

To the end of his life, Humboldt enthusiastically supported du Bois-Reymond's electrophysiological work.

Du Bois-Reymond, in turn, respected Humboldt's love of experimentation. He asked Ludwig, "Who ever saw an eighty-year-old man visit a young man of science in an apartment over the stables, himself set to work on experimenting, . . . then go trundling off in a splendid court equipage to his daily dinner party at the king's side?"[104] Humboldt's letters suggest that he was drawn to du Bois-Reymond partly out of nostalgia. As a companion to the Prussian king, he lacked control over his time, having to leave appointments and rush off to court at short notice. Although he gladly used his influence to advance science, he missed his younger days as an experimenter. In 1850 he wrote, "The daily pressures . . . [and my] penchant for work, at which I'm constantly interrupted, give me a joyless life! So that all the more, my dear Du Bois, I long to see *you*."[105] For Humboldt, du Bois-Reymond embodied his earlier life: that of a bright, inquisitive young man with no administrative duties and all the time in the world to experiment. In 1850 Humboldt told Müller: "I know of no one so well suited for the Academy due to the breadth of his knowledge, his talent, the fineness of his experiments, and the achievements he has already made, as Du Bois!"[106] Together, his mentors worked to get du Bois-Reymond elected to the Academy of Sciences, which would ensure his scientific success.

Although du Bois-Reymond became an Academy member in 1851, he did not receive any more space in which to conduct experiments. In the early 1850s, he worked mostly at the Anatomical Museum. With his usual sarcasm, he wrote to Helmholtz in 1851:

> Supposedly, some sort of physiological institute is going to be set up at the museum this summer. Together, Müller and I have posted an announcement for *Exercitationes physiologicas* [a laboratory course]. . . . These facilities have arisen thanks to a step from our enlightened ministry.[107]

Du Bois-Reymond's new "Institute," located on the third floor of the university's main building, offered a far better space for dissections than for physiological experiments. Though reasonably clean and well lit, the high rooms trembled with the vibrations of passing traffic. Until the Berlin Physiological Institute was completed in 1877, du Bois-Reymond had to make do with cramped spaces in, around, and above the museum, which became officially known as the Physiologisches Laboratorium. He called it the "musty hole-in-the-wall."[108]

The Current and the Spear

Du Bois-Reymond lived as intimately with his instruments as with his laboratory animals. Despite his cultural elitism—or perhaps because of it—he loved building setups with his own hands.[109] To solve technical problems, he consulted mechanics in machine shops. Viewing the manual labor of physiology as he did the gymnastics at which he excelled, he claimed that exercise refined the body and mind simultaneously.[110] In his determination to incorporate technology into scientific research, he differed greatly from Müller.

In Berlin, du Bois-Reymond worked directly with the city's leading artisans to design his electrophysiological equipment. In 1851, he bragged to Henry Bence Jones, "I have mechanics on hand, who are always ready to work for me at the slightest notice."[111] At the Physiological Institute's opening ceremony, he looked back at the days when

> We rolled our own coils [of copper wire], soldered our own elements, even glued our own rubber tubes together, for at that time you couldn't

> buy rubber hoses. We sawed, planed, drilled, filed, turned, and sharpened. Our need for advice and help in mechanical matters drove us to the workshops, where we learned all kinds of maneuvers through exchanges with highly skilled artisans. They made us as familiar with the construction of instruments, right down to the last screw, as with the anatomy of an animal.[112]

Here again, du Bois-Reymond was defying theorists who believed that a mysterious force distinguished animals from machines. Preparing his own instruments improved du Bois-Reymond's self-confidence, but he never struggled alone. In Gustav Magnus's physics lab, du Bois-Reymond met other young scientists who shared his interest in the anatomy of instruments, which supplanted Müller's less exciting anatomy of fishes.

As an experimenter, du Bois-Reymond tried to construct the most functional equipment possible, comparing the movements of an efficient machine to those of an expert gymnast.[113] A devotion to streamlined instruments characterized his aesthetics of experimentation, a sense that he would claim Müller lacked. During the 1840s, while Müller was admiring the rococo forms of sea urchin embryos, an anti-ornamental movement dominated German machine construction.[114] Du Bois-Reymond joined it enthusiastically. Describing his studies of nerve currents to Ludwig in 1848, he asked, "Can one hope with this [work] to get through to the physiologists, before whose coarsely woven visual matter the most that the word electricity conjures up is the rococo image of a colossal electrostatic machine on feet of sealing wax?"[115] In his eyes, rococo swirls and cherubs became a visual symbol of eighteenth-century ignorance. On the one hand, du Bois-Reymond's objection to unnecessary decorations illustrates the ebbs and flows of opposing artistic ideals in nineteenth-century Berlin. Magnus's physics lab, where young experimenters designed beautifully functional setups, had been built in a rococo house.[116] But du Bois-Reymond's 1890 reference to "horrible memories of the impure forms of rococo architecture and furniture" suggests that to the end of his life, he was exasperated by Müller's love of fragile, elaborate structures.[117]

Besides the frog, the laboratory "animal" with which du Bois-Reymond most often struggled was the galvanometer. Since 1820, when the Danish physicist Hans Christian Ørsted (1777–1851) had shown that the flow of electrical current deflected a magnetized needle, physicists and physiologists had used galvanometers to detect electricity. The most sensitive detector was still a frog leg, which would start twitching in the presence of current. But a galvanometer—a free-swinging needle mounted within a coil of wire—offered a quantitative measurement because the angle of deflection varied with the intensity of the current. The more turns the coil contained, the more sensitive the instrument.[118] Until the mid-1830s, most galvanometers could not reliably detect the electrical activity of muscles and nerves. Matteucci had used one with 2500 turns but had obtained ambiguous results.[119] Soon after starting the animal electricity project, du Bois-Reymond realized that he would have to build his own instruments.

Between 1841 and 1848, he tried to construct the most sensitive galvanometers ever made, winding his coils by hand. By the spring of 1842, he obtained reliable readings of muscle and nerve current with a coil of 4650 turns. Eventually, with 24,160, he detected electrical activity in nerves in the absence of any external electrical stimulus.[120] Du Bois-Reymond's instruments must have impressed Müller, for he wrote to Retzius that during his 1846 summer research trip to Helgoland, "I had an extremely sensitive galvanometer with me" to study the organs of electric rays.[121] He did not say who had built it. Du Bois-Reymond's collaborator, the instrument maker Halske, later recalled:

> I saw in my mind how a galvanometer housing was fitted with 33,000 turns with the greatest feeling and care, only—on completion—to have no current. Horrible! The wire was unwound, and in its overblown condition occupied the workshop personnel as a source of entertainment for a long time, as a curly wig.[122]

Under these conditions, it is not surprising that images of coils and needles permeated du Bois-Reymond's unconscious, affecting the way that he read and wrote. In May 1841 he confided to Hallmann, "Last night in a dream I saw G. rolling a galvanometer."[123]

When du Bois-Reymond republished the introduction to *Animal Electricity* as the essay "On Life Force," he added an epigraph:

> And now I will unclasp a secret book,
> And read you matter deep and dangerous,
> As full of peril, and advent'rous spirit,
> As to o'erwalk a current, roaring loud,
> On the unsteadfast footing of a spear.[124]

To introduce his argument, du Bois-Reymond quoted Shakespeare's *King Henry IV, Part I.*[125] With these references to "peril" and "secrets," he was mocking life force, exposing its alchemical roots. At the same time, however, Shakespeare's spear and current suggested the "adventurous spirit" of animal electricity. A drawing from a letter to Helmholtz in March 1862 hints why du Bois-Reymond may have chosen this epigraph. In the sketch, two spearlike, precariously balanced electrodes wobble over a muscle that resembles a small stream. In 1597, when Shakespeare wrote the play, "current" did not mean the flow of electricity; he was referring to the force of a river. Du Bois-Reymond's English was excellent, and it is unlikely that he misread the lines. He read Shakespeare—as he read British scientific journals—in the original, but probably he could not resist the suggestive power of the word "current." To du Bois-Reymond, the attempt "to o'erwalk a current roaring loud, on the unsteadfast footing of a spear" brought to mind electrophysiological recordings.[126] Possibly Hamlet's "to be or not to be" speech ("their currents turn awry, and lose the name of action") produced the same associations.

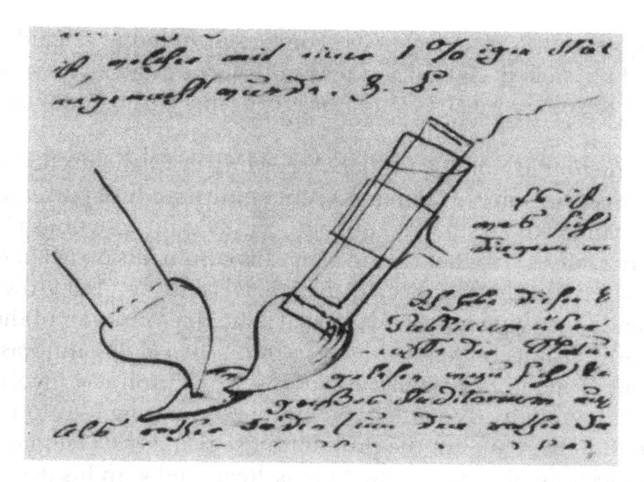

Figure 3.3 Two electrodes probing a muscle. Source: letter of du Bois-Reymond to Helmholtz, 25 March 1862. Emil du Bois-Reymond and Hermann von Helmholtz, *Dokumente einer Freundschaft: Briefwechsel zwischen Hermann von Helmholtz und Emil du Bois-Reymond 1846–1894* (Berlin: Akademie Verlag, 1986), 202, printed with kind permission of the Akademie Verlag.

The Seduction

Like his dedication to cutting-edge technology, most of du Bois-Reymond's relationships did not last. As soon as he suspected that a friend did not share his scientific values, he would grow disappointed and allow the connection to dissolve.[127] He could not withdraw from Müller, since his scientific career depended on their troubled relationship, but he did withdraw from Karl Reichert, who replaced Henle as Müller's prosector in 1840. Although du Bois-Reymond at first looked up to Reichert and asked for his advice, he lost respect for him when he realized that Müller's senior assistant did not think biological phenomena could be explained by physical laws. "All Reichert really does is hunt water bugs," he wrote to Hallmann disgustedly.[128]

Du Bois-Reymond's attachment to Eduard Hallmann followed the same pattern. In early 1839, the two met at a gym. During the years in which *Burschenschaften* were banned, gyms became places of political activism, but du Bois-Reymond seems to have been more interested in developing his muscles.[129] He and Hallmann liked each other immediately. Five years older and from a well-to-do family, Hallmann had studied theology in Göttingen, but in 1834 had left for Berlin, partly because he wanted to study science with Müller, and partly because of a political incident.[130] He must have impressed Müller, for a year after his transfer, the professor made him a museum helper. It was Hallmann—a friend of Theodor Schwann—who convinced du Bois-Reymond to study medicine, outlining a three-year course schedule that included Müller's anatomy and physiology classes, and his public lectures on anatomy of the sensory organs. "Reading Müller's *Handbook* is extremely tiring but indispensable," he advised.[131]

But Hallmann did not like Müller. According to Estelle du Bois-Reymond, the two split over "irritable" Hallmann's comparative study of the temporal bone, leaving Hallmann with a lasting resentment.[132] Estelle did not publish Hallmann's letters with her father's, so that we can gauge his comments on Müller only from Emil's responses. Clearly, he expected an ogre. His reluctant but growing respect for Müller and his willingness to challenge Hallmann's opinion of their mentor suggest his slow disillusionment with his friend. When Hallmann was forbidden to practice medicine in Prussia because of his former political involvement, du Bois-Reymond wrote to Alexander von Humboldt, urging him to use his connections to undo the ban, but the prohibition remained.[133] Hallmann followed Schwann to Belgium, disappointing du Bois-Reymond with his decision to give up experimentation for clinical practice.[134] After 1841, his letters to Hallmann—his most vivid and intimate writing—grow shorter, less communicative, and more infrequent.

Müller did not warm to du Bois-Reymond as he had to Henle and Schwann. On the contrary, he had to be won over. In du Bois-Reymond's letters to Hallmann, he described his progress with senior scientists as ironically as an eighteenth-century libertine might narrate his conquests. In September 1839, Hallmann had urged him to meet Matthias Schleiden in Jena, and du Bois-Reymond's story of their encounter reads like a tale of seduction:

> When I spoke, I turned toward [Schleiden]. At last the guests begin to disband. We are almost alone. I get more and more involved in what I am saying. I tell him, cautiously, selectively, what I am finding, speaking supplely and determinedly. I sit opposite him, leaning forward, elbows resting on the table, looking intently at him with all of my strength. The *ingénu* [*Der Unbefangene*]. His thoughts must have been far from my youthful enthusiastic delirium.[135]

Written in the present tense to give a sense of immediacy, du Bois-Reymond's description reads much like Count Valmont's reports of his "campaign" against virtuous Madame de Tourvel. In Choderlos de Laclos's *Dangerous Liaisons*, seduction is narrated as a hunt or military conquest, and du Bois-Reymond's account—in which the *ingénu* is the aggressor—owes much to eighteenth-century French displays of wit and narrative power.

In his letters to Hallmann in 1839–1845, du Bois-Reymond creates Müller as a literary character, developing him as their relationship progresses. For Hallmann's entertainment, he reports his impressions as they occur, so that his reader can experience them simultaneously. Du Bois-Reymond thus makes himself a character in an epistolary novel, but it is crucial to remember that he is also the author.[136] By telling Hallmann about Müller, he was writing to please his friend, who was trying to establish a medical practice in Belgium and recalling their mentor with mixed emotions. By turning Müller into a fictional personage, du Bois-Reymond shows off his wit, but more important, he takes control. His career as a scientist depended on pleasing a physiological ogre, and writing about his conquest eased his fears.

In a letter to Hallmann in September 1839, he describes his first meeting with Müller:

> I had to wait around for a while; then the grim-faced man appeared and led someone out. When he came back, he turned to me and withdrew politely into his room, inviting me to follow. I followed in his footsteps as closely as I could (for I have to say, I found his presence imposing) and said what was necessary in a cold way, without making real eye contact. But in fact he was quite friendly and kept saying repeatedly, "Good, good, I'll take care of it."[137]

Du Bois-Reymond freely admits his anxiety over the first encounter, making the initial conditions seem unfavorable.

With a beginning like this, his wooing of Müller becomes even more impressive.[138] By mid-October, he boasts:

> It seems as though I've tackled him in the right way, since everyone I've told about my meeting with him has been astonished by his politeness. In fact, he still knew my name and accompanied me to the hall door in a very friendly way. You know, Müller seems to me like a person with a weak character who hangs that imposing air over himself like a lion's skin.[139]

In this letter, his intimidation is mixed with resentment that he should have to win Müller's confidence in order to become a researcher.

As the wooing progressed, du Bois-Reymond found himself pulled in two directions. To please Hallmann, he needed to make Müller look bad, but to prove that he was gaining ground, he had to show that Müller was treating him better. In the first week of December, he reported:

> Your Müller is by far the most unpleasant person I've encountered behind a podium in a long time. Certainly he's always very polite to me, so that if any fine or interesting preparations turn up, I go to his room, where he's pacing around and mumbling to himself, humming to himself in all sorts of ways, and I look at the things more closely. Then, to be friendly, he starts making the strangest faces.[140]

Here Müller becomes a bizarre figure, eccentric and socially challenged. Yet in the process of the seduction, the mumbling professor began winning over his seducer. In mid-January, du Bois-Reymond wrote to Hallmann:

> I can't deny that for me Müller has become the daily theme. I know you're still mad at him [*du bist über ihn blasiert*], but sometime you must also have felt the power of his personality. In the end he has certainly

> imposed himself on me, [and] I've found him outwardly gruff, as only he can be to me. [But] I have [given] a definite sign that I would like him to show me special favor.[141]

In early February, Müller responded to this sign, granting the eager student the ultimate token of his trust. Du Bois-Reymond reported triumphantly, "I have the keys to the cabinets in the [Anatomical] Museum and manage things there as I please."[142] He also mentioned that Müller had begun showing him his drawings, perhaps because he recognized the student's artistic ability.

As he grew closer to Müller, du Bois-Reymond began to distance himself from Hallmann, even as he assured his reader of his continued confidence. About two weeks after winning the coveted keys, he wrote:

> Naturally I still haven't said a word about you to Müller. I'm now almost too much indebted to this man. I don't know if I've told you that I have in my hands the keys to the museum. But in no case will I offer myself to him as an assistant [*Famulus*] as I had planned for a time. . . . Nor do I have the least desire to work for the museum, as far as messing around with skeletons . . . I think it's a waste of time. . . . But since New Year's I've gotten a whole different notion of anatomy and have developed much more respect for it.[143]

To maintain the friendship with Hallmann, his experienced adviser, and to impress Müller, his potential mentor, du Bois-Reymond had to practice some delicate diplomacy. He needed to criticize Müller to show Hallmann that his views still mattered, and to convince Müller that he needed the keys to the museum cabinets because of a nonexistent, all-consuming interest in anatomy.

That spring and summer, du Bois-Reymond reported his continued progress in quick, ironic "dispatches."[144]

> 27 May 1840. Relationship to Müller: I only go to him when I need him, then impress him as much as possible with my firmness, if not to say crudeness, and also with sparkling, brand-new scientific items like the Wheatstone stereoscope, of which he still knew nothing.
>
> 27 July 1840. Relationship to Müller: Has only gotten better. Concessions much more on his side than on mine. An offer to have tea with him and his clever wife. My judgment of him remains . . . . [145]

We can only guess what du Bois-Reymond thought about Müller that July, since his daughter Estelle omitted the next words when she edited his letters in 1918. Her edition of his correspondence with Hallmann contains many such ellipses in places where he apparently insulted other scientists. Considering the passages that Estelle left in, his judgment must have been harsh, delighting his reader in Belgium. But as he worked with Müller, his opinion continued to change.

When summer vacation began in August 1840, du Bois-Reymond reflected:

> The men who seem ingenious [*geistreich*] to me are those who, without achieving anything precise or really extraordinary in a field, without

> bringing to light any new general idea, are endowed with a richness of ingenious [*ingeniös*] and enlightening thoughts and give them a form that is not necessary for the whole but is complete, appropriate, plastic, and resilient only in individual cases. . . . Clearly, Müller fits my definition of the ingenious man. . . . Müller's mighty head, that dynamic lion's face . . . often strikes me as ridiculous, but I always feel small before such naturally talented individuals. While I in no way defend his outward behavior, I praise him with my whole heart.[146]

Like Valmont, du Bois-Reymond made his conquest, but in the process, he changed his attitude toward the person whose confidence he was trying to win. In spite of his desire for scientific independence, in spite of his initial scorn for anatomy, du Bois-Reymond became Müller's assistant and found himself dissecting animals in Müller's museum.

When Müller left for his yearly research trip in August 1840, he gave du Bois-Reymond a list of things to investigate. The new assistant passed it on to Hallmann verbatim, hoping that, like Steffens's nature philosophy, it would speak for itself:

> Analysis of the liver of crustaceans and snails.
> Examination of the ears of insects.
> Organization [*Bestimmung*] of some fossilized amphibian skulls in the Mineralogical Museum.
> Determination [*Bestimmung*] of the effect of peripheral illumination on the human eye by widening the iris with belladonna extract. (A hideous experiment! You dab the inner eyelid with the extract, then the iris stays unusually wide open for several days, and you can't read or see anything close up. And this is not supposed to be risky?)
> The digestion of birds.[147]

Even knowing that he wrote to entertain Hallmann, one can see that for du Bois-Reymond, Müller was a means to an end. The young physiologist had no interest whatever in the digestion of birds, and wanted the professor for his workroom and instruments. In October 1840, he told Hallmann, "I'm working as Reichert's assistant. That way I have the advantage of living without Müller's good graces, while being allowed to gain some experience working in his room."[148] Still, he longed for the approval of the eminent scientist and felt real pleasure when he won it. A year later, in the fall of 1841, he wrote to Hallmann, "I'm starting to believe that my relationship with [Müller] can mean more than simple exchanges with an honored teacher."[149]

The Oppressive Mentor

From the time that Hallmann began advising du Bois-Reymond, Müller's presence hung over the younger student. In March 1840, he wrote to Hallmann that his friend Henry Smith had hung Müller's portrait up in his (du Bois-Reymond's)

Figure 3.4 The 1837 portrait of Müller that hung over du Bois-Reymond's head, a color lithograph by Louisa Corbeaux based on a drawing by S. Lawrence. Source: Wilhelm Haberling, *Johannes Müller: Das Leben des rheinischen Naturforschers* (Leipzig: Akademische Verlagsgesellschaft, 1924), plate IV, 193.

room.[150] Thirty-seven years later, at the Physiological Institute, a bust of Müller graced the main portal, and when the Institute opened, du Bois-Reymond referred dramatically to a time when he, Brücke, and Helmholtz had "sat here at the master's feet."[151]

As an adviser, however, Müller practiced laissez-faire politics. He was not authoritarian or controlling. Du Bois-Reymond recalls that although Müller offered his books, instruments, even his museum collection, to promising students, "just as he stood on his own two feet, he demanded that his students learn how to help themselves. He gave them tasks and encouraged them; but as for the rest, he contented himself—to use a chemical analogy—with exerting a kind of catalytic action."[152] In this capacity, he gave du Bois-Reymond his life's work.

As du Bois-Reymond's adviser and mentor, however, Müller offered much more than intellectual matchmaking.[153] He supported his student, even rescued him on a political, material level. In the fall of 1842, when du Bois-Reymond had been experimenting for a year and a half, Matteucci wrote to Müller that he was on the brink of "discovering" the electrical phenomena for which du Bois-Reymond had been gathering evidence. As a result, Müller decided to publish du Bois-Reymond's findings in the physics journal *Poggendorff's Annals* rather than his own *Archive,* so that du Bois-Reymond could get his work into print faster and claim priority for his discoveries.[154] In March 1843, du Bois-Reymond wrote to Hallmann, "Here I really got to know my Müller. For the delicacy of mind, the readiness to help, and the warmth that he showed me on this and many other occasions, I'll be indebted to him all my life."[155]

The student was not always so appreciative. Between 1840 and 1849 he worked for Müller as an unpaid assistant, for although he had received his medical degree in 1843 and completed his *Habilitation* in 1846, he chose not to practice medicine or teach at the university so that he could devote all of his time to his electrophysiological experiments. Normally, Müller gave his two paying jobs, prosector and museum helper, only to his best and neediest assistants. When the prosector Reichert accepted a professorship of anatomy at Dorpat in 1843 (the one Henle had scorned), du Bois-Reymond hoped that he might become a paid assistant, but the museum helper position went first to Ernst Brücke and then to Hermann von Helmholtz in 1848. When du Bois-Reymond's father lost his job in 1848, however, Müller did help the struggling young physiologist. In the summer of 1849, the museum assistantship and a position teaching anatomy at the Art Academy became available, and Müller saw to it that du Bois-Reymond got both. In 1851, he joined Humboldt in nominating his student for membership in the Academy of Sciences. Du Bois-Reymond wrote to Ludwig, "Müller . . . is giving me tremendous backing, whatever his intentions may be."[156] Although intensely aware of his adviser's support, he never quite trusted Müller, viewing him as an enigma at best, and an exploiter at worst.

Du Bois-Reymond's misery in Müller's museum can be seen in his letters to Carl Ludwig, who shared his distaste for comparative anatomy. In August 1849 he complained, "Müller has kept me occupied at the museum, carrying out what is in his opinion the highest activity of the human intellect, namely, classifying fossil vermin."[157] After Müller's shipwreck, Ludwig seems to have hoped the compulsive collector had been scared straight, writing:

> Thus you have at last been delivered from the drafting of descriptions and labels of the unfortunate objects incarcerated by Müller's insatiable covetousness. I hear that the marine animals he is now after primarily have recently taught him a serious lesson and all but got their own back on him.[158]

Ludwig knew Müller mainly from du Bois-Reymond's descriptions, and in the early 1850s, they must have been bad indeed. Spurred on by Müller's unwilling assistant, Ludwig imagined the senior professor swimming for his life, surrounded by vengeful sea cucumbers and electric rays.

In 1854, when du Bois-Reymond began teaching physiology at the Berlin University, he saw Müller as a competitor. On a material level, this was an accurate view, since professors and lecturers received fees according to the number of students enrolled in their courses. But from this regard, Müller's willingness to let du Bois-Reymond teach physiology seems even more generous, since lecture fees paid for his research trips and museum specimens. In late 1854 du Bois-Reymond wrote to Ludwig, "My sole hope of competing successfully with Müller is based on showing many experiments in the lectures."[159] A year and a half later he told Helmholtz, "Here I have the task of winning over the listeners, who have been used to Müller for twenty-three years. I can do this only with experiments."[160] In the spring of 1857, the feeling of competition must still have been strong, for he confessed to Henry Bence Jones, "The only way in which I can hope to detach the students from Müller's and to lead them over to my lectures is by making a great many experiments."[161] To survive academically, du Bois-Reymond had to promote his experimental approach to physiology over Müller's anatomical one, which continued to attract Berlin students.

Because du Bois-Reymond respected Müller's early work as a physiological experimenter, he regarded his anatomical studies with all the more contempt. He knew that Müller had the potential to do greater things. In 1857 he raged:

> Will the rabble never be satisfied? Didn't they complain about Müller just the same way when he taught physiology from his scientific standpoint? Now they can take pleasure in listening to *him* lecture about it, if endless expositions on sea urchin development seem more practical to them than waveforms, Regnault's respiratory apparatus, or the ophthalmometer.[162]

Whatever the students and public thought of du Bois-Reymond's experimentally oriented lectures, Müller must have admired them. In 1854, he recommended that du Bois-Reymond be made an associate professor, ensuring his success at the university.

Even as a junior faculty member, du Bois-Reymond remained frustrated by his lack of power. At the age of thirty-five, he was still fully dependent on Müller and could order laboratory equipment only by persuading his mentor to buy it for him. In June 1854, he wrote to Helmholtz that he wanted a myograph, a rotating drum that graphically recorded muscle movements over time. "I'd like to induce Müller to purchase one for the museum, that is, for me," he explained. Helmholtz replied that a myograph cost 150 taler. In despair, du Bois-Reymond confessed, "Your apparatus is really too expensive, then, for me to prevail upon Müller to buy it for the museum. If I were made professor of physiology, in time, I would obtain a fund for experiments and apparatus and fight for expenditures like that."[163]

In 1858, he suddenly got the chance. When Müller died and his chair was divided, du Bois-Reymond passed from academic purgatory to the blessed state of full professorship, becoming Berlin's professor of physiology. Du Bois-

Reymond's mixed reaction to Müller's death reveals his complex feelings for his mentor. He reported to Helmholtz, "On Saturday night, with torches and sad countenances, we bore Johannes Müller to his grave in the old academic way. Today I opened his physiology lectures." Helmholtz replied, "Now, finally, the fulfillment of your long-standing wish is starting to look possible: the attainment of a full professorship in Berlin."[164] In November 1858, in his first letter to Ludwig since Müller's demise, du Bois-Reymond wrote, "My fate here has finally changed for the better . . . strokes of luck, like sorrows, do not come alone."[165] But to Henry Bence Jones he expressed real regret, saying about his new professorship, "My only grief is that poor Müller was to die for it, when he was so kindly disposed in my behalf."[166] Deep down, he knew that Müller admired the physiology he had ceased to practice.

Du Bois-Reymond's mixed feelings had their roots in Müller's ambiguous response to animal electricity. While he praised his student's work to the Cultural Ministry and recommended him enthusiastically for any available job, Müller seems not to have expressed this admiration to his assistant's face. "Müller seldom utters praise, you know," he wrote to Bence Jones in 1854.[167] If he had communicated his support to du Bois-Reymond, the physiologist might have seen him as the ally that he was. But by the 1850s, Müller was communicating little to anyone.

Even as a full professor, du Bois-Reymond was haunted by Müller's presence. Müller had assigned him his life project, supported his academic advancement step by step, and finally bequeathed him his position, though perhaps not voluntarily. Du Bois-Reymond may have objected to Müller's sea urchin embryos so strongly because he knew how intricately his own identity was wrapped up in his adviser's. To establish himself as a scientist, he needed to declare independence, which proved nearly impossible. In 1849, trying to win du Bois-Reymond government support, Alexander von Humboldt described him to the cultural minister as "a very talented young man, Johannes Müller's pupil."[168] No matter how many experiments he performed, he would always be Müller's student.

The Memorial Address

By the 1850s, the Prussian government had recognized science's military and technological value and was encouraging top scientists to present their work to the public.[169] As one of the leading experimenters in the German-speaking world, du Bois-Reymond offered many popular lectures, providing insights into the history and philosophy of science as well as ongoing research.[170] As a young student, the physiologist had considered himself a poor public speaker and had worked hard to improve his ability.[171] In time, he became a respected orator, holding thirty-four popular lectures between 1851 and 1894.[172] The Berlin public appreciated du Bois-Reymond's talks, and he attracted very large audiences. Complaining to Helmholtz about the English people's low enthusiasm for his work, he claimed, "If I were to announce popular lectures here in Berlin, where I could give them with a tenth of the effort I would have to exert [in England], I could have a thousand listeners and, objectively speaking, surely accomplish more than I would in London."[173] Once the German Empire was founded in 1871, holding popular lectures became part of a Berlin scientist's job. In the spring of 1870, du Bois-Reymond wrote to Helmholtz (with a certain glee) that Helmholtz was being considered for a full professorship of physics in Berlin, but would probably not get it because (1) he was not a professional physicist, (2) his demands were too high, and (3) he was "no avid teacher, no teacher for the rabble."[174] This implied, of course, that du Bois-Reymond was. For nearly twenty years, he had been attracting thousands of eager Berlin listeners.

Du Bois-Reymond's memorial address for Müller must be read in the context of these popular science lectures. It was a social event, a public performance, with much more at stake than the accurate depiction of a scientist's achievements.[175] While the eulogy illustrates a scientist's quest to establish his identity in contrast to that of a powerful precursor, it also represents a clash of worldviews, with the spokesman for the surviving outlook assessing his dead forebearer's way of doing science. As du Bois-Reymond told his audience, "Writing Müller's story [*Geschichte*] means writing the history [*Geschichte*] of the anatomical-physiological sciences during the time of his activity, which is interwoven with it in every point."[176] In the address, as in many of his public lectures, du Bois-Reymond consciously crafted a history of science, describing how physiology had evolved toward its highest form, his own electrophysiological experiments.[177] The real aim of the speech is to bury Müller, not to praise him. Presenting his mentor as a vitalist who left philosophy for physiology after a mental breakdown, du Bois-Reymond reduces Müller's career to a vector pointing toward his own.[178] In this lecture, he tells a story so good that it becomes history.

The written address that we have inherited is not the one that du Bois-Reymond read on 8 July 1858. A few days afterward, he wrote to his close friend Henry Bence Jones, "I have just completed an awful piece of work, Müller's Éloge, which I have read last Thursday before an imposing audience sitting next to Humboldt at the public meeting of the Academy of Sciences."[179] Du Bois-Reymond worked for weeks on this "awful" speech, telling Helmholtz that he was "so absorbed [in it] that [he] didn't have time for anything else."[180] As the probable inheritor of Müller's physiology chair, du Bois-Reymond had a lot to gain from his performance, so that it was well worth the time that he invested. The Berlin professorship he had coveted for twenty years was now at hand, and he wrote to Bence Jones, "It is all but certain that I will get the physiological place."[181] Always self-confident, he was thinking ahead. If he became the Berlin physiology professor, he would need funding for a state-of-the-art physiological institute. To win the position and the money, he needed to highlight his own achievements and show Prussian officials the value of laboratory science, all while appearing to praise his fallen mentor.

Even more important to du Bois-Reymond, however, was the chance to control the written record. During the next year and a half, using all of the best sources available to him, he turned his speech into a 190-page biography. In March 1860 he wrote to Bence Jones:

> I shall soon be able to send you the result of my work, Müller's biography, which gradually and against my will has grown into a book. . . . It was a foolish undertaking to which I have been prompted by the sense of gratitude and devotion to his memory, and in which I afterward persisted as it is my habit to do, although the result for myself of course is quite out of proportion with the time and labor it has cost me.[182]

A perfectionist, du Bois-Reymond hated to release any project with questions unanswered, and he investigated Müller's life with the same energy he devoted to animal electricity. It is very unlikely, however, that the speech became a biography against du Bois-Reymond's will.

Between July 1858 and late 1859, he collected and read all the major obituaries for Müller, including those of Ernst Brücke, Hermann von Helmholtz, Rudolf Virchow, and the Royal Society of London.[183] He went through Müller's file at the Cultural Ministry, carefully noting the funding Müller had received each year.[184] He contacted Müller's former mentor at the Cultural Ministry, Johannes Schulze, and solicited the letter from Theodor Schwann quoted in chapter 2. To bolster the information he was getting from academic sources, he approached Müller's family. He wrote at least five times to Nanny, who asked him to seek English buyers for Johannes's library. From Müller's son Max, he received a year-by-year enumeration of his father's research trips and estimates of his father's salary. In a letter of December 1858, Max frequently mentions Virchow and 1848, suggesting that he wanted du Bois-Reymond, not his politically active colleague (whose memorial address had already been published), to write the definitive account of Müller.[185] With the support of the family and colleagues, this is just how du Bois-Reymond presented his published text. In the first footnote (out of 209), he wrote, "As a factual representation [*quellenmässige Darstellung*] of the life and work of a man like Johannes Müller, [my account] may perhaps claim a valuable place in the history of science."[186] The most widely quoted primary source on Müller was keenly aware that he was one.

Du Bois-Reymond structured his address around Müller's scientific development, dividing it into nineteen short sections. Moving rapidly through his mentor's early life, the physiologist describes Müller's father as a successful shoemaker who wanted his son to become a saddle maker, ruling out any "great advancement beyond his present circumstance."[187] While he tells of Müller's desperate poverty at the Bonn University in the 1820s, he avoids creating a rags-to-riches story, which might not have appealed to his eminent audience.

Instead, du Bois-Reymond builds a narrative of scientific growth, with an early climax and a crucial turning point. He emphasizes that in later life, Müller was ashamed of his 1822 philosophical essay on animal movement. The physiologist presents his mentor's year and a half in Berlin as a critical period in which the anatomist Rudolphi turned him away from nature philosophy. Although Müller attended Hegel's lectures, he claims, where history was presented as the realization of abstract ideals, he was already too well developed as an empirical researcher to be influenced by them. Du Bois-Reymond dismisses Müller's 1826 book on fantasy images as a youthful work dissociated from the "real" Müller. In it, he says, "the future founder of the experimental physiological movement in Germany goes so far as to mock" the French physiologist François Magendie's (1783–1855) studies of rabbit eyes.[188] The "real" Müller valued animal experiments.

Here du Bois-Reymond introduces a motif that recurs throughout the address, calling Müller the hero who freed German science from the "dragon" of nature philosophy. According to his student, Müller described the specific energy of the senses with "the power of a reformer."[189] Whereas Virchow would compare Müller to a Catholic priest, du Bois-Reymond makes him a Protestant warrior, the Luther of German physiology. Müller's *Handbook of Human Physiology*, whose "penetrating power of reform" extended beyond German borders, sounds a lot like Luther's Bible.[190]

To shape his narrative, Müller's successor uses his teacher's 1827 breakdown as a rock-bottom point from which the scientific hero soars upward. Worn out by his "subjective-physiological-philosophical" self-experiments, Müller collapses, but he quickly recovers, thanks to a generous and understanding cultural minister. "Thus," proclaims du Bois-Reymond, "he was restored to science, but not as the earlier Müller. . . . Here begins the Johannes Müller that we knew."[191] According to his student, Müller's experimental physiology began in 1831, when he reintroduced the frog as an experimental animal. He asserts that until Müller demonstrated the different roles of spinal nerve roots in the frog, Bell's hypothesis was merely an interesting thought. Plugging the frog, du Bois-Reymond writes that since Galvani's animal electricity studies in the 1790s, the animal had been largely forgotten. He mentions that Müller made many "vain attempts" to study the electrical activity of nerves, leaving it to the audience to recall his own successful ones.[192]

In du Bois-Reymond's narrative, Müller wins the Berlin chair of physiology because he is "the most outstanding representative of a new movement."[193] Müller's *Handbook* created a new physiology, introducing a superior way of doing science so that, in an image that must have pleased some Prussian politicians, scientific quandaries "fell like medieval castles before the new war machines."[194]

But here a problem arises, for the narrative of the heroic reformer converted to physiology gives way to a second, conflicting tale. Up this point, du Bois-Reymond has associated Müller with experimental physiology, winning acclaim for his own science by associating it with that of his already respected mentor. But to ensure future funding, he needs to show that Müller's science has been surpassed. He must present his teacher as a reformer who didn't go far enough.

This narrative splitting creates a contradiction that undermines the address.[195] Whereas the first half praises Müller, the second demolishes him, leaving readers wondering what sort of scientist Müller was.[196] "Everyone knows," writes du Bois-Reymond, "that Müller was a decided vitalist from the start, and he remained so to the end."[197] Sadly, his teacher believed that living things contained a force different from any operating in inorganic matter and that they possessed a unique *Zweckmässigkeit*, through which the function of each part was subordinated to that of the whole. Here du Bois-Reymond deplores the "insufficient theoretical grounding of Müller's education," arguing that if he had ever worked with machines—as his students did—he would have seen that in them, too, each element forms an integral part of a whole.[198] Led by Schwann, the new physiological school "drew the conclusion for which Müller had provided the premise."[199] Du Bois-Reymond's second narrative reaches a high point when he declares:

> Müller never possessed the art of grasping and analyzing tasks mathematically or the familiarity with mechanical tools that are as necessary for physiologists as for physicists today. . . . What we call the aesthetics of experimentation was foreign to him.[200]

What du Bois-Reymond most valued—functional instruments specially designed to reveal the physical forces operating in living organisms—meant little or nothing to Müller. It took "an almost moral compulsion" to get him to look at new laboratory equipment.[201]

Consequently, Müller withdrew from the "universal monarchy" of physiology in 1840 when he was still its king.[202] Given the rate at which the field was growing, he realized that he could never master the new techniques, and retreated like Milton's Satan to a place he knew he could rule: the hell of comparative anatomy.[203] Given his contempt for anatomy, du Bois-Reymond describes Müller's studies of marine life quite aptly. Müller sought "characteristics which, more than any other relationships between their forms, would divide fish into groups according to their inner affinities."[204] The physiologist comments sarcastically, though, that once the cartilaginous fishes were "forever assigned to their proper places," it remained to sort the bony fishes into their orders and families.[205] Du Bois-Reymond is particularly critical of Müller's echinoderm studies, pointing out that because of his love for pelagic fishery, he never got to see every stage of embryonic development. Rather than raising sea urchins in the museum, Müller restricted himself to observing the embryos he could skim off the waves, so that he could only infer the intermediate stages from the ones that he captured.[206] Without comment, du Bois-Reymond mentions that Müller's last lecture at the Academy dealt with fossilized echinoderm embryos from chalk cliffs. Although he doesn't say so, the implication is clear: by 1858 his teacher had become a fossil.

Du Bois-Reymond leaves Müller's museum for the end of his speech, suggesting that his activity there best summarizes his science. When his mentor came to Berlin, he inherited a collection of 7197 specimens, which he united with his own private collection of 418 prepared and 385 unprepared items. The day before he died, Müller entered preparation number 19,577 in the catalog. Under his guidance, then, the anatomical collection grew at a rate of thirteen to fourteen specimens every ten days. During the academic year he spent much of his time there, "deliberating, ordering, rearranging, comparing, determining, and making entries."[207] In discussing Müller's life "outside of science"—a very short section—du Bois-Reymond mentions his "famous sense of order," a quality that characterized his science as well.[208] The discovery of slugs growing in sea cucumbers disturbed Müller so much because "he saw in his mind the house of systematic zoology, which he had so eagerly helped to build, shaken and split by deep cracks."[209] Sarcastically, the student comments that Müller "sought a general plan in the system of animals, . . . apparently fully convinced that there was one."[210] At fifty, when he found the slugs that he could not explain, the reformer must have felt too old to be "a smasher of the old order."[211]

In his discussion of Müller's politics, du Bois-Reymond emphasizes this same anxious quest for order, the quality that kept his teacher from creating experimental physiology. Like Virchow, du Bois-Reymond calls Müller a conservative, an "aristocrat of intelligence" with "an almost childlike respect" for the government. "To a man of Müller's rigid sense of order," he writes, "anarchy in the state machinery or total chaos on the street was no less horrifying than disorder among his museum preparations or in his library."[212] In the physiologist's description of 1848, the heroic reformer becomes a pathetic figure:

> Now, in his mind, Müller saw . . . his worst horrors realized, the flames coming out of the arched windows of the anatomical collection and irreplaceable treasures destroyed. With his sword girded, his arms folded, and a dark look, he personally stood watch day and night at the university gate.[213]

Home under the covers, du Bois-Reymond would not have had the chance to see Müller defending the museum, so his account is secondhand at best. Probably it got a laugh from the Academy members, as they uneasily recalled the events of ten years past.

It is significant that when du Bois-Reymond sums up Müller's scientific achievements, he does so in printed pages. A skilled writer himself, he knew how deeply science depended on communication and publishing. Between 1822 and 1858, his mentor produced twenty books and about 250 articles, which amounts to three and a half printed pages every five weeks. He describes the way Müller wrote, investigating everything three times: once before writing it up, once while writing it up, and once while the article was in press, so that his manuscripts and page proofs were full of corrections.[214] As an experimenter trying to secure an academic job, du Bois-Reymond knew the central role that writing played in science.

In creating the character of Müller, du Bois-Reymond was writing for his life. By describing the history of science, he was trying to determine its future. Reaching out to a worldwide audience, he asked Henry Bence Jones, in July 1860, "What do you think of having [my work on Müller] translated into English[?] Would not Lewes, the author of Goethe's life, be the proper person for it[?]"[215] George Henry Lewes's (1817–1878) *Life and Works of Goethe* (1855) had enjoyed tremendous success, creating nationwide interest in German culture. If Lewes had translated the Müller biography, English readers might have viewed Müller as another Goethe, another German genius worthy of international acclaim. And if du Bois-Reymond had contacted Lewes directly in 1858, he might well have agreed, since the self-taught physiologist was deeply interested in marine biology. Both Lewes and his partner, George Eliot (who translated D. F. Strauss's *The Life of Jesus* in 1846), were proficient in German and followed the latest developments in German science. In 1854, while writing his Goethe biography, Lewes had met du Bois-Reymond and Müller in Berlin.[216] But Bence Jones did not share du Bois-Reymond's confidence that Lewes—or English readers—would care about Müller's science. In his reply several weeks later, he spoke only of electric fish, saying nothing about the translation for four months. Then, in November, Bence Jones wrote, "As to the translation of your Müller by Dr. Baly, I do not ask anything of him. He is not a man I am intimate with."[217] Unless some intervening letters have been lost, Jones's memory altered du Bois-Reymond's request into one that made more sense to him. William Baly had translated Müller's *Handbook of Human Physiology*, and this was the way that Bence Jones saw Müller: not as an immortal genius, but as a physiologist. The very fact that du Bois-Reymond thought of Lewes—not Baly—as translator, suggests that in du Bois-Reymond's mind, Müller remained a towering figure, a predecessor who could be rendered less threatening only through writing.

Active Words

Du Bois-Reymond's relationship with Müller evolved considerably between 1840 and 1858. At first, he respected and feared the famous physiologist, hoping to impress him in order to advance his career. Once he got to know Müller, he began to admire him, although he never shared his teacher's passion for comparative anatomy. Müller exerted the greatest possible influence over his science, offering him his life's work and the living system in which he conducted it. Du Bois-Reymond's association with Müller also determined the spaces in which he could work and his means of economic support. From the mid-1840s onward, du Bois-Reymond began viewing Müller's influence as restrictive. Although his mentor eventually won him a junior faculty position, he remained financially dependent on Müller for research funding. As their scientific values diverged, du Bois-Reymond felt smothered and crippled.[218] He could have left Berlin, but his work demanded close collaborations with instrument makers that would have been much more difficult from distant cities. And with no shortage of self-esteem, du Bois-Reymond believed that he deserved to teach and experiment in the German cultural capital. He held on, expressing his frustration in writing.

Du Bois-Reymond's memorial address for Müller suggests that he suffered "anxiety of influence." He genuinely admired Müller's physiological studies of 1828–1831, seeing in them a model and inspiration for his own investigations of the nervous system. To assert his identity as a scientist, however, he needed to show that he had "swerved away" from Müller, and did so by claiming that Müller's science "went accurately up to a certain point, but then should have swerved, precisely in the direction that the new [science] moved."[219] By using Müller's words (e.g., *Zweckmässigkeit*) in new ways, du Bois-Reymond hints that his teacher had the right ideas but failed to execute them, relying on his student to perform electrophysiological experiments that his belief in life force prevented him from doing (though not from delegating). Du Bois-Reymond's scientific writing also shows signs of what Harold Bloom has called "daemonization," "open[ing] himself to what he believes to be a power in the parent-[science] that does not belong to the parent proper."[220] This power would be animal electricity, of which the scientists in Müller's generation were becoming aware but only du Bois-Reymond was fit to define with words and quantitative experiments. Finally, du Bois-Reymond's science—and his "poem" to Müller—involve what Bloom has called "the return of the dead." Because of du Bois-Reymond's skill as a writer, "the new [science's] achievement makes it seem to us . . . as though the later [scientist] himself had written the precursor's characteristic work."[221] As a physiologist, Müller was so innovative that his work approached the level of du Bois-Reymond's.

In his memorial address for Müller, du Bois-Reymond skillfully negotiated among potentially contradictory demands. Simultaneously, he showed that Müller's experiments made him a great scientist; his anatomical studies, a misguided one; and that he, du Bois-Reymond, resembled Müller the physiologist (not Müller the anatomist) and would return Müller's work to the course from which it had so tragically veered. Both his resemblance to and his difference from Müller made him the best qualified candidate for the Berlin physiology chair.

In the play that du Bois-Reymond admired, Hamlet instructs the actors:

> Suit the action to the word, the word to the action, with this special observance: that you o'erstep not the modesty of Nature: for any thing so overdone, is from the purpose of playing, whose end both at the first and now, was and is, to hold as 'twere the mirror up to Nature.[222]

Du Bois-Reymond would have heard in this speech a similarity between good "playing" and good science. Theater worth seeing—like experiments worth doing—did not impose pre-set forms upon nature. In his research, he tried to follow Hamlet's advice. He aimed principally to act, but at times acting meant writing. Whether wielding electrodes or phrases, he tried to promote his empirical outlook on reality. A literary as well as a scientific creator, the student asked to define Müller's life knew the power of narrative.

4

Physiological Bonds

The Training of Hermann von Helmholtz

Whatever you do, don't let yourself be forced to do anatomy.
—Carl Ludwig to Emil du Bois-Reymond, 10 July 1849

Like Müller, du Bois-Reymond spent much of his life longing for friends who shared his scientific outlook. In his twenties, he was far luckier than his mentor, finding in Berlin a group of talented physicists and physiologists with whom he could work. Among them was Ernst Brücke (1819–1892), a painter's son who had hoped to become an artist but after meeting Müller had shifted his allegiance to experimental physiology.[1] In March 1841, du Bois-Reymond organized a group of several young scientists who supported active experimentation and opposed the notion of life force. In January 1845, this group became the Berlin Physical Society, meeting weekly at the physicist Gustav Magnus's house to exchange ideas about experiments.[2] Later that year, they were joined by Hermann von Helmholtz and the engineer Werner von Siemens (1816–1892), who designed Prussia's telegraph network. Because all of these scientists built their own innovative equipment, the Physical Society provided a forum for technical discussions among physicists, physiologists, and mechanics. In addition to university lecturers, the society included instrument makers and technicians from the Prussian military who wanted to collaborate with physicists.[3] As they shared technical tips, these young experimenters inspired each other and formed passionate scientific bonds.

Du Bois-Reymond and Helmholtz had a great deal in common. Both thought that physical laws could explain life functions, and both worked with mechanics to design their experimental setups. Both took Müller's courses and talked to him during 1839–1842, although, oddly, the young physiologists did not become acquainted until 1845. In their representations of Müller, however, they differed greatly. This chapter will explore why that was. Helmholtz delivered no memorial address, and his retrospective comments on Müller were rarely critical. In evaluating the differences between these two scientists' portraits of Müller, one must consider their backgrounds, their career choices, and perhaps also their personalities.

Helmholtz as a Student of Müller

Three years younger than du Bois-Reymond, Helmholtz came from a similar background, but with one crucial difference. Like du Bois-Reymond's family, his relatives belonged to the educated elite, but they were not particularly wealthy. The oldest of four surviving children, he was baptized and confirmed a Protestant.[4] His father, a classics teacher at Potsdam's excellent *Gymnasium*, had studied at the Berlin University and retained scholarly ambitions. Unfortunately, this cultured man did not earn enough to pay for his son's education.

Like du Bois-Reymond's father, Ferdinand Helmholtz greatly respected Immanuel Kant's philosophy, believing that people could best know the world by contemplating the mental structures that shaped human perceptions. Hermann resisted this view, suspecting that science would reveal the relationship between sensory perceptions and external events. Eventually, his program of physiological experiments would challenge the notion that knowledge comes from within.[5]

Helmholtz attended the Potsdam *Gymnasium,* which was among the best in the German-speaking realm. There he studied Homer's poetry with his own father. In addition, he learned Hebrew, Latin, English (including Shakespeare), French, German, Italian, geography, history, logic, mathematics, physics, psychology, and religion.[6] Unlike du Bois-Reymond, Helmholtz knew from the beginning that he wanted to study physics. His father, however, could not afford to send him to the Berlin University and thought it unlikely that Helmholtz would be able to support himself as a physicist. In the 1830s, this conviction did not indicate any lack of confidence in his son, since there were almost no academic jobs available. Helmholtz agreed with his father that the best way to learn science was on a military medical scholarship.[7]

In the fall of 1838, he began studying medicine at the Friedrich-Wilhelms-Institut, or Pépinière. Although the Institute had been founded to train bright, impoverished young men as military surgeons, many of its students came from the lower middle class.[8] In its "magnificent building" on Friedrichstrasse, about a hundred students lived two to four in a room, ate their meals, and memorized the structures and disorders of the human body.[9] Although their tuition and accommodations (including heat and light) were free and they received 6 to 8 eight taler per month for living expenses, their parents had to pay heavy fees for exams: 10 taler for a qualifying exam after a year of coursework, 130 taler for a predoctoral exam after four years of instruction, and 80 taler for the Prussian state medical exams.[10] After four years of classes and one year of clinical rotations at the Charité Hospital, all graduates had to practice medicine, repaying the Prussian army with eight years of service. Most were also expected to complete an independent research project and publish an illustrated thesis of about thirty pages, for which they earned a Ph.D. This cost another 200 taler.[11] In return for these fees, the students received a grueling, first-rate medical education.

Helmholtz's letters home from medical school show a great deal about his expectations. A visual thinker, he carefully sketched his dorm room for his parents, numbering and identifying each piece of furniture. "To make it all clear for you, I'll draw a little map," he explained. Item number 10 on the list of his furnishings is "my instrument"—not a galvanometer, but Helmholtz's piano, which he had brought with him from Potsdam. His parents worried about whether he was keeping up with his music, and he complained about people wandering into his room and playing his instrument when he wanted to study.[12] A recurring theme in his letters home is his dirty laundry—or more particularly, how to get it to his mother. Although Helmholtz learned medicine on a military scholarship, he and his parents expected that he would not have to do menial labor and that he would maintain the accomplishments of an educated man.

Even more than du Bois-Reymond, Helmholtz benefited from his family connections. He did extremely well on his Pépinière entry exam, but even if he had made a mediocre showing, he might have been accepted because his great-uncle Christian Ludwig Mursinna had directed the gynecological clinic at the Charité Hospital and had taught at the Friedrich-Wilhelms-Institut. Helmholtz's father used Mursinna's name liberally to get him in.[13] When Helmholtz arrived in Berlin, he found a network of relatives and family friends to support him, not all of whom were terribly pleasant. His aunt, Frau von Bernuth—who lived at am Kupfergraben 5, between Müller and Magnus—persistently corrected his table manners. When he complained to his father, the classics teacher replied, "Behind the forms of finer dealings there is usually a deeper meaning."[14] Although he could study the natural sciences only on a military scholarship, Helmholtz was not allowed to forget his social standing.

Pépinière cadets took many of the same classes as Berlin University medical students, including most of Müller's lectures: general human anatomy, comparative anatomy, pathological anatomy, physiology, and anatomy of the sensory organs.[15] In May 1839, Helmholtz wrote to his parents, "Müller's physiology is excellent."[16] In the summer semester of 1840, he carefully copied out his lecture notes from Müller's comparative and pathological anatomy classes.[17]

Helmholtz's record confirms Bidder's and du Bois-Reymond's remarks about Müller's comparative anatomy. Inspiring and visually rich, it presented anatomy from the standpoint of physiology. Müller began with embryology and development, then moved through the skeletal, circulatory, nervous, respiratory, excretory, and reproductive systems. Throughout the lectures he referred to microscopy and to the bony and cartilaginous fishes that so fascinated him. Helmholtz filled the margins with beautiful, meticulously labeled drawings, probably reproductions of Müller's sketches on the board. The extreme care with which Helmholtz copied the drawings shows his respect for the speaker and the

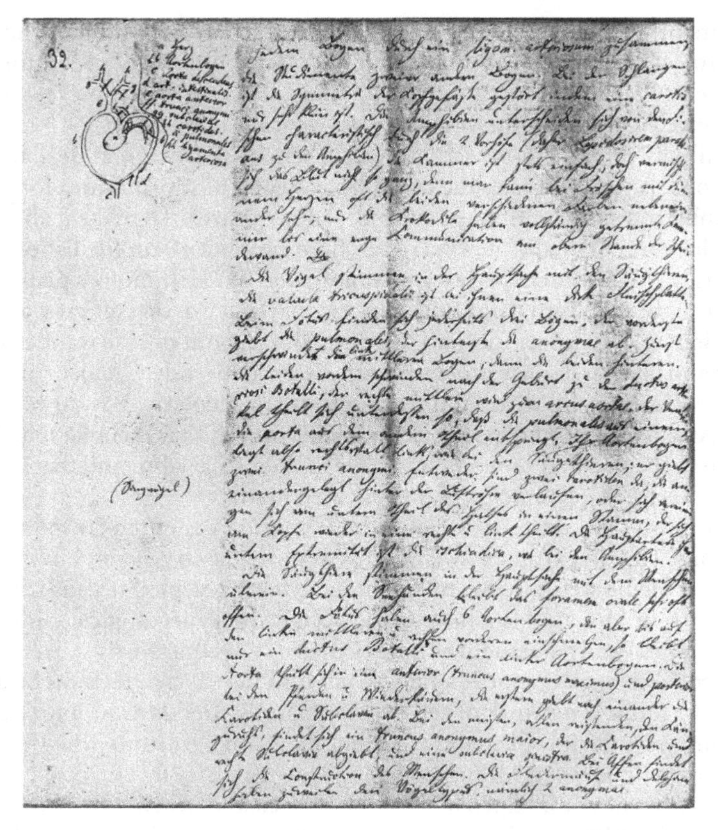

Figure 4.1 Helmholtz's notes from Müller's summer 1840 comparative anatomy course describing circulatory systems. Compare with Virchow's notes from the same course a year later, in chapter 5. Source: ABBAW, NL Helmholtz, nr. 538, 32, printed with kind permission of the ABBAW.

material discussed. In contrast, his pathological anatomy notes contain relatively few sketches, and one senses that neither Helmholtz's nor Müller's heart was in the class.

While Müller's physiology and anatomy lectures inspired Helmholtz, he owed his first scientific break to a good health insurance policy. In August 1841, a month after he had begun independent research for his dissertation, he came down with typhus and had to spend five weeks in the hospital. As an Institute student, he received free care at the Charité, so he saved enough on living expenses to buy a microscope.[18] "The instrument was not beautiful," he recalled in 1877, "but it allowed me to recognize the nerve processes of ganglion cells in invertebrates which I described in my dissertation, and to pursue the vibrios in my work on fermentation and putrefaction."[19]

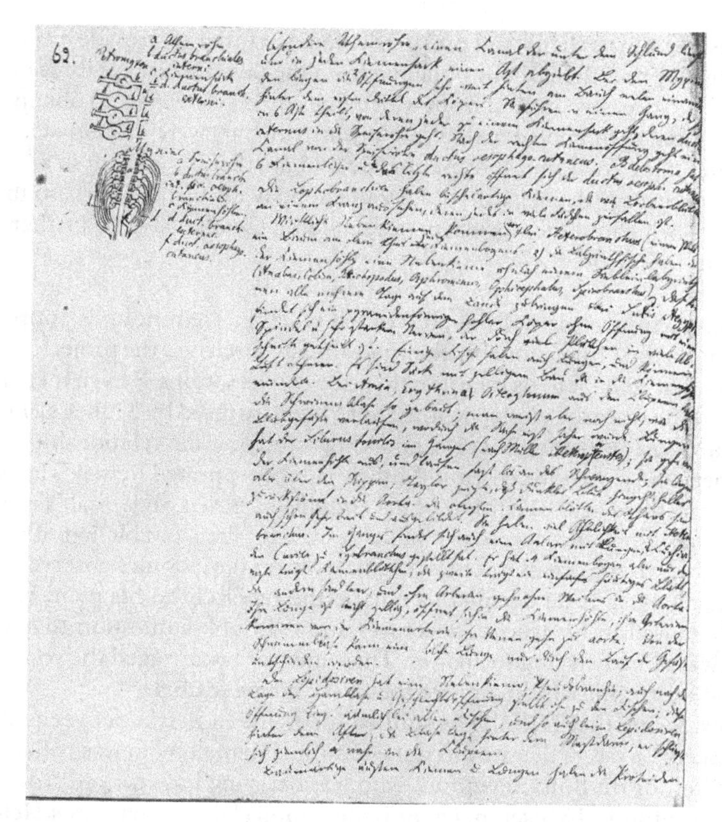

Figure 4.2 Helmholtz's notes from Müller's summer 1840 comparative anatomy course illustrating the respiratory systems of fish. Source: ABBAW, NL Helmholtz, nr. 538, 62, printed with kind permission of the ABBAW.

It is not clear when Helmholtz began talking to Müller about his own research, a microscopic quest to determine whether nerve fibers emerge from ganglion cells (today known as nerve cell bodies) in some invertebrates. At the time, it was not known that axons and cell bodies belonged to the same cells, so that by showing their association, Helmholtz provided some early evidence for the neuron theory, which would not be developed until the late 1880s.[20] In a letter home on 1 August 1842, he described how Müller had advised him on his doctoral study:

> Today I went to Professor Müller with my dissertation. He received me in a very friendly way, and after I had gone through the main results and the evidence for them, he explained that [the project] was certainly of great interest, since it demonstrated an origin for the nerve fibers that had been hypothesized in higher animals but could not be proven. He advised me, however, to investigate the matter first in a more complete

> series of animals than I had previously done, in order to provide more stringent proof than can be obtained from the investigation of three or four. He suggested several from which one could best expect to obtain results and even invited me, in case my instruments were inadequate, to use his at the Anatomical Museum. He told me that if I was not in a hurry to get my degree, I should use the holiday for further work, so that I could bring into the world a perfect child that would have no further attacks to fear.[21]

From this account, it is not possible to tell whether Helmholtz is approaching Müller for the first time. Possibly they had already discussed the project, although the letters that remain to us mention no prior conversations.[22] Even if Helmholtz conducted the research alone, he must have been inspired by Müller's *Handbook* and lectures. At the time, Robert Remak was studying the relationship between nerve fibers and ganglion cells, and Müller had incorporated Remak's latest findings into the third (1838) edition of his *Handbook*. Based on Remak's results, he wrote, it was "to some degree probable or at least a reasonable hypothesis that the gray fibers of the organic nerves originate" from the globules observed in sympathetic ganglia.[23] If he decided to pursue the research on his own, Helmholz probably sensed in his teacher's lectures and textbook a question that a senior student was already trying to answer. Their mentor encouraged the younger one to look at more animals, in a move that was vintage Müller.

Considering that Helmholtz studied medicine in Berlin between 1838 and 1842, it seems strange that he did not encounter Remak, who was collaborating with Müller, or du Bois-Reymond, who was carrying bags of iced frogs to and from the Anatomical Museum. Du Bois-Reymond claims to have met Helmholtz in Gustav Magnus's home physics lab in December 1845, when Helmholtz returned to Berlin on a five-month leave to prepare for his state medical exams.[24] Since the statement comes from a contemporary letter rather than a memoir, there is no reason to doubt du Bois-Reymond's word. If he is right, their near misses suggest the differences in their personal styles as well as their medical backgrounds.[25] Because of du Bois-Reymond's interlude in Bonn and his hesitancy to choose a major field, he began his medical studies at the Berlin University a year after Helmholtz started at the Pépinière, even though he was three years older. Both men attended Müller's lectures, but Helmholtz took physiology in the spring of 1839, before du Bois-Reymond had returned to Berlin. In the summer of 1840, they spent four hours a week together in the same lecture hall, listening to Müller teach comparative anatomy.[26] Even in a class of 200 students, they must have known each other by sight, but the differences in their lives outside of class may have kept them from socializing. Until 1845, du Bois-Reymond lived with his parents at Potsdamerstrasse 36A, then quite far from the city center. Meanwhile, between 1838 and 1842, Helmholtz was locked up at the Institute, where he was forced to rise at five and retire at ten. He could leave its main building only to attend university lectures or with special written permission.[27] Still, one wonders why, in the summer of 1842, Müller did not simply introduce the experimental physiologist and the gifted young medical student interested in nerves. Apparently, as the young scientists' mentor, he didn't see this as part of his job.

Helmholtz's Clinical Work and Early Research, 1842–1848

Once Helmholtz completed his doctoral dissertation in the fall of 1842, he had much less opportunity to cross paths with du Bois-Reymond. In October, he moved to the Charité Hospital, where he lived for a year while performing his clinical rotations. Founded in 1710 as a facility for plague victims, the Charité had long served as a teaching hospital for army doctors, and it had outgrown its eighteenth-century role as a poorhouse, but not its military flavor.[28] Located on the city's northwest side, the hospital treated patients from Berlin and Potsdam who could not be cared for at home, but turned away those who were incurable.[29] In 1842, it could hold a thousand patients, having acquired a new, 300-bed building in 1835. With a staff of 230, the hospital had an annual operating budget of 76,000–110,500 taler. For efficiency's sake, patients were divided among surgery, obstetrics, internal medicine, syphilis, pediatrics, psychological, ophthalmological, and wounds clinics, each headed by a distinguished physician. On the wards, patients were strictly separated by sex and, to a certain degree, by social class. A stay in the Charité did not come cheap. The minimum cost per day was 7.5 *Silbergroschen*, but the hospital also offered first-, second-, and third-class service, with costs of 1, .50, and .33 taler, respectively. As a Pépinère intern, Helmholtz served briefly in each clinic, working daily from 7 A.M. until 8 P.M.[30]

In October 1843, when Helmholtz had finished his clinical training, he obtained the enviable position of staff surgeon to the regiment of Royal Hussars in Potsdam, which left him a lot of time for research.[31] While resigned to eight years' military service, Helmholtz wanted only to conduct physiological experiments.

Probably Müller had stimulated Helmholtz's interest in invertebrate nervous systems, but the research he began during his service in Potsdam had a different impetus. Like du Bois-Reymond, he was seeking a system in which he could demonstrate that physical laws explained physiological phenomena. While the two did not begin sharing ideas until late 1845, Helmholtz developed many of the same attitudes independently. He appreciated Müller's anatomical knowledge but believed that studying animal forms for their own sake was a waste of time. A few comments in his letters home suggest that although he valued microscopic studies of cellular connections, he disliked the memorization of gross anatomy. During his first year of medical school, he complained, "We often have to sit through the evening learning one muscle after another till our heads split."[32] In an 1848 lecture at the Art Academy, he admitted, "Medical anatomy is . . . in its essentials a collection of dry concepts."[33] Unlike Henle, Helmholtz was unmoved by the beautiful purposefulness of muscles. He wanted to know how they worked.

Like du Bois-Reymond, Helmholtz admired Schwann's studies, and he began his postdoctoral research by following up on Schwann's fermentation work. Schwann had shown that fermentation could occur only in the presence of microorganisms, but the chemist Justus Liebig had challenged his results, presenting fermentation as an inorganic chemical process. In a setup in which he carefully controlled access to boiled culture media, Helmholtz confirmed Schwann's finding that fermentation required a living organism. Helmholtz claimed that putrefaction (the breakdown of living matter) did not, however, although it attracted microorganisms because of the nourishment it provided.[34] Even though both processes were closely tied to living organisms, Helmholtz viewed their chemical reactions as no different from those occurring in inorganic matter. It was not life force that drove yeast to make alcohol.

Suspecting that chemical reactions produced muscle motion, Helmholtz began investigating the chemistry of muscles. By comparing the aqueous and alcohol extracts of exhausted frog legs, he showed that intense contractions produced chemical changes.[35] While conducting these studies, he became interested in animal heat, which he believed could also be explained by chemical processes in organisms' bodies. As one might expect, Helmholtz found that exercise increased muscle temperature, but he proved that this warming was caused by chemical processes in the muscle, not just by increased blood flow.[36] Once he met du Bois-Reymond, the two collaborated closely to design innovative equipment for these studies. While creating the instruments to study physiological heat, Helmholtz proposed the theory for which he is best known today, the Law of Conservation of Force. It states that when energy changes forms, none is lost or created, thus ruling out any ephemeral life force that emerges at conception and vanishes at death. Since Helmholtz was correlating mechanical movements, temperature changes, chemical reactions, and electrical currents in his day-to-day work, it is not surprising that his quantification of life functions led him to this general physical law.[37]

In his papers on muscle chemistry and animal heat, Helmholtz argued that no unique forces existed in living organisms, but Müller seems to have been more impressed by his experimental designs than threatened by his results.[38] Seeing Helmholtz's commitment to physiological research, he gave the military doctor a crucial boost. In the summer of 1848, he recommended Helmholtz for the positions of museum helper and anatomy teacher at the Art Academy, which had opened when Ernst Brücke accepted a professorship of physiology in Königsberg (today Kaliningrad, Russia). "Helmholtz is one of the rare great talents that I would pick out as exceptional," wrote Müller to the Cultural Ministry.[39] To make Helmholtz eligible for these jobs, Müller helped release him from his three remaining years of military service. Du Bois-Reymond could not have been pleased to see Helmholtz receive these positions when he had been assisting Müller without pay for nine years. A year later, when Brücke moved to Vienna, Helmholtz became associate professor of physiology at Königsberg, enabling him to marry Olga von Velten, the niece of his regiment's physician in Potsdam.[40] Müller also played a key role in the awarding of this job, as will be discussed in the next section. When Helmholtz left Berlin, du Bois-Reymond took over his abandoned posts. Academically, he would never catch up with Helmholtz again.

Figure 4.3 Hermann von Helmholtz. Source: Leo Koenigsberger, *Hermann von Helmholtz* (Brunswick: Friedrich Vieweg, 1902), daguerrotype of 23 March 1848, 1: 55.

The Königsberg Job Search

In the 1849 Königsberg job search, a gifted physiologist came into play who shared du Bois-Reymond's and Helmholtz's commitment to quantitative experiments. Carl Ludwig (1816–1895) never attended the Berlin University or took a course with Müller, but he played a crucial role in the relations among Helmholtz, du Bois-Reymond, and their adviser. As an innovative experimenter who shared the students' views, he encouraged them, listened to their complaints—and competed with them for chairs of physiology.

In his 1877 lecture "On Thought in Medicine," Helmholtz called Ludwig a *Studiengenossen* [comrade] in Berlin, which has led to considerable confusion.[41] Although Ludwig didn't study with Müller, he became a "comrade" of Helmholtz and du Bois-Reymond because of their shared interest in designing equipment to study life functions. The three met when Ludwig visited Berlin in 1847, which perhaps prompted Helmholtz to place him there in his memory.[42]

Except for a brief expulsion for dueling, Ludwig had worked at the University of Marburg since 1834, first as a medical student, then as a lecturer, and after 1846 as an associate professor.[43] In his physiological studies at Marburg, Ludwig sometimes collaborated with the chemist Robert Bunsen (1811–1899), who invented the Bunsen burner. For his quantitative studies of fluid circulation, Ludwig was trying to build instruments that would measure the volume of blood in an organ or the amount of gas that blood carried.[44] In 1846 he had designed the kymograph, a device for monitoring blood pressure over time.[45] Like du Bois-Reymond and Helmholtz, Ludwig believed that life functions involved no forces unique to living beings, and could be explained through well-conceived experiments.

Almost all of Ludwig's letters to du Bois-Reymond ask for equipment—or at least advice about instruments—and urge his Berlin friend to come and visit. Ludwig had spent thirteen years in the sleepy college town of Marburg, and after a taste of the Berlin Physical Society, he wanted to join a community of experimentalists in a major city. His loneliness drew him to the cosmopolitan du Bois-Reymond, two years his junior, and their common thoughts about how to do science produced an intellectual and emotional bond. "How I long to have a closer, personal relationship with you," wrote Ludwig in February 1849. "It would be hard for you to comprehend the feeling of longing that draws me to you, Brücke, and Helmholtz because you have never had Diogenes' lantern in your hands."[46] Like the Cynic philosopher, Ludwig felt that he was wandering the scientific world in search of honest men, and had found them only at the Berlin Physical Society. During a trip to Halle, Ludwig wrote, "The great proximity from which I write you is not able to make me content, because for me it is the same as being far away; it does not bring me closer to you." He asked du Bois-Reymond to visit him there, adding ironically that "to complete the cozy get-together, we are to live in the same room."[47] That fall du Bois-Reymond did visit Ludwig, but in Marburg, where he could talk to Bunsen as well. When he returned, he told Helmholtz, "I would have to count the eight days in Marburg with Ludwig and Bunsen as among the few good ones I've had in my life so far."[48]

While Ludwig, Helmholtz, and du Bois-Reymond were drawn together as researchers, they also competed for academic jobs. Their response to the 1849 opening in Königsberg shows the complexity of their relationships. On the one hand, du Bois-Reymond knew that he should apply, since the allowance from his father had ended. On the other hand, he didn't want to start teaching until *Animal Electricity* was finished, and the thought of leaving Berlin for East Prussia filled him with horror. While Ludwig longed to escape Marburg, he at first refused to apply, feeling that the job should go to du Bois-Reymond.[49] But the Berlin physiologist urged him to contend, offering advice on how to solicit Humboldt's help.

After ten years with Müller, du Bois-Reymond enjoyed playing mentor himself, and he sometimes used Humboldt's influence to impress his friends. In early 1849, when Ludwig was thinking of applying for the Königsberg job, du Bois-Reymond asked, "Can Humboldt be of any use to you? . . . I associate with him a great deal. . . . I am on particularly good terms with Humboldt, so that I am in touch with him on one point or another almost every two or three days."[50] Du Bois-Reymond's advice to Ludwig about how to approach Humboldt shows his academic savvy as well as his willingness to manipulate:

> With him measures must always have a certain form. It is necessary first of all that you write to him yourself. You can approach him in a general sort of way. After telling him you want something from him, give him as

> drastic an account of your story and of your present situation as you did me, but without going into the details of which you informed me relative to the faculty. . . . Then say that I had promised you that Humboldt would do something for you. Do not fail to say something complimentary about him, too; for Humboldt is susceptible to appreciation. . . . Include a couple of choice tables with curves as a gift, for he loves to have such things in his possession. . . . Be sure to write your letter to him in large and very meticulous handwriting, for he is a presbyope.[51]

Du Bois-Reymond's own letters to Humboldt—written in large, unusually neat handwriting—contain beautiful, detailed diagrams of frogs and electrical currents. In Humboldt, he had a crucial mentor and sympathetic reader.

Although two years younger than Ludwig, du Bois-Reymond assumed the role of politically skilled adviser, a part that he played with relish. He had not yet taught at a university, but his closeness to Humboldt and Müller, and his presence in Berlin, made him an insider. Like Humboldt, he sent periodic reports on everything he was doing for his friend. In early April he told Ludwig, "The day before yesterday Müller stepped in here most cheerfully and wearing a truly democratic beard. . . . I have told Müller that you should go to Königsberg *à tout prix* [at any price] and explained to him and Humboldt that I want to remain here *à tout prix*."[52] But by mid-May, du Bois-Reymond reported, "The affair does not look very promising for you after all."[53]

The reason was Müller. "Hear what I did in order to clear the field for you," boasted du Bois-Reymond. He explained that since the Prussian authorities suspected Ludwig of harboring "democratic convictions," he had written to Johannes Schulze (Müller's old mentor) at the Cultural Ministry, describing Ludwig's true political character. Like Henle, Ludwig had been involved in *Burschenschaften*, as his dueling scar revealed. "But Müller then went and spoiled everything," complained the would-be mentor. "Not content to praise you (which he really did do, by the way), he also at the same time recommended Helmholtz and probably Remak, whom he declares also to be a physiologist."[54] Here is the letter that Müller actually wrote:

> Recently, the experimental method in physiology has developed a very close relationship with the physical sciences, namely physics and chemistry, and the same demand for exact methods is being made in physiological investigations as in physical ones. Physiology can expect to achieve the most progress in this direction from those men armed with all the knowledge necessary for a doctor or physiologist in the organic field but who have also attained a basic education in physics and can apply it confidently to their experiments. In Germany, the most promising young talents in this direction are Brücke, du Bois-Reymond, Helmholtz, and Ludwig. The lecturer Dr. du Bois has acquired the best-laid claims to a professorship of physiology with his classical studies of animal electricity but does not seem inclined to accept the position in question until he has completed his present work. I have already had the honor of reporting to

> the High Ministry in full about Helmholtz, the current helper at the anatomical facility and anatomy teacher at the Academy of Arts. I regard him as one of the most significant talents in physiology. . . . Of his qualification as teacher there cannot be the slightest doubt. If he receives and accepts the professorship in Königsberg, I would then suggest to Your Excellency that Dr. du Bois be made helper [at the museum] and anatomy teacher at the Academy of Arts. . . . Professor Ludwig of Marburg is on par with Brücke, du Bois, and Helmholtz and would also be fully qualified for the position in question. He has distinguished himself through his physiological investigations in a decidedly recognized way, is a very beloved teacher, and has already educated many pupils and caused them to carry out substantial works. . . . Given this opportunity, I must also mention the achievements of the lecturer Dr. Remak with respectful recognition. His direction is not that of the aforementioned men but is oriented mainly toward anatomical-physiological microscopy with a regard for pathology and development, which constitutes a significant part of physiology. The direction of Dr. Remak in the field of physiology is as justified as the experimental one, but it is not as exclusively physiological as that of the other candidates.[55]

It is a particularly revealing letter. First, it shows that communication between Müller and du Bois-Reymond was relatively good. Müller does mention him, Helmholtz, Ludwig, and Remak—in alphabetical order—just as du Bois-Reymond thought he did. Müller also tactfully conveys du Bois-Reymond's unwillingness to accept the position. Second, the letter illustrates something that his student did not recognize: Müller's tremendous respect for experimental physiology and his determination to obtain jobs for rigorous experimenters. Third, the recommendation demonstrates that to the Prussian Cultural Ministry, effective teaching was as important as respected research. In making the appointment, they wanted not just a scientific star but also someone who would create future investigators. Finally, the letter shows who Müller really wanted to receive the Königsberg job. Remak is an afterthought, and du Bois-Reymond does not want the position. This leaves Helmholtz and Ludwig, but Helmholtz is praised more highly, and Müller suggests promoting du Bois-Reymond if Helmholtz accepts the job. He does not need to mention that Helmholtz is the grand-nephew of a renowned Prussian surgeon and Ludwig is a notorious democrat. Overall, the recommendation makes clear how much Müller knew about physiological politics and how much he appreciated his students'work.

Helmholtz's Physiology and du Bois-Reymond's Mentorship, 1849–1870

Perhaps du Bois-Reymond learned scientific mentoring, as well as science, from Müller. As soon as Helmholtz and du Bois-Reymond met, the Berlin physiologist began counseling his like-minded friend.[56] In Helmholtz's letters to du Bois-Reymond, which begin in July 1846, the military doctor first calls his friend "Sie," but in February 1847 changes to "Du." Their correspondence in the 1840s and 1850s is full of drawings and discussions of instruments, particularly the details of other scientists' equipment. Certainly Helmholtz learned a lot from du Bois-Reymond about how to conduct electrophysiological experiments, but as the younger student's career advanced, du Bois-Reymond's mentorship became less and less valuable.

Although du Bois-Reymond could probably have had the Königsberg job if he had wanted it, he told Müller that he wished to finish his animal electricity research before he began teaching, a miscalculation for anyone determined to advance. Probably du Bois-Reymond did not want to leave Berlin for the remote East Prussian city, meaning to stay close to the instrument makers with whom he collaborated, but his choice also shows a lack of self-confidence. It had never occurred to Helmholtz that he could not teach and experiment at the same time. Years later, du Bois-Reymond wrote, "I lack your enviable and to me incomprehensible gift for working on ten projects at once and at the same time teaching two or three courses."[57] It is not hard to hear the envy in this praise.[58]

When Helmholtz left for Königsberg, du Bois-Reymond told Ludwig, "Helmholtz's departure from Berlin naturally leaves me a void that cannot be filled."[59] To console himself, du Bois-Reymond mentored his younger friend diligently, emphasizing his crucial role in publicizing Helmholtz's experiments. In late 1849, while investigating the mechanical properties of frogs' muscles, Helmholtz discovered that the time between stimulation and muscle contraction increased if he stimulated farther out along the nerve. Systematically, he studied the relationship between this delay and the distance the nerve impulse had to travel, and calculated that it must be moving at 24.6–38.4 meters per second.[60] On 15 January 1850, Helmholtz sent du Bois-Reymond a report of his finding, asking him to read it to the Berlin Physical Society.[61] For the rest of that year, du Bois-Reymond eagerly promoted the study, for like Helmholtz, he realized its implications. If the velocity of nerve impulses was measurable and constant, then they must be electrical and/or chemical in nature, governed by discernible physical laws.

But du Bois-Reymond doubted Helmholtz's writing and marketing skills, insisting that he was better qualified to communicate Helmholtz's work to the scientific community. Before Alexander von Humboldt could announce Helmholtz's results in Paris, du Bois-Reymond had to explain them in Berlin:

> First [I will discuss] your work which, I tell you with pride and sorrow, was understood and appreciated here in Berlin only by me. Namely—now don't hold this against me—you have represented the matter with such boundless obscurity that your report could pass at most for a brief guide on the rediscovery of the method. The result was that Müller did not rediscover it . . . I had to explain it to them one by one, Riess, Dove, Magnus, Poggendorff, Mitscherlich, and finally Müller himself, who

> hardly wanted to hear about it. . . . Humboldt was very much put off and refused to send your article to Paris, whereupon I offered to rework it until it was comprehensible. This I have done on my own responsibility.[62]

Du Bois-Reymond offers Helmholtz the *Dangerous Liaisons* reproach: Helmholtz has written to express himself, not *to* and *for* his readers. The *ingénu* must be rescued by a self-appointed editor and literary agent who always writes with his readers in mind. Hoping to appear invaluable, du Bois-Reymond presents himself as Helmholtz's only sympathetic listener in a crowd of uncomprehending scientists centered around Müller.

But Müller's response belies du Bois-Reymond's narrative. Less than a week after Helmholtz sent his first report, Müller read it to the Academy of Sciences, and two weeks after that, he wrote to Helmholtz, "I am convinced that . . . a great stride has occurred, the first in the art of measuring the nerve effect." He encouraged Helmholtz to measure the time required for reflex actions and the velocity of impulses in the autonomic nervous system.[63] Du Bois-Reymond may have had to explain Helmholtz's setup to Müller, but he exaggerated their teacher's resistance.

If Müller couldn't understand Helmholtz's work—according to du Bois-Reymond, at least—then Helmholtz couldn't always understand Müller's. The quest for a grand plan manifested in all living forms seemed to him fully misguided. Helmholtz lacked du Bois-Reymond's sarcasm, but his comments to his Berlin friend make it clear what he thought of comparative anatomy. When Helmholtz began his new job in October 1849, he wrote that on the Königsberg medical faculty, the only "productive mind" was Martin Rathke, "but like J. Müller, he's consumed by zoological interests."[64] In the spring of 1852, Helmholtz remarked, "I presume that at present, J. Müller is fishing for *Synapta* in Trieste."[65] Even with his teaching and his ever more interesting studies of nerves, Helmholtz longed for a like-minded physiologist, asking du Bois-Reymond in 1851, "Can't you travel with me just a little bit?"[66] Like Ludwig, he related best to experimentalists.

In Königsberg, Helmholtz began studies that linked physiology to physics. He was particularly fascinated by the ways that living bodies gathered, stored, and interpreted information about the world. While designing a lecture demonstration on the eye's ability to glow, he got the idea for the ophthalmoscope, an instrument that uses reflected light to produce a detailed image of the retina.[67] This device quickly became one of the most essential tools of clinical practice.

Although Helmholtz thrived academically, the cold East Prussian weather hurt his wife's health, so with Alexander von Humboldt's assistance, he obtained a professorship of anatomy and physiology in Bonn.[68] Here again, du Bois-Reymond represented himself to Helmholtz as an indispensable mentor, bragging that he had instructed Humboldt (who had really wanted to recommend du Bois-Reymond), "Support Helmholtz as if I played no role in the matter."[69]

When Helmholtz assumed the Bonn professorship in 1855, he became responsible for teaching anatomy as well as physiology. Possibly he began to understand Müller for the first time. Among his duties were managing the museum collection and obtaining cadavers for his medical students. Helmholtz found the Bonn Anatomical Institute "in ghastly condition," its few physiological instruments having "died from the filth."[70] By the spring of 1857, he was already hoping to win a position in Heidelberg, remarking bitterly, "For now I am stuck in this old dirt-hole." He confessed, "The most revolting thing is these endless negotiations about the delivery of cadavers, against which the Catholic clergy are always secretly scheming."[71] As a Prussian Protestant, he was not used to such obstacles. Since moving to Bonn, he wrote, he had had to experiment almost entirely in his own apartment, and the Anatomical Institute was "a pig sty, where there's no escape from the disorder and dirt."[72] In 1858, Helmholtz gladly accepted the Heidelberg job. When he negotiated his offer, he saw to it that his chair of physiology included the directorship of the new Physiological Institute that gave him and his students space and equipment to explore sensory physiology.[73]

Unlike the Königsberg and Bonn positions, the Heidelberg job had been one that du Bois-Reymond wanted and was sure he was going to get.[74] From the time that Henle had planned to leave Heidelberg for Göttingen in 1851, he had been seeking a successor. In early 1852, Carl Ludwig told du Bois-Reymond that Henle had just written to him asking whether to recommend Helmholtz or du Bois-Reymond as his replacement. Ludwig claimed that he had lobbied Henle for years to bring a good physiologist to Heidelberg, but, convinced that du Bois-Reymond did not want to leave Berlin, he had "urged [Henle] to look at Helmholtz personally and to talk to him." When Henle hesitated, Ludwig "resolved to opt for [du Bois-Reymond.]"[75] The job went to Helmholtz. Du Bois-Reymond, who had yet to be offered a full professorship, wrote to Helmholtz, "Your letter, which I received instead of the hoped-for call to Heidelberg, was no pleasant surprise."[76] Later he claimed that he was not angry about the Heidelberg job, although he had good reason to be. He asked Helmholtz to suggest him as his successor in Bonn.

In Bonn and Heidelberg, Helmholtz continued his studies of the human senses, publishing the three volumes of his respected work *Physiological Optics* between 1856 and 1867. In the first volume (1856), Helmholtz carefully described the anatomy of the eye and explained how it created images. In the second (1860), he analyzed the way that the eye registered light, and argued that three different types of receptors made color vision possible, a theory originally proposed by the British physicist Thomas Young (1773–1829) in 1801. In the third volume (1867), he reported his studies of eye movements, depth perception, and binocular vision. Here he discussed the relationships between external objects and visual representations of them, assessing the reliability of human knowledge.

In his acoustical as well as his optical studies, Helmholtz relied on Müller's Law of Specific Sense Energies, which states that each sensory receptor can register inputs only in a particular way. He represented the inner ear as a system of specially tuned resonators, each of which responds only to a given frequency. To convey his model to readers, Helmholtz compared the resonators to the strings of a piano. Because of the central role that music played in middle-class European society, his book *The Sensations of Tone* (1863) became a great popular as well as scientific success.

By early 1858, du Bois-Reymond could not help noticing that the younger student he had once advised had been promised his own physiological institute, while he still had to pester Müller for equipment. In the 1860s, Helmholtz began taking over the role of mentor. He told his Berlin friend in 1864 that on a trip to England, he had heard debates about whether to elect du Bois-Reymond to the Royal Society. Opponents were backing Matteucci, whose experiments had led the Berlin physiologist to the subject of animal electricity. Helmholtz wrote that du Bois-Reymond's experiments were considered "too subtle," and that he had tried to help him by demonstrating some of them in his public lectures.[77] As a member of the Royal Society since 1860, Helmholtz felt that he could now advise his friend on how to present his work.

In 1870 Gustav Magnus, the experimenter in whose home laboratory du Bois-Reymond and Helmholtz had met, died. This meant that the Berlin chair of physics was available, and Helmholtz, whose visual and acoustical studies had made him internationally known, became the leading candidate for Magnus's position. Well aware of how badly Prussia wanted him back, Helmholtz set high demands, asking for an unprecedented salary of 4000 taler and a Physical Institute under his personal control.[78] Knowing his value, the Cultural Ministry met his terms, and for the rest of his career, Helmholtz studied heat and electricity as du Bois-Reymond's colleague in Berlin. In 1877–1878, they set up their physical and physiological laboratories back-to-back in the same complex on Dorotheenstrasse.[79]

Like Helmholtz, Ludwig thrived from 1849 onward, despite losing the Königsberg job. That same year he received an offer from Zurich, and from there he moved to increasingly prestigious positions in Vienna (1855) and Leipzig (1865).[80] In Leipzig, he designed an institute for experimental physiology that served as a model for all those built after it, including du Bois-Reymond's in Berlin. As Ludwig's career progressed and du Bois-Reymond's languished in the 1850s, both men became aware of the growing distance between them. "I do not understand why you do not write to me at all," complained du Bois-Reymond in 1851. "Your letters and those from Brücke and Helmholtz are virtually my only pleasure in life."[81] By 1857, Ludwig reflected, "I feel such a need to look you in the eye again. We have gone very different ways."[82] He perceived a growing difference between them that du Bois did not want to recognize.

The more academic power du Bois-Reymond acquired, the less research he performed. Considering the amount of time consumed by teaching and administrative duties, this was to be expected, but du Bois-Reymond differed significantly from his mentors, Müller and Humboldt, and his peers, Helmholtz and Ludwig, in devoting his scientific life to a single project.[83] From the time that he became the Berlin professor of physiology, inheriting a third of Müller's chair, he steadily gained academic power, serving as a secretary of the Academy of Sciences from 1876 onward and rector of the university in 1869–1870 and 1882–1883.[84] His activities in the 1870s and 1880s make one wonder about his dedication to math, physics, and mechanics. Once he became director of the Berlin Physiological Institute in 1877, he showed less and less interest in developing new technologies, clinging to techniques he had perfected in the 1840s.[85] In contrast,

Ludwig's research diversified from the 1850s onward as he explored new aspects of circulation—the nervous stimulation of the heart, for instance—and developed new instruments to quantify diffusion and flow.[86] Meanwhile, du Bois-Reymond kept pursuing his animal electricity work, a task that would occupy him until his death in 1896. By the 1870s, the onetime mentor regarded Ludwig's and Helmholtz's achievements with envy. Still, the competition among the three physiologists never undermined their emotional bonds.

The scientific and emotional ties among Müller, du Bois-Reymond, Helmholtz, and Ludwig show the degree to which science is rooted in personal relationships. While none of their achievements can be explained by references to some vague "personality," their mutual support—both intellectual and political—helped them to design experiments, build equipment, and obtain academic jobs. Viewing Müller's work as a foundation, sometimes as a physiology manqué, they defined their own scientific lives, reinforcing their choices as they characterized their mentor.

Helmholtz Re-Creates Johannes Müller

Unlike du Bois-Reymond, Helmholtz never delivered a funeral address for Müller. Probably the Academy did not consider him as a speaker since he had studied only a few years with Müller and had been working outside of Berlin for the past ten. In April 1858, Helmholtz was planning his Physiological Institute in Heidelberg and extracting himself from his anatomical "dirt-hole" in Bonn. But like most of Müller's students, he wrote well and wanted to shape the historical record of his teacher. Since he eulogized Müller only briefly, in a public lecture in 1877, it is important to consider his representation of Müller in his major scientific works. His comments in his *Handbook of Physiological Optics* (1856–1867), in particular, show his ability to emphasize what he most respected in Müller's science while tactfully denying what he rejected.

When he was promoted from associate to full professor at Königsberg in 1852, Helmholtz delivered an inaugural lecture titled "On the Nature of Human Sensations." At the time, he was beginning his optical studies and trying to understand the mechanism of color vision. Arguing from analogy that both sound and light consisted of waves and must be perceived in related ways, he invoked Müller's Law of Specific Sense Energies, which would become the point of departure for his investigations of the visual system.[87] "Certainly," he told his audience, "the fundamental features of this relationship [between our sensations and the objects perceived] have long since been determined by one of the sharpest thinkers and most reliable observers among the new physiologists, J. Müller."[88] He cited "the conclusion drawn by Müller, that the quality of the sensation of light comes not from the special condition of light, but from the special activity of the optic nerve."[89] As Müller had before him, Helmholtz proposed to his mixed academic audience that the human sensations of light and color were mere symbols of the world outside our bodies.

It is significant that Helmholtz, a thirty-year-old scientist, called Müller one of the "new physiologists." His description suggests that in 1852, he saw fifty-year-old Müller as an ally whose shared assumptions about physiological experiments mattered more than their philosophical differences. In Helmholtz's *Optics*, his frequent references to Müller's studies three decades earlier show his tremendous respect for experiments that he viewed as fundamental to the field. In his specialized bibliographies—on eye movements, for example—Müller's *Comparative Physiology of Vision* and *Handbook of Human Physiology* appear often, as well as his 1826 study *On Fantasy Images*. Helmholtz also mentions Müller's equipment, such as his stroboscopic disk to study wave motion, and his optical models, such as the horopter circle, to illustrate binocular vision. These references make his own research seem like a continuation of Müller's work.

If imitation shows respect, it is revealing that Helmholtz followed Müller not just in his setups but also in his style. Müller's optical studies of the 1820s incorporated his own experiences, which Helmholtz occasionally quoted, adding his own and those of other scientists. In a book that is mathematically rigorous in its analysis of visual fields, Helmholtz saw no problem including the experience of vision, which he viewed as essential to the understanding of optics. Citing Müller's *Handbook of Human Physiology*, Helmholtz wrote, "During congestion in the cranial region, or when he bent over and then suddenly rose up, Johannes Müller saw something that looked like black bodies with tails to them, jumping and flying about in the most manifold directions." He specified that "so far, the author has not seen anything like this."[90] But Helmholtz offered his own perceptions when explaining "streamers," images that appear as floating debris in the visual field: "The author's experience is that they generally look like two sets of circular waves gradually blending together."[91] While both scientists used their own visual memories to help their readers see what they saw, it is striking that Müller, a comparative anatomist, described his images as microorganisms (or perhaps gentlemen in 1820s frock coats), and Helmholtz, a physicist, as interfering waves. Yet by copying his teacher's use of personal perceptions, Helmholtz showed respect for the way that Müller understood science.

In the late 1850s, Helmholtz continued to present Müller's Law of Specific Sense Energies as a fundamental rule of the visual system, but as he experimented in his new Physiological Institute in Heidelberg, he looked for ways to escape Müller's understanding of the nerves. In the second volume of *Optics*, he wrote, "Müller's law of specific energies was a step forward of the greatest importance for the whole theory of sense perceptions, and it has since become the scientific basis of this theory. In a certain sense, it is the empirical fulfillment of Kant's theoretical concept of the nature of human reason."[92] In his *Critique of Pure Reason* (1781), Kant had explained space and time as innate forms of human perception which would inevitably shape any understanding of external reality. This view dissatisfied Helmholtz, who believed that eyes showed people more than the structure of the mind and retina. In the third volume of *Optics*, he tried to explain how the inputs to two retinas could be combined to create an image of three-dimensional space. Müller had proposed an inborn mechanism of corresponding retinal points, but Helmholtz disagreed:

> Under the influence of Kant's theory, that space is an innate form of our apperception, Johannes Müller [argued] . . . that feeling and seeing depend on the same fundamental apperceptions of the extension of our own organs in space. Thus, he starts with the assumption that we come into the world with an innate knowledge of the dimensions in space of the sensitive portions of the retina. . . . But external vision, judgment of distance and material form of objects are supplied by experience. . . . [Gradually] we learn to combine in the idea the two localizations by means of the sense of touch of the skin and the sense of sight. . . . Yet Müller realizes that this must seem queer from his point of view, and he compares it with the perceptions that may occur as the result of the action of the sense of touch along with that of looking at a reflected image of ourselves (as in shaving).[93]

In this historical review, Helmholtz emphasized that Müller's concept of spatial perception differed from Kant's, leaving a role for learning from experience. His own studies of depth perception were showing that very little was inborn. Vision was a learned skill, not an innate faculty.[94]

Helmholtz's and Müller's contrasting explanations of spatial perception reflect their different notions of science and knowledge.[95] For Helmholtz, saying that an ability was inborn was a non-explanation, offering no information about how it worked. It was not good science, and it was not physiology. Describing forms and capacities given by nature no longer qualified as research.

In *Optics*, Helmholtz presented Müller as du Bois-Reymond did in his memorial address: as an investigator who moved in the right direction but failed to go far enough. Helmholtz differed strikingly from du Bois-Reymond, however, in the way that he conveyed this idea. Rather than characterizing his teacher's science as a whole, he restricted his comments to particular issues. In the third volume of *Optics*, Helmholtz wrote:

> Some physiologists have accepted J. Müller's theory, that the retina has the faculty of perceiving its own dimensions in space. In this case the tangential lines near the periphery would not have to be apparently too large, as they are, but rather apparently too small.[96]

When necessary, Helmholtz stated that the facts were against Müller, and his writing indicated his self-confidence as an experimenter. Of course, Helmholtz was writing a textbook for physiologists, not a speech for a diverse academic audience. Still, when he might have attacked Müller, Helmholtz defended him. In the conclusion of *Optics*, he stated:

> J. Müller, especially, is not to be blamed, if, before any observations whatever had been made concerning the law of ocular movements, and

> when nothing but absolutely vague deductions could result from trying to explain localization by means of them, he was not inclined to proceed further in his efforts of explanation.[97]

If Müller had had all the facts at his command, Helmholtz implies, he would have supported Helmholtz's view that vision was learned. Like du Bois-Reymond's Müller, Helmholtz's is retarded by eighteenth-century ideas, but for very different reasons. Lacking the facts, Helmholtz's Müller declines to speculate, and the reader pictures an innovative experimenter who gave the field its first impetus. From reading Helmholtz's *Optics*, one would never know that Müller believed in life force.

When Helmholtz did eulogize Müller in a public speech ten years later, he again emphasized Müller's methods over his philosophical approach. On this occasion, he was describing the progress of science and medicine at a celebration to mark the founding of the Pépinière. In "On Thought in Medicine" (1877), written more than three decades after he had worked with Müller, Helmholtz recalled:

> It was chiefly one man who gave us the enthusiasm to steer our work in the right direction, namely Johannes Müller, the physiologist. In his theoretical views he still favored the vitalistic hypothesis, but in the most essential point he was an investigator of nature [*Naturforscher*], firm and immovable. To him all theories were just hypotheses that had to be tested against the facts and about which only the facts could decide. Even his opinions on those points that are most easily petrified into dogma, such as the functioning of life force and the activities of the conscious soul, he persistently tried to define more precisely, to prove, or to refute, all by means of facts. While the techniques of anatomical studies were most familiar to him and he preferred to rely on them, he also worked himself into the chemical and physical methods that were more foreign to him.[98]

Here, in a speech to army doctors, Helmholtz characterized Müller as a vitalist, but his teacher came across as a rigorous investigator.

In assessing Müller's achievements, Helmholtz highlighted the studies that most resembled his own, such as Müller's investigation of the propagation of sound in the eardrum. He proposed that Müller's greatest contribution to neurophysiology was his Law of Specific Sense Energies, but he also praised his experiments showing the different functions of spinal nerve roots. According to Helmholtz, Müller's way of demonstrating these concepts proved even more important than the concepts themselves. Rather than simply proclaiming a general law, Müller would investigate many possible applications, then try to explain any exceptions that arose. "What until then had been suspected from the data of everyday experience," wrote Helmholtz, "and expressed in a vague way, the true mixed up with the false, . . . left Müller's hands in the form of classical perfection, a scientific achievement whose worth I am inclined to equate with the discovery of the law of gravity."[99] Helmholtz concluded that "[Müller's] spirit and chiefly his example continued to work in his pupils," encouraging his audience to think of his own accomplishments as a continuation of Müller's best work.[100]

In his "Autobiographical Sketch" in 1891, Helmholtz again stressed Müller's role as an inspirer. When he began studying medicine, Helmholtz recalled:

> I quickly came under the influence of a profound teacher, the physiologist Johannes Müller. . . . In the great puzzling questions [*Räthselfragen*] about the nature of life, [he] still struggled between the old, essentially metaphysical view, and the new, still developing scientific [*naturwissenschaftlich*] one, but the conviction that nothing could replace knowledge of the facts came to him with increasing certainty, and because he still wrestled with these issues, his influence over his students may have been that much greater.[101]

In this belated elegy, Helmholtz presented his teacher much more favorably than du Bois-Reymond had thirty-three years earlier. Both described a scientist who struggled between belief in an ambiguous life force and rigorous, quantitative experiments. But whereas du Bois-Reymond's Müller failed, Helmholtz's succeeded, inspiring students because of his own inner conflict. Like du Bois-Reymond, Helmholtz presented Müller as an initiator and an enabler, the founder of a field in which his students would shine. When Helmholtz occupied Berlin's new Physical Institute, he redesigned Müller as one of its founding fathers.

In comparing du Bois-Reymond's and Helmholtz's representations of Müller, one must consider their different experiences of him. Du Bois-Reymond worked under him in Berlin for almost twenty years, whereas Helmholtz interacted with him for much less time. In 1849, du Bois-Reymond chose to stay with Müller in Berlin, and Helmholtz struck out on his own. Their career decisions, shaped by institutional factors and personal values, determined their interactions with Müller, and consequently the feelings they had about him. Neither accepted his notion of life force, and neither could stomach his comparative anatomy. But their depictions of him were motivated by more than their scientific views. Helmholtz never had to beg Müller to buy a myograph, and his writings praise his teacher rather than burying him.

5

Rudolf Virchow's Scientific Politics

Science and Social Advancement

Rudolf Virchow differs from all of Müller's other students discussed in this book in that he studied medicine because he wanted to practice it. For the others, a medical degree offered a way to learn science, but clinical practice was an undesirable fallback. Although Virchow became as renowned a scientist as any of the others, he grounded his work in practical medicine. His clinical outlook affected his experiences with Müller and the way that he represented his teacher for posterity.

As with du Bois-Reymond and Helmholtz, Virchow's economic background and scientific goals shaped his interactions with Müller, but in Virchow's case, another factor played a role. In 1848, Müller and Virchow found themselves on opposite ends of the political spectrum. Virchow's liberal politics, inseparable from his science, also influenced his depiction of Müller.

Born in 1821, Virchow came to the Berlin scientific community in 1839 from rural Pomerania, a region of poor farmers ruled by aristocratic Junkers.[1] Of all of Müller's students considered in this book, he had the humblest origins, comparable to those of Müller himself. Virchow's father owned a forty-acre farm and served as the treasurer of Schivelbein, an agricultural town of about 2000 inhabitants. Neither the treasurer's job nor the potato crop brought in much money. Virchow's parents, both of whom came from families of butchers, married late in life, and he was their only child. As a scientist and politician, Virchow identified with working people and never lost his sense of closeness to the land. In 1846, when he had been in Berlin for seven years, he wanted to know how his father's grain and potatoes were doing.[2]

Prussia, to which Schivelbein belonged, offered poor boys some opportunities for advancement. During the first half of the nineteenth century, 25–30 percent of its university students were social climbers from lower middle-class families like Virchow's.[3] In Berlin, at the Friedrich-Wilhelms-Institut where Helmholtz studied, bright, diligent young men could receive free medical training in return for eight years of military service. As early as 1837, Virchow wrote in a high school essay that he planned to dedicate his life to medicine; he declared the same intention in his application for his final high school exam in 1839.[4] From a very young age, he had loved science and wanted to learn how living bodies worked.

As the top student at the Köslin *Gymnasium*, he had a reasonable chance to enter the Pépinière, and he greatly impressed the officer who examined him. But the farm boy also had some family connections.[5] His father's brother, Major Johann Christoph Virchow (1788–1856), was modernizing the Prussian army's equipment, including its medical facilities, and he helped to ensure his nephew's acceptance to the Pèpinière. His mother's brother, the architect Ludwig Ferdinand Hesse (1795–1876), worked for the Prussian court and designed the Charité's new veterinary school. In his letters home, Virchow frequently mentioned his uncles. He called the new Charité building "uncle's work."[6] In 1845, he wrote that he was having lunch with the major each Sunday and that Prussian soldiers would soon be carrying backpacks of his uncle's design.[7] With his talent, Virchow could probably have entered the Pépinière without any help, but his uncles' influence offered a much-needed boost into the Berlin medical community and cultured society.

Despite Virchow's success in medical school, relations with his father remained emotionally tense. His father had wanted him to go into business, doubting that a medical career would turn his son into a gentleman. In February 1842, during his third year of medical school, Virchow protested hotly when his father called him an egotist:

> You wanted to make a fine gentleman of me; even today that means very little to me. You explained to me during every vacation that without this, all of my knowledge would be nothing, though I could still be proud of it. If you had just criticized me less and praised me more, even if only a little, maybe that would have helped create a bond between us that was outward as well as inward.[8]

To challenge his father, Virchow quoted a saying he had read in the *German Yearbook for Science and Art*: "If your son is good, he'll tear himself away from your authority; if he's a weakling, he'll abandon your discipline only to become the slave of someone else."[9] Still, on several occasions, Virchow apologized to his father for being emotionally inexpressive and assured him that he loved and admired him, although he rejected his father's attempts to shape his future: "Maybe until now my invincible drive toward independent activity and my hermetic nature have hidden my love for you; but it will always be as great as my admiration for your inexhaustible and prudent action."[10] Virchow's letters reveal a complex negotiation between his desires to please his ambitious father and to succeed on his own terms. If Müller's wealthier students had trouble convincing their fathers to let them do research, Virchow had to struggle even harder.

Like Henle's letters home, Virchow's were largely about money, but unlike Henle, Virchow never joked about his expenses. His requests were earnest, self-justifying pleas for sums he knew his father could ill afford. In March and again in June of 1843, he wrote that he urgently needed to take his doctoral exam, since most of his classmates had already done so, and his lack of a doctorate was hurting him in the competition for good hospital internships. In July he offered a detailed breakdown of the costs, perhaps at his father's request: 15 taler to print the dissertation, 3 taler to correct it, 5.5 taler to bind it, 13 *Friedrichsd'or* for the awarding of the degree, and 20 taler for a party: about 200 taler in all.[11] Two years later Virchow asked for 80 taler to take the Prussian state medical exams, so that he could finally write "what I want and the way I want." [12] Advancing socially through medicine required a considerable investment, which Virchow's father reluctantly made. In the summer of 1845, Virchow replied to an angry letter that had accused him of asking for money every time he wrote. "It's certainly true that I demand money almost every time I write," he responded, "but it's not really my fault." He blamed the patriarchal Prussian monarchy and the police state that maintained it, in which a hardworking person could grow rich only through private enterprise. Virchow pointed out that a young man of his age could make as much in a day working for the railroad as he did in a month at the hospital, but insisted that he would never trade places, since he "loved medicine now more than ever." [13] In the end the investment paid off, for as soon as Virchow began receiving the 300 taler salary of Charité prosector in early 1847, he started sending his father money. On the day of Müller's death in April 1858, Virchow's letter home contained 400 taler.

Virchow Takes Müller's Classes, 1839–1842

As a Pépinière student, Virchow took Müller's anatomy, physiology, comparative anatomy, and pathology courses, just as the Berlin University students did. In his letters home, he sometimes referred to these "other students," whose families could pay their lecture fees and support them while they studied. Although Virchow claimed that relations with them were good, he was well aware of the differences between them. To his father, he described the distances the Pépinière students had to run between classes. Mitscherlich taught in a hall on Dorotheenstrasse, five minutes from the Pépinière, but Müller and Schlemm lectured at the Anatomical Institute, much farther away. [14] Other classes were held in the main university building, and the endless races between lectures must have reminded the Pépinière students of their status. Virchow was also deeply conscious of the Pépinière's military atmosphere. Senior doctors inspected classes to make sure that the students were attending, and if too many were cutting, the offenders' entire section could be punished. [15]

Considering their rigorous schedule, it is not surprising that Virchow's fellow students missed classes. In the winter semester of 1839–1840, they attended fifty-four hours of lectures per week, distributed over six days from 7 A.M. until 6 P.M., and they were expected to study at night. Six mornings a week, from seven to eight, a military doctor named Meyer ran an anatomy lab, using the same preparations that Müller had lectured on the previous afternoon. Four times a week, from nine to ten, Friedrich Schlemm taught them about the bowels.[16] On six days, from eleven to twelve, they learned chemistry from the university professor Mitscherlich in a hall crowded with 200 students. Six days, from two to three, Müller taught them general human anatomy; and on three days, from three to four, the anatomy of the sensory organs. Virchow reported that both of Müller's lectures were packed, attracting over 200 students. Twice a week, from three to five, he and the others learned physics; and on four days, from four to six, they were subjected to "a horribly boring lecture" on logic and psychology.[17] If Müller's lectures stood out in the avalanche of classes that first semester, Virchow didn't say so at the time.

He gave his contact with Müller emotional value only in retrospect, when he began to think back on what he had learned and from whom he had learned it. In 1852, in a brief memorial essay for his friend and co-editor Benno Reinhardt (1819–1852), Virchow wrote in the *Archive for Pathological Anatomy*:

> For me, the winter of [1839–1840] is still the source of the most pleasant memories. Back then we sat with some ambitious classmates on the benches in Johannes Müller's lectures: the fine, pale face of Brücke; the robust form of Dubois-Raymond [sic].[18]

In his memory, the fellow students who had distinguished themselves as scientists emerged as his cohorts, but he didn't describe them that way in his letters of 1839–1840.

Like Helmholtz, Virchow kept his notes from Müller's lectures, and his records show a good deal about the ways the anatomist taught and his student learned. In the summer semester of 1840 (4 May–14 August), Virchow took Müller's physiology course.[19] Virchow seems not to have recopied his notes, as Helmholtz did. Instead, he studied by underlining key terms in black pen. He wrote at breakneck speed, sometimes in complete sentences, apparently wanting to take down every word that Müller said. In the first lecture, Müller listed the most important literature in physiology, including works by Albrecht von Haller (1708–1877), François Magendie, himself, and Lorenz Oken. The juxtaposition of Magendie, an experimental physiologist, and Oken, a speculative nature philosopher, shows Müller's double commitment to experimentation and philosophy. In his introductory section on the nature of life, Müller taught the students about life force, then went on to consider the body's functions systematically, from the simplest to the most complex. He began with the circulatory and lymphatic systems. In the margins of these pages, Virchow copied diagrams illustrating the principles of pressure and flow. In the next section, on the digestive system, Virchow began drawing cells in the margins, for Müller was discussing events at the cellular level. The professor then explained the functions of muscles and nerves, and on 13 July began teaching about the senses: vision, hearing, smell, taste, and

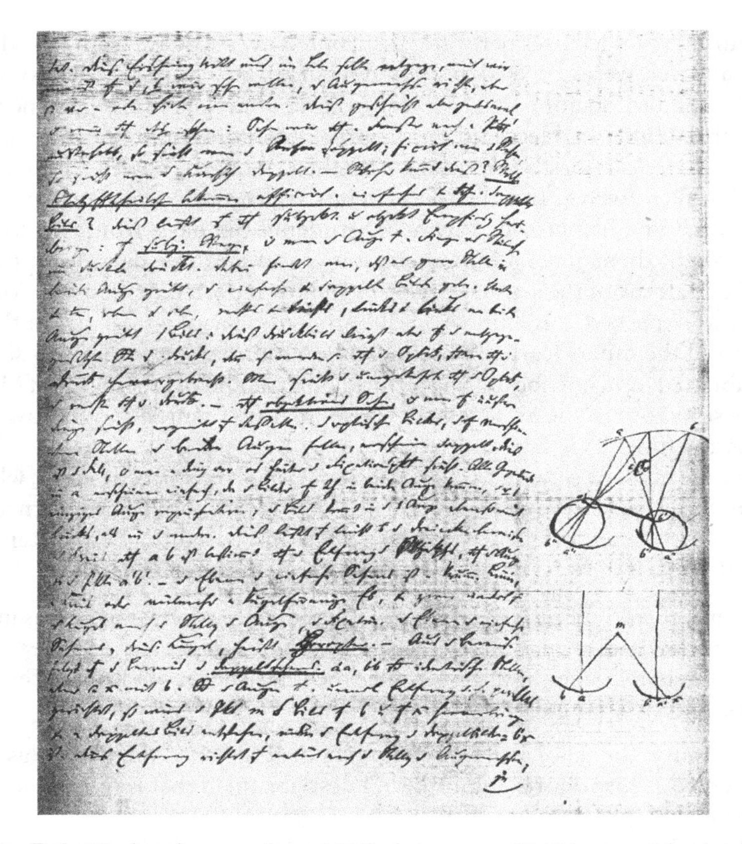

Figure 5.1 Virchow's notes from Müller's summer 1840 human physiology course describing the mechanism of binocular vision. Source: ABBAW, NL R. Virchow, nr. 2803, 141, printed with kind permission of the ABBAW.

touch. Judging from Virchow's drawings, the section on vision must have been one of the high points of the course. Müller explained the principles of optics and the mechanism of binocular vision, making sure that students saw for themselves how the inputs of two eyes could be combined.

Virchow's notes include Müller's embryology lectures, which used the same systematic approach. They begin with a list of great embryologists, including Caspar Friedrich Wolff (1733–1794), then outline spontaneous generation, asexual, and sexual reproduction. In the margins, Virchow made superb drawings of developing embryos, probably copied straight from Müller's blackboard. Artistically, Virchow lacked the talent of du Bois-Reymond and Haeckel, but embryology seems to have inspired him greatly.

A year later, from 3 May to 6 August 1841, Virchow took Müller's comparative anatomy course, the one for which his teacher was most famous.[20] In the introductory literature list, Müller mentioned Aristotle, Carl Gustav Carus (1789–

1869), and Georges Cuvier, again offering classical, German Romantic, and more recent French approaches in parallel. Significantly, Müller organized the class as he had designed his physiology course: by life function rather than individual organism. Instead of considering animals from "lowest" to "highest," Müller taught them system by system, comparing the ways that they solved the problems of circulation, sensation, movement, and digestion. Just as he had with humans, Müller included a section on the sensory organs, emphasizing the differences and continuities from animal to animal. This strategy reflects his efforts to organize his museum, where he endlessly studied the relationships among living forms.

The next summer (25 May–30 July 1842), Virchow attended Müller's pathological anatomy lectures, the ones most significant for his future work.[21] Here again, Müller began with a history of the field, telling students of Giovanni Battista

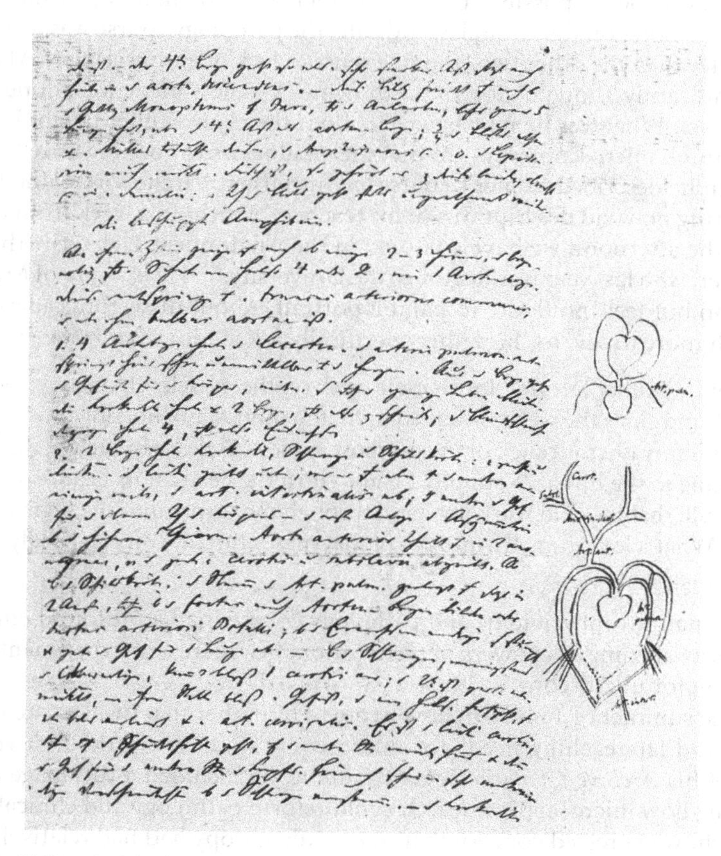

Figure 5.2 Virchow's notes from Müller's summer 1841 comparative anatomy course describing the circulatory systems of amphibians. Source: ABBAW, NL R. Virchow, nr. 2804, 65, printed with kind permission of the ABBAW.

Morgagni's (1682–1771) and Xavier Bichat's (1771–1802) achievements and offering them a bibliography complete with places of publication and dates. On 15 June, Virchow jotted down a reference to one of Henle's works in the margin. Apparently he had begun following Müller's invitations to learn about recent or ongoing research. One sees references to the Charité's chief surgeon, Johann Christian Jüngken, whose studies Müller must have been recommending. Müller's course included a large section on tumors, showing his interest in offering his own research experience to the students. During the past four years, he had been applying Schwann's cell theory to the study of tumors, a move in which Virchow must have taken a keen interest. Even if Müller didn't give medical students microscopic preparations to look at, he aroused their interest in microscopy by describing recent and ongoing studies.

Although pathology would become Virchow's field of choice, he made fewer drawings in his pathology notebook than in his physiology and comparative anatomy notebooks, possibly because Müller was not making as many. As Helmholtz's notes on the same class indicate, the pathology course was considerably shorter than the other two, and one senses a lack of enthusiasm on Müller's part. Significantly, though, most of the drawings Virchow made were done at the cellular level. Whatever he may have felt about the class, Müller taught his students to think microscopically—to investigate illnesses by looking at cells.

When he described Müller's courses to his father, Virchow remained neutral, offering no vivid descriptions of his teacher: "Six times a week from two to three in the afternoon we have anatomy in the Anatomical Theater with Prof. Dr. Müller, who last year was Rector of the University."[22] He is aware of Müller's reputation but feels no desire to paint a portrait. About dissecting cadavers he has much more to say, for his father specifically asked him what it was like:

> Recently Müller lectured for a whole week on the muscles in one severed head, and since the weather was so mild, the thing stank horribly. In lab, the military doctor called on me to show some muscles way back in the opening to the throat. Naturally I had to turn the head on its head—on its skull, that is—and work my hand through the throat into the pharynx. What a lovely smell came out of that man's throat, you can hardly imagine![23]

Müller's anatomy, physiology, and pathology courses had a profound effect on Virchow, awakening his powers of observation, his interest in experiments, and his passion for microscopy.

In the summer of 1844, Virchow wrote to his father that the Charité had an "urgent and far-reaching need" for microscopic studies.[24] In the first volume (1847) of his *Archive for Pathological Anatomy*, he included a fifty-page article explaining how microscopic research could reform pathology and clinical practice. Virchow expressed concern that so far, microscopy had had relatively little impact on practical medicine, a situation that he hoped to remedy.[25] In the prospectus for the journal, however, he wrote that he was excluding everything not directly linked to "real medicine (pathology and therapy)."[26] While Virchow became an excellent microscopist, inspired partly by Müller, he always remained close to the clinic. He worked to cure patients, not to discover the great plan of life.

Medical Politics: Virchow on the Wards, 1843–1848

As a Pépinière student and young clinician, Virchow never worked with Müller as intimately as Henle, Schwann, or du Bois-Reymond did. Because of Virchow's clinical outlook, he also experienced Müller quite differently than his fellow Pépinière student Hermann von Helmholtz. Based at the military institute and later at the Charité clinics, Virchow knew Müller as a powerful leader in the institutional conflicts between the Berlin University and the Charité Hospital. Although he admired Müller's science, he had to deal with him as a political opponent. During the years that Virchow studied and taught in Berlin (1839–1849, 1856–1902), he was never personally close to Müller. All his thoughts about his teacher's aura were developed in retrospect.

When Virchow began working on the Charité wards in 1843, the hospital contained 1036 patients. With its doctors, attendants, and administrators, it resembled a small town of 1500 inhabitants.[27] In June 1843, Virchow wrote to his father that he rarely left the Charité except sometimes in the evenings, after supper.[28] Two years later, he reported that he worked every day from 8 A.M. until 8 P.M., then went out until eleven or twelve, came back home, and worked for a few hours more.[29] Virchow so rarely left the Charité because he also lived there: first in a room with two colleagues; then with just one; and finally in his own room (heating fuel included) when he became prosector.[30] Like Henle and Schwann, he would have been hard pressed to distinguish his living space from his working space.

Because Virchow wanted to combine research and practice, he stayed at the Charité as long as he could. Both his investigations and his plans for political reform were rooted in the most practical aspects of health and pathology. During Virchow's last semester of coursework at the Pépinière in 1843, a surgeon's position opened up at the Charité, and the military doctors invited him to apply. Although eager to enter the wards, he did so only on the condition that he receive his full year and a half of clinical training.[31] Soon after accepting the surgical position, Virchow wrote to his father that so far he had had no opportunity to regret his choice. Working for Jüngken in the ophthalmology clinic, he sometimes had to get up at five to bleed patients, but he proclaimed that "this activity, exhausting as it may be, still makes me happy, because for the first time one steps out of school theorems into real life and replaces clinical mirror-dancing with real, creative action."[32] He quickly became a valued clinician.

On the wards, Virchow wrote, the senior military doctors were like gods, issuing orders but "maintaining an awe-inspiring distance." The attendants, on the other hand, stuck too close to the patients, "like the devil to faithful Christians." As a result, young doctors like himself made most of the practical decisions regarding patients, such as what they should eat, when they could take walks, and when they should be given sleeping powders, leeches, or a good bleeding. [33] Virchow's patients liked him, calling him "the little doctor."[34] By the spring of 1844, although he had begun his own microscopic research, he still expected to be stationed with a military unit and knew that he wanted the cavalry.[35] Two years later, however, he was released from military service, with the support of the senior military doctors and the Prussian cultural minister, so that he could become interim prosector of the Charité. Virchow did not give up clinical practice for research, although he gladly relinquished his service to the Prussian army. By assuming responsibility for the hospital's autopsies, he got exactly what he wanted.

Virchow had begun his own research in 1843, receiving his doctorate that October for a thesis on the cornea.[36] As the university's anatomy professor, Müller acted as his official adviser, but it is unclear how often they met or how actively he supervised the research. To expand his studies with investigations of blood clots, tumors, and leukemia, Virchow needed access to the Charité's cadavers. By becoming prosector, he increased his pickings but by no means gained free dominion. Rather, he entered an ongoing battle between Berlin's hospital and university for control of the city's dead.

In the eighteenth century, Prussia's kings had worked to make Berlin a center for science and medicine.[37] Part of this effort had been the founding of the Anatomical Institute in 1713 as a place where military doctors could study the human body. When the Berlin University was founded in 1810, the Anatomical Institute, then located on Dorotheenstrasse, was turned over to the university's medical faculty.[38] From the time that the Anatomical Institute was founded, its instructors had struggled with the Charité's doctors for cadavers, since both wanted them for their teaching and their research.[39] By the late eighteenth century, they had developed an agreement that bodies would go to the Anatomical Institute in winter and stay at the hospital in the summer, when time was of the essence.[40] Until the mid-nineteenth century, the Charité directors maintained their control over the flow of cadavers, but as the Prussian government realized the prestige—and military improvements—that a top research university could bring, the balance of power began to shift. In his early microscopical studies, Virchow was making observations he hoped would have an immediate effect on the treatment of patients, and he needed cadavers to provide diseased tissue. He took charge of the morgue just as the university anatomists were gaining the upper hand, and his scientific development can be understood only in light of Berlin's medical and academic politics.

The plans to establish a Charité prosectorship dated back to 1829. Johannes Schulze, who had mentored Müller and since 1818 had worked for the Cultural Ministry developing Prussia's high schools and universities, helped to get the position funded.[41] In May 1831, the Berlin-trained physician and pharmacologist Philipp Phoebus (1804–1880) became the first prosector, and that October, the Cultural Ministry outlined his duties: (1) maintaining the morgue, (2) conducting autopsies, (3) teaching medical students how to do autopsies, and (4) creating a collection of anatomical specimens for instructional purposes.[42] From the beginning, the prosector moved in an unstable orbit, pulled in different directions by competing authorities. He had to answer not just to the Charité administration and the senior military doctors but also to the Cultural Ministry and the university's medical faculty.[43] Conflicts erupted immediately over the rights to dissect cadavers, since many senior doctors wanted to autopsy their own patients. Intimidated, Phoebus drafted a new outline of his perceived duties, writing that autopsies were to be directed by senior doctors and that the prosector would act "only as their organ" in the process.[44] Phoebus also deferred to the university's medical professors, writing that the prosector would not teach university students. To make matters worse, a doctor supporting a family could not live on the prosector's salary of 300 taler a year, and when Phoebus left the morgue to teach or practice medicine privately, both senior and junior doctors took advantage of the situation.[45] Frustrated, Phoebus quit in 1833, but many candidates applied for the job.

As soon as Müller joined the Berlin University's medical faculty in 1833, he began using all of his shrewdness and energy to increase his department's power. One of his first acts was to back for prosector Robert Froriep (1804–1861), an anatomist three years his junior who had studied in Bonn and had just completed his *Habilitation* at the Berlin medical faculty. In 1833 Froriep, with his strong inclination for microscopic research, would become an associate professor at the Berlin University, so that—Müller hoped—a man sharing his own interests would take charge of the Charité's cadavers and anatomical collection.[46] In September 1833, Froriep got the job, which by then included a fifth duty: sketching pathological preparations and anatomical specimens.[47] Artistically inclined, Froriep soon began teaching anatomy at the Art Academy, and later published a detailed anatomical atlas. As the new prosector proudly accumulated the Charité's anatomical collection, Müller realized he had backed a scientist a little too much after his own heart.

Froriep fared no better than Phoebus in the battles to control cadavers and anatomical preparations. In October 1833, the Charité administration drafted a list of duties that weakened the prosector's position relative to that of the senior doctors, so that Froriep was even less "master of the morgue" than Phoebus had been.[48] Froriep struggled for research space. In his letter of application, he asked the Charité administration for a "spacious, well-lit, heatable room with two windows, if possible," and for an assistant to help him with dissections.[49] In 1843, he appealed to the Cultural Ministry to establish an independent pathological anatomy institute, but the minister turned him down.[50] Froriep never received the work space he desired, but he did get an assistant: young Rudolf Virchow took on the job in 1844. Worn out by medical politics, Froriep gladly assumed his father's position as regional industrial accountant and personal physician to the duke of Weimar in 1846, proposing his assistant Virchow as his successor. With his growing interest in tumors and cellular pathology, Virchow was eager to become prosector, writing to his father that "pathological anatomy is completely undeveloped in Berlin, and there is a universal, urgent wish for it."[51] In May 1846, Virchow took over as prosector.

Froriep had failed to develop pathological anatomy as an independent field partly because of aggressive opposition from Müller. In his efforts to teach pathological anatomy and to create a Charité-based anatomical museum, the prosector found himself up against the university's senior anatomist. By installing Froriep as prosector, Müller had hoped to obtain droit du seigneur over the hospital's anatomical specimens, not to create a rival museum. During Froriep's thirteen troubled years as prosector, Müller used all of his influence with the Cultural Ministry to restrict the clinical anatomist's independence, particularly with regard to his collection.

From the time of the Anatomical Institute's founding in 1713, Berlin's anatomists had accumulated interesting specimens: enlarged tumors, deformed babies, and curious growths. When it joined the Berlin University in 1810, its collection came with it, one of the grains of sand around which Müller's pearl would grow.[52] Probably Johannes Schulze at the Cultural Ministry shared Müller's passion for museums, for not only did he support Müller's candidacy, with his declared intention to make Berlin an anatomical Paris; Schulze also considered the creation of a clinical collection to be the prosector's first duty. In his letter defining the prosectorship, Schulze argued that the position was needed so that a doctor could "build a pathological anatomical collection appropriate for clinical instruction which, so that it can be used better, will be kept separate from the main collection in the university building and housed in the Charité-hospital itself."[53] In his preliminary sketch of his duties, Phoebus, like Schulze, had listed the creation of a "practical anatomical collection" first. The Charité administration placed it last.[54]

When Froriep took over, he gave the anatomical collection top priority, calling attention to it in all his reports. In his application letter, he wrote that the collection should be open for two types of use: "first for clinical instructors, so that they can come by at any time and pick up the necessary preparations for their lectures—secondly for students, so that they can enter the collection daily and consult individual preparations in the available catalog."[55] In his report after one year as prosector, Froriep wrote proudly that the collection contained 900 specimens. In a letter to Virchow a year after he left, Froriep explained that he had planned to organize the specimens according to pathological processes and the affected organ systems.[56] By 1842, the voluminous correspondence between Froriep and Müller was annoying the cultural minister, who had to settle their conflicts like a deus ex machina.[57] Generally he sided with Müller, who was winning Berlin a reputation as a center for medical research. The university anatomist demanded free access to Froriep's preparations, and gained assurance from the cultural minister that Froriep's collection could not exceed 1500–2000 specimens and that each year he must receive a list of all new preparations so that he could take the most intriguing ones for his Anatomical Museum.[58]

When twenty-four-year-old Virchow became prosector, he found himself facing his former teacher and Germany's foremost anatomist. Ten years later, Müller would lend crucial support to the establishment of pathological anatomy as a university subject, but in 1846, he saw the young prosector's collection as a threat. Soon after taking the job, Virchow wrote to his father:

> My position still remains more or less precarious and demands a certain caution. A lot needs to be changed, and some things fully abolished, in the institutional situation that has existed up to now. Right now I'm fighting to preserve our collection of pathological-anatomical preparations as a self-sufficient, independent institution against the demands that Geh. Rath Müller is making for the University's Anatomical Museum. The Ministry sees the matter from Müller's point of view, and I'll have to shift it somewhat if my future position in this respect is not to be fully dependent.[59]

Two months later, Virchow wrote to Froriep that the Cultural Ministry had granted him 200 taler for the clinical collection—contingent on Müller's right to inspect it personally and take any specimens that he desired. In March 1847, Virchow told Froriep that he had again received 200 taler—and that Müller had just carried off a hundred preparations.[60] While Virchow admired Müller's science, he experienced him as a powerful predator. Müller was never a scientific confidant or a partner in research.

Although Müller's classes offered numerous examples of microscopic studies, it was Froriep who convinced Virchow to start his own investigations. Rather than turning to his thesis adviser for guidance, Virchow looked to his immediate clinical superior, the Charité prosector. Besides suggesting research topics, Froriep helped integrate him into the European scientific community, encouraging him to publish his findings and to read French and English journals.[61] Froriep shared Virchow's conviction that medicine would improve if doctors would learn more science, and Virchow looked up to him as a physician who set an example with his own work.[62] In November 1844, Virchow wrote to his father that "Froriep, the prosector of the Charité, especially supports me and guides me in my anatomical-microscopic investigations; on his advice, I've picked out a special theme for more exact study."[63] This topic was the cause of blood poisoning, which at the time killed many patients who had undergone surgery. Froriep urged Virchow to read the reports of the French pathologist Jean Cruveilhier (1791–1874), who had attributed these deaths—and most of pathology—to phlebitis, the inflammation of vein walls, which supposedly allowed pus to enter the blood and produce deadly clots. This view conformed to that of Europe's leading pathologist, Karl Rokitansky (1804–1878) of Vienna, who attributed most diseases to imbalances in blood chemistry. In Virchow's systematic, microscopic studies of the blood during 1844–1847, he demonstrated that clotting and inflammation were independent processes and that the whole notion of "pus in the blood" was fictional.[64] Virchow might never have carried out this research if Froriep had not encouraged him to read foreign scientific literature.[65]

Froriep also helped to get Virchow's findings into print. Along with his father, whose job as Weimar's regional industrial accountant focused on publishing, Froriep edited his own small journal, *Scientific and Medical Notes*, in which he offered to print Virchow's work. In late August 1845, Virchow wrote excitedly to his father that his "lovable boss" and bureaucrat father were about to publish "a series of my own observations."[66]

Once Froriep left, Virchow found himself alone, sometimes longing for counseling. In March 1847, he wrote to Froriep:

> I often think about how much more peacefully and circumspectly I could develop myself if I had your experience, your friendly and wise counseling, at my disposal, and I would give a great deal if I could have you back in your old role. . . . I am my own adviser, since so far I have discovered no one in Berlin who can give me any good advice.[67]

Like Müller, Virchow longed for a close friend with whom he could discuss his scientific ideas, and from 1843 to 1847, Froriep played this role. Their relationship could not have been that close, however, since just six months later, Virchow stopped writing to Froriep and did not contact him until two years later. As a mentor, Froriep did not meet all of Virchow's needs, so the young scientist sought out others.

During 1847–1849, Virchow became increasingly politically active, sharing ideas with liberals such as his future father-in-law, the obstetrician Carl Mayer (1795–1868), and the physician Benno Reinhardt. His memorial article on Reinhardt shows how their friendship inspired his scientific thinking: "Interactions that occurred almost daily, an endless exchange of opinions and experiences, soon brought us to the idea of joint activity and the grounding and development of a medicine driven by the natural sciences."[68] Yet when Virchow told Froriep that he could get no good counseling, he was already talking to Reinhardt, so he must have had a senior scientist in mind. Müller, who respected bright young researchers, was readily available, but his advice must not have been good enough. Müller did not mentor Virchow, although each respected the other's research.

Considering how poorly Froriep fared in the power struggle with Müller, if the prosector been Virchow's only adviser, his assistant would not have survived. He could have wooed Müller, as du Bois-Reymond so carefully did, but Virchow saw himself as a clinical researcher and wanted to work within the Charité's medical community. Luckily, he won the support of almost all the senior military doctors, who saw in the gifted, energetic researcher a chance to gain some ground at the university. When it appeared that a surgical position for microscopic research would be funded, Heinrich Grimm (1804–1884), one of the highest-ranking military doctors, urged Virchow to apply because he didn't want a civilian to hold the job.[69] In late 1844, Virchow wrote to his father that the chief of surgery, Johann Christian Jüngken, had taken him under his wing, asking him to stay on at the surgical clinic and busying him with chemical and microscopic studies.[70] The following summer, Grimm asked Virchow whether he would like to become a lecturer at the university; Virchow replied that he still planned to become an army doctor.[71]

When Froriep retired as prosector, Virchow won the position not just because of Froriep's support but also because of the military doctors' backing.[72] There was one key dissenter: Johann Lukas Schönlein (1793–1864), Berlin's most illustrious clinician, who wanted his assistant, Robert Remak, to have the job. Since Schönlein had come to Berlin from Zurich in 1840, he and Müller had been building Berlin's reputation as a center for rigorous, scientifically grounded medicine. Schönlein urged doctors to perform percussion (thumping patients' chests and abdomens to assess what was going on inside of them) and auscultation (listening to their breathing through stethoscopes), techniques that were emerging in France but were not yet practiced in Germany.[73] He also advocated the use of microscopic studies and chemical analyses to diagnose patients. From the time of his arrival, Schönlein had exerted great influence, acting as personal physician to the Prussian king and medical adviser to the cultural minister. With no link to the Prussian army, Schönlein held the post of university professor, and he supported his gifted assistant Remak, an anatomist and Orthodox Jew, who had no more connection to the military than he did. When it became apparent that the cultural minister would not appoint a Jewish candidate, no matter how talented, Schönlein changed his backing to Virchow.[74] Virchow wrote to Froriep, "Schönlein no longer finds it fashionable to take any other course; he called me to his office, declared his protection of me and remarked that I would learn from this matter that what counted for him was not individuals but the matter at hand."[75] The staff physician Johann Lohmeyer (1776–1852), who oversaw the Pépinière students' personal matters, wrote to Cultural Minister Eichhorn that Virchow had almost unanimous support from the Charité staff, and asked that Virchow be released from his military service obligation so that he could become prosector.[76] The minister granted his request. Like Helmholtz, Virchow got out of the army because of his talent for research, but his backer was a military doctor, not Müller.

Lohmeyer and his staff did not release Virchow from his duty just because they knew he was worth more in Prussia's finest hospital than he could have been at any military outpost. They were also preparing him, as one of their own, for an invasion of the Berlin University, where he could demonstrate the skill and intelligence of military doctors and increase their chances of teaching university students. After describing Schönlein's conversion, Virchow wrote to Froriep: "Now they're demanding that I do my *Habilitation* at the University and prove myself as a competent worker in this 'vineyard.'"[77] Two weeks later, Virchow wrote to his father that Lohmeyer and Gottlieb Eck (1795–1848), associate director of the Pépinière, were urging him to complete his *Habilitation* and that Friedrich Hecker (1795–1850), dean of the medical faculty, was willing to let him try. Virchow hoped that with his support at the Cultural Ministry, he might soon be appointed associate professor. The minister had just granted him 150 taler for a study trip to Vienna and Prague.[78] But in March 1849, Virchow lost his job as prosector. Branded as a political activist, he was summarily fired, and his powerful friends scrambled to save him.

Virchow in 1848

For Virchow, science and politics were inseparable. He hoped to reform Prussia through improved education and public health, and to accomplish this goal, he needed to free science and medicine from the blind submission to authority that he believed was crippling society.[79] In May 1848, when hope for liberal reform reached its peak, Virchow wrote to his father, "As a scientist I can only be a republican, since the realization of the demands determined by natural laws, which come from human nature, can be achieved only in a republican state."[80] Even in December 1848, when the monarchy and military had regained power, he remained hopeful that "history will keep moving forward; its task is the mental and bodily liberation of humankind."[81] Virchow never saw political or scientific freedom as inevitable, however. They could be achieved only through intelligently planned campaigns. Self-confident and pragmatic, Virchow argued aggressively for individual liberty, always bearing his audience in mind.

On 3 May 1845, when the Friedrich-Wilhelms-Institut celebrated the birthday of its founder, Johann Görcke (1750–1822), its directors asked Virchow to present a "formal medical declaration of faith," and Virchow took advantage of the opportunity to demand reforms and attack medical conservatives.[82] In his speech, he declared that modern medicine must base itself on physical laws, regarding the cell as chemists did the atom. Medical knowledge must build itself upon close clinical observation, animal experiments, systematic autopsies, and microscopic analysis.[83] Eck, the Friedrich-Wilhelms-Institute's associate director, preapproved Virchow's speech—although he complained that the young doctor sounded like a French Academician—and the talk was well received by the military doctors.

When the Pépinière celebrated its fiftieth anniversary on 2 August 1845, Eck asked Virchow to make another speech "of the same type as the last one," for the scheduled speaker had canceled on short notice. Virchow wrote to his father:

> I'm glad to do it, for one seldom has the opportunity to address the conscience of ministers, generals, and privy counselors, and I hope to make myself clear to them. I am not deceiving myself in this matter; nowadays, with some real knowledge, one can impose his ideas on anyone, even the most highly placed, for everything is hollow and worm-eaten from top to bottom.[84]

His father grew concerned that the young man's criticism would jeopardize his career, but in 1845, Virchow's eloquent demands for reform only increased the respect of his military superiors. Virchow wanted a renewal of medicine from the inside out—from the morgue and the microscope to the wards—and he focused on clinical practice. To him, the bottom line of any epistemological strategy was its value to the suffering patient.

In February 1848, Virchow gladly accepted Cultural Minister Eichhorn's order to investigate a typhus epidemic in Silesia (today southwestern Poland). After three years of negligible potato crops, the region's peasants were starving and succumbing to disease at an alarming rate. In the area that Virchow investigated, 14.5 percent of the people had typhus, and 20 percent of those who fell ill died—2.9 percent of the total population.[85] Very likely, Eichhorn sent Virchow and Privy Counselor Stephan Barez (1790–1856) because he hoped to avoid scandal by placing the matter in the hands of an energetic young scientist and a respected pediatrician.[86] Explaining the trip to his father, Virchow wrote that Barez was to "oversee and carry out practical measures," whereas Eichhorn had specified that Virchow "was to occupy himself with the disease in a scientific way" and "study its nature and origin more closely."[87] The young reformer gladly accepted his charge. The week before, he had told his father that the thousands of deaths were a disgrace to the Prussian government.[88]

Virchow's letters from Silesia read much like his 150–page report. In both, he uses detail and vivid imagery to make readers see the suffering peasants:

> The houses consist mainly of piled-up beams—log cabins; the rooms are quite small, the animals mixed in with the people, the windows small and incapable of being opened, the greater part of the rooms occupied by the stoves and the beds. And the people—horrible, miserable forms who walk through the snow with *naked* feet, the feet mostly swollen, the faces pale, the eyes dull. And they are so submissive, they kiss your arm, the hem of your coat, your knee, all in one breath. Enough, it is repulsive.[89]

Virchow blamed the epidemic on failed government policies, declaring that Prussia was denying its people's right to health. He pointed out that the Silesian aristocracy lived in the cities, sucking their wealth out of the land and the peasants' bodies.[90] To prevent future famines and epidemics, he recommended economic investment, improved education (including instruction in Polish), new roads, and the withdrawal of the Catholic Church from teaching and health care.

Years before he went to Silesia, Virchow had identified Catholicism as a force that blocked the growth of knowledge. In an 1843 article on a Schivelbein monastery, he had described priests as opponents of free inquiry.[91] While stationed at the Charité's mental ward, he had lamented the number of minds that religion was destroying. Religious mania ranked third as a cause of internment, after masturbation and alcoholism. (In Berlin, however, the maddening religion would probably have been Protestantism.)[92] When characterizing the Silesian epidemic, both to Minister Eichhorn and to his father, Virchow specifically blamed the Catholic Church for keeping people ignorant.[93] "One sees here quite clearly," he wrote to his father, "what can become of a people enslaved by the Catholic hierarchy and the Prussian bureaucracy. This apathy, this animal-like enslavement is *horrifying*."[94] In science and politics, Virchow valued unbiased observation and independent thinking above all. Recalling how he and Reinhardt had founded the *Archive for Pathological Anatomy*, he told readers, "It was enough to begin a struggle for principles and methods against schools and authorities, simply trusting in one's own powers."[95] Any system in which people were expected to believe and obey without rational justification outraged his sense of human dignity, and he opposed the Church's administration of hospitals as a remnant of feudal times.[96] In the 1870s, Virchow would join Chancellor Otto von Bismarck's (1815–1898) *Kulturkampf* against Catholicism, one of the few times he would back the hawkish arch-conservative on any issue. As a twenty-six-year-old medical reformer, he attacked the Church and the Prussian government together, yet even his scathing report did not ruin his medical career. To lose his job, he had to take up arms and distribute leaflets denying the legitimacy of the Prussian government.

When Virchow returned from Silesia on 10 March 1848, Berlin was about to explode. Of all of Müller's students, Virchow participated most actively in the 1848 revolution. He wrote openly to his father about the barricades, and his account can be confirmed by one outside source. Carl Ludwig Schleich (1859–1922), who worked for Virchow in the 1880s, recalls a story his father told him about his supervisor's past:

> It was 1848. The revolution broke out. In rushing haste, Virchow stormed into the room of my father, who was then an assistant doctor.
>
> "Schleich, do you have any weapons?"
>
> "Nothing but this old rifle and a rusty saber!"
>
> "Give them here! To the barricades!"
>
> And he was off.[97]

Schleich's story has the ring of a tale polished in the retelling, so we should judge Virchow's eagerness only after hearing his own account. At 11 A.M. on 19 March, Virchow wrote his father a long, dramatic chronicle of the past night's fighting:

> Everyone gathered before the palace, cheering and calling out. The King appeared, and everyone shouted "hurrah!" The citizens had only one more wish, that the military be withdrawn. That was too much for the King. He told General Möllendorf he wanted order established; the Prince of Prussia ordered an attack, and suddenly the unsuspecting people were dispersed at the points of the dragoons' sabers. . . . From this moment the revolution began. Everyone cried betrayal and revenge. In a few hours all of Berlin was barricaded, and anyone who could get weapons armed himself. Unfortunately the number of large rifles was extremely small, since the dealers had been forced to turn in their stock, and Berliners only very rarely possess rifles or guns. . . . My part in the uprising was relatively insignificant. I helped to build a few barricades, but then, since all I could get was a pistol, I wasn't of much further use, since the soldiers were mostly shooting from too great a distance, and with the small number of citizens [*Bürger*], at least at my barricade, hand-to-hand combat was out of the question.[98]

Both Schleich's and Virchow's accounts indicate that if Virchow could have shot Prussian soldiers, he would have. The discrepancy is the senior Schleich's rifle, which the scientist may have given away because his eyesight was not the best. Virchow manned the barricade at the corner of Friedrichstrasse and Taubenstrasse, just a block from the boardinghouse where Henle and Schwann had lived ten years earlier.

In the year that followed, Virchow proved much more useful to the revolution as an orator, writer, and political organizer. On 1 May, he wrote to his father that he had been elected as a delegate for Berlin's eighty-seventh district, both to the Frankfurt National Assembly and to the Prussian Assembly, but he refused to go to Frankfurt, suspecting that this congress would not realize his dream of a German republic.[99] Since at twenty-six he was too young to vote in the Prussian Assembly, he focused his efforts on organizing a democratic party and electing delegates who would fight for a republic.[100]

As a political activist, Virchow identified with Berlin's workers, something his father never understood. After announcing his election, Virchow explained to his father:

> I can see very well how hard it must be for you as an old property owner and solid citizen [*Bürger*] to appreciate fully the role of the workers in this revolution, since you're unable to see these relations in a larger context, but I can assure you that we have here among these workers a great number of men who without exception are far ahead of all your solid citizens [*Bürger*]. You're right to say that it was really the workers who decided the revolution, but I believe that you in the provinces don't think enough about the fact that this revolution is not simply a political, but essentially a social one. Everything political that we're now doing, the whole constitution, is only the structure through which social reform will come into being, the means by which the conditions of society will be transformed right down to their foundations.[101]

During the summer of 1848, Virchow met with representatives of all of Berlin's districts, including groups of doctors, craftsmen, and machinists. His letters home often take the plural form "we," his term for all reformers, regardless of economic background. Less than a week after the worst fighting, he noticed a split between the workers and the bourgeoisie, although he described it without committing himself to either class:

> For the present no one is in control: neither the government nor the people, or as one might say for the latter, neither the bourgeois [*Bürger*] nor the workers have seized power. It will be lucky for us if this division of power lasts, but that doesn't look likely. Already it's starting, the reaction of the citizenry [bourgeoisie] against the workers [the people].[102]

Virchow differed from his father—and from Müller—by including workers in the collective "we," but he sometimes revealed his allegiance to the middle class. Describing the constituents of his democratic Berlin district, Virchow wrote, "Certainly many of them are proletarians, but the spirit of our time has made men [*Menschen*] of them."[103]

Between July and November 1848, Virchow focused his energy on revitalizing the medical profession, editing the weekly journal *Medical Reform* with the liberal psychologist Rudolf Leubuscher (1821–1861). But when troops returned to Berlin that November and the liberal reforms seemed endangered, Virchow reassumed an active role.[104] When eighty university professors sent the king an affirmation of loyalty approving his adjournment of the Prussian Assembly, eighteen lecturers—including Leubuscher, Virchow, and Robert Remak, his rival for the Charité prosectorship—signed a dissenting document in support of the National Assembly. All eighteen risked their academic futures, but for Virchow there were still no direct consequences.[105] These would come with the next electoral campaign.

In January and February of 1849, Virchow worked tirelessly to get a democrat elected to the new Prussian Assembly. Since he was known for writing and distributing campaign literature, government troops searched his apartment at the Charité for revolutionary leaflets.[106] By early March, he wrote to his father that two surgeons had been dismissed for supporting the political opposition and that he expected a decision about his own case any day. When a representative of the Cultural Ministry had asked him how he could reconcile his government-backed position with his distribution of inflammatory literature, he had answered, "My official position has nothing to do with my political activity, and I have not abused the former to conduct the latter."[107] The Cultural Ministry thought otherwise. On 30 March, the hospital administration informed Virchow that as of 15 April, he would no longer be prosector, and as of 1 May, he would have to leave his Charité apartment. Virchow assured his father that "several influential people" were working to get him reinstated and that the protest among doctors and students was so great that the cultural minister would probably yield.[108] Sure enough, Virchow received word that he could keep his job, but not his heated apartment, as long as he stayed out of politics. Eight months later, he would write to Froriep,

Figure 5.3 Rudolf Virchow in 1849, the year he moved to Würzburg. Source: Rudolf Virchow, *Rudolf Virchow: Briefe an seine Eltern 1839 bis 1864*, ed. Marie Rabl (Leipzig: Wilhelm Engelmann, 1906).

"I was thus immediately reinstated as prosector, once I had sworn not to democratize the doctors and bureaucrats at the Charité, but I was banished forever from the Charité apartment and the free meals."[109]

Virchow did not despair, however, because since mid-February he had known that he was being considered for a full professorship of pathology in Würzburg, Bavaria. He feared that he might not get the position, since the conservative Bavarian government would communicate with the Prussian Cultural Ministry about his political activity. During that anxious summer, he asked his mother several times for bedding for the new Berlin apartment he would need to rent. In early August, Virchow obtained the Würzburg position, but only after assuring the Bavarian Senate that he would not turn Würzburg into "a playground for [his] known radical tendencies."[110] While Virchow was not eager to leave Berlin, the Würzburg job offered a large hospital, a good salary, and scientific independence. He departed for Würzburg at the end of November 1849, just after his betrothal to Rose Mayer.

Of all of Müller's students considered in this book, Virchow made one of the shrewdest marriages. As had been the case with Müller, his scientific work allowed him to rise in society, and his choice of a mate facilitated that climb. Through Mayer's family, Virchow established permanent bonds to Berlin's most influential liberals and doctors, and secured a place in its upper middle class.[111] In 1846, at meetings of the Berlin Obstetrical Society, Virchow had met Rose's father, Carl Mayer, the city's most respected obstetrician and one of the founders of modern gynecology. Mayer shared Virchow's determination to make medicine more scientific, and he supported the young doctor's political goals, introducing him to prominent liberal and radical writers.[112] The extended Mayer family included politicians, lawyers, manufacturers, court gardeners, artists, and several physicians, and by 1848, Virchow was spending two or three evenings a week in their company.[113] The aristocratic branch of the Mayer family grumbled that Virchow would "spoil" their clan, but Rudolf and Rose's marriage proved a successful partnership. Between 1851 and 1873 they had six children, all of whom survived to adulthood.[114] Carl Ludwig Schleich recalls that in the mid-1880s, Rose seemed to have taken on all of Virchow's mannerisms and gestures. He did not let his large family draw him away from his work, however, since he regularly read until one in the morning and rose at six at the latest.[115]

In Würzburg the Virchows had the first three of their six children, and Rudolf conducted his most important cellular research, establishing an international reputation as a pathologist. By 1856, the power struggle between the university and the Charité had focused on this field, since Berlin was losing gifted medical students to Würzburg and Vienna because it lacked a leading pathologist.[116] Whatever grief Virchow may have caused Müller in the spring and summer of 1848, Müller led the campaign to bring him back to Berlin.[117] When the Charité prosector and associate professor Heinrich Meckel von Hemsbach (1822–1856) died, Müller suggested replacing him with a full professor of pathology. By 1856, Müller was exhausted, and knew that he could not continue teaching physiology, comparative anatomy, and pathology to the university's medical students each

Figure 5.4 Johannes Müller in 1856, the year he helped bring Virchow back to Berlin, as painted by Oscar Begas for the portrait collection in the Prussian Royal Palace. Source: Wilhelm Haberling, *Johannes Müller: Das Leben des rheinischen Naturforschers* (Leipzig: Akademische Verlagsgesellschaft, 1924), plate VIII, 441.

year. Meckel had been teaching pathology at the Charité. Müller knew that pathology had developed to the point that it warranted its own professorship and that Virchow was the best pathologist in the German territories. By bringing in Virchow, who was no longer a military doctor, Müller would secure pathology as a university field. By yielding part of his own chair, he could win crucial territory—and cadavers—for Berlin University.[118] Müller's energetic backing of Virchow reveals a great deal about the aging anatomist's values. More than he feared radical politics or disliked Charité doctors, Müller respected Virchow's science. Seeing his own energies dwindling, he wanted to fill the medical faculty with scientists committed to close observation and microscopic studies. Virchow, who knew hospital politics, wrote that he would come only if the Cultural Ministry would guarantee him a newly constructed pathological institute, his own department at the Charité, and a salary of 2000 taler.[119] The minister met his demands. In the 1850s, the Cultural Ministry was spending very little on science, but its members still listened to Müller.[120]

Without his conservative teacher's help, Virchow might never have become the scientist that he did. Müller shaped Virchow's career not just by introducing him to microscopy, but by creating the position from which Virchow would establish microscopy as the key investigative tool of pathologists.

Virchow the Writer

Virchow's tenure of the Berlin pathology chair did not require a promise to stay out of politics, and within five years of his return, he was at the center of Berlin's political as well as its scientific life.[121] In 1859, he was elected to the Berlin City Council, on which he served until the end of his life. Two years later, he helped to found the German Progressive Party, and in 1861 won a seat in the Prussian Assembly, where he remained until 1902. In Assembly debates, he emerged as a leading opponent of the militaristic Chancellor Otto von Bismarck, demanding that tax money be used for education and public health instead of an arms buildup. Virchow enraged Bismarck so greatly that in 1865 the chancellor challenged him to a duel, but the pathologist refused to fight with any weapon but his own intelligence. Newspaper editorials throughout the German territories praised Virchow for confronting the conservative chancellor and depicted him as a champion of free speech.[122] In 1880, Virchow ran for a seat in the German Reichstag, the imperial parliament that had existed since the country had unified in 1871. The pathologist won the right to represent Berlin's second district by attacking Adolf Stoecker's (1835–1909) anti-Semitic policies, and he remained in the Reichstag until 1893.[123]

During the second forty years of his life, Virchow used his city, state, and national offices to realize the health and educational reforms for which he had argued in the first forty: a modern sewer system and a clean water supply for Berlin, the reorganization of hospitals, meat inspection, and an overhaul of school curricula. As a politician, he commanded tremendous popular support, and Germans respected his medical background. After the duel challenge, the *Silesian News*—whose readers lived in the region once devastated by typhus—proclaimed that "a man who through his profession risks losing his health and life every day cannot be accused of cowardice if he refuses to risk them in a single moment."[124]

Virchow launched his political career as he had built his scientific one: by writing. Knowing that his family connections led only into the Prussian army, he used writing to create a reputation and spread his views. One cannot emphasize strongly enough how much and how well Virchow wrote. In the fall of 1845, he told his father, "The next task I've set for myself is to win general recognition through a series of literary publications."[125] Virchow was referring to the articles on blood that Froriep would publish, but he saw the printed word as the best way to shape people's vision, whether the subject was cell structure, military spending, or medical history.

In his eighty-one years, Virchow published over 2000 articles and books.[126] He spoke Dutch, English, French, German, and Italian, and read Arabic, Greek, Hebrew, and Latin.[127] He maintained an extensive correspondence with some of the best-known writers and editors of his day, so that he knew a great deal about the practical aspects of publishing as well as literary production.[128] In the preface to his widely read textbook *Cellular Pathology* (1858), Virchow called readers' attention to the ability of language to shape their thoughts about the body:

> Even the language of medicine is gradually assuming another appearance; well-known processes to which the prevailing system had assigned a certain place and name in the circle of our thoughts change with the dissolution of the system, their position and their denomination. When a certain action is transferred from the nerves, blood, or vessels to the tissues, when a passive process is recognized to be an active one . . . then it becomes absolutely necessary to choose other expressions whereby these actions, processes and products shall be designated; and in proportion as our knowledge of the more delicate modes, in which the processes of life are carried on, becomes more perfect, just in that proportion must the new denominations also be adapted to this more delicate groundwork of our knowledge.[129]

Virchow knew that to change scientists' and the public's understanding of disease, he would have to change language, and he chose every word consciously, with his readers in mind.

Virchow shaped the course of pathology through his writing and editing as much as through his discoveries in the morgue. In the spring of 1847, he and his friend Benno Reinhardt began producing the *Archive for Pathological Anatomy and Physiology and for Clinical Medicine*, which came into being largely because Virchow persuaded the liberal publisher Georg Reimer to print it. Virchow wrote to Alexander Frantzius (1821–1877) in 1850 that he and Reinhardt had had to start their own *Archive* since "no respectable journal existed in which we could write."[130] The name Virchow and Reinhardt selected echoes that of Müller's own journal, the *Archive for Anatomy and Physiology and for Scientific Medicine*, which the young editors may have used as a model.[131] Since Müller's *Archive* had been appearing for thirteen years, Virchow must have regarded it as either insufficiently clinical or insufficiently respectable. He wrote most of the 583 pages of his journal's first volume himself, and in the *Archive*'s early years, his articles ran a hundred pages or more. After Reinhardt's death from tuberculosis in 1852, Virchow edited their *Archive* for the rest of his life. It became one of the most "respectable" journals in nineteenth-century Europe, and one of the most widely read.

Virchow illustrated his articles with detailed case histories and autopsy reports that gave readers vivid pictures of his patients' bodies. In 1853, in an eighty-five-page article on the relation of splenetic and lymphatic disorders to blood composition, he included nine long case histories which showed readers where he was getting his ideas. Each opens with the patient's name, age, geographic origin, and profession, then recounts his or her medical history and a detailed chronology of the illness in question. None of these nine was cured, since each account concludes with an autopsy report that reads like an adventure story. Virchow's cases make absorbing reading, and his descriptions are deeply engrossing, as in the following history of a hemorrhaging engineer:

> George Seelos, forty-eight years old, from Friedberg in upper Bavaria, foreman in the Würzburg Army Corps of Engineers. . . . He was a man of middle height, black hair, eyes, etc., dark yellow skin tone, lean build,

> very irritable, choleric temperament, presented on the whole more the image of a Southerner. . . . During the objective examination I found an enormous tumor in his abdomen which began in the spleen and stomach area and extended back to the *Spina anter. oss. Ilei*. . . . On the night of 23–24 September, after he had already been feeling unwell for several days, Seelos was suddenly seized by a violent headache, dizziness, and vomiting. . . . Eight ounces of blood were removed from his veins and twenty leeches applied to his forehead and temples. . . . The death occurred at six that evening. . . . I conducted the autopsy myself at the military hospital. When the cranial cavity was opened, all parts of the brain and its meninges were brimming with blood, and even from the surface a series of large extravasal foci were visible in the substance of the brain. I cannot recall ever having seen so many and such proportionately large hemorrhages in the brain.[132]

Virchow's writing demonstrates why, for him, clinical findings and theories of disease were inseparable. In his view, patients were the source of knowledge just as they were the reason for its creation. He could publish his work only in a journal for "clinical medicine," not in Müller's *Archive* for "scientific medicine."

Founding and editing the *Archive for Pathological Anatomy* was in itself a great achievement, but it was only one of Virchow's many writing projects.[133] Between July 1848 and June 1849, he and Rudolf Leubuscher published *Medical Reform*, a weekly paper that applied their political views to the medical profession. Virchow wrote many of its finely crafted articles. In 1851 he began co-editing the *Yearly Report on Achievements and Progress in Medicine*, and he continued to do so until his death. In 1863 he produced a popular book on trichinosis and began publishing an 1800-page textbook on tumors. Starting in 1869, after co-founding the Berlin Anthropological Society, he began editing its periodical, the *Journal for Ethnology*. As anthropology occupied more and more of his time, this journal became his passion, and he edited it personally until 1902. His best-known work, the textbook *Cellular Pathology*, surprised its publishers because it sold so well. Virchow wrote with great facility; according to his son, he whipped out articles "in one sitting" and did not revise them once he had finished.[134] A gifted orator, he wrote as he spoke, producing a narrative that entranced his listeners or readers.

Although Virchow may have written quickly, no one was more aware of the structures and meanings of words. Carl Ludwig Schleich recalls his supervisor's "intense philological passion which betrayed itself through an inclination for linguistic hair-splitting." According to Schleich, Virchow was obsessed with etymologies and terrorized medical students who mangled Greek or Latin terms in their oral exams.[135] Virchow's assistant of the 1880s may have been right, for volume 11 of the *Archive for Pathological Anatomy* contains an article by Virchow on the correct way to spell the word *Epithel*. As a politician, doctor, and anthropologist, Virchow made full use of his sensitivity to language.

Even in Virchow's letters home one can see his awareness of his readers. When he explains his work to his father, he writes for a person whose intelligence he respects but who has no knowledge of medicine. Describing his first microscopic studies with Froriep, he emphasizes the practical value of his work: "After major surgical interventions, amputations, for instance, there is often a so-called inflammation of the veins, a condition that manifests itself through the entry of pus into the blood and blockage of the blood vessels, usually leading to the death of the patient."[136] Virchow's letters to his mother have an entirely different tone. In May 1848, when she worried—with good reason—that her only child was about to get shot, he reassured her that by rebelling, Berlin was doing Pomerania a favor:

> This aversion to Berlin is really provincial. People should be happy that Berlin is looking out for them, otherwise they would stay sitting on the old dung-heap for centuries. . . . I'll give you an example. When a house is in flames, and a whole city is in danger of burning, one tears down the neighboring house in order to stop the fire from spreading. . . . One sacrifices the smaller, in order to obtain the larger end. Politics is just like that.[137]

Apparently Virchow expected little knowledge or intelligent thought from his mother, for all of his letters to her are equally patronizing. The difference between them and those to his father shows his awareness that to convey his perspective to different readers, he has to adjust his rhetoric.

Among the pictures that Virchow most wanted to paint for his listeners were those of the history of science. Like du Bois-Reymond and Helmholtz, Virchow took a passionate interest in teaching science and its history to the public.[138] In 1868, he and the lawyer Franz von Holtzendorff began publishing *Collected Popular Scientific Lectures*, which Virchow edited each year until 1901 to make sure they would be comprehensible to general readers.[139] He was equally committed to informing scientists and doctors about the history of their fields, since he often included historical articles in the *Archive for Pathological Anatomy*.[140] When he wrote about scientific or medical history, Virchow promoted his political views: that medicine should be used to reform society, and that it had been created—and should be run—by the practical, hardworking middle class.[141] Virchow's comments on Müller must be read in the context of all of his writing, especially his attempts to bring science to the public.

Virchow's Müller

When Virchow began writing about Müller in 1843, he did not think that he was composing the history of science. Müller appears in his letters home as the "famous physiologist" with whom he is privileged to study.[142] Materially, the main purpose of these letters was to obtain support for his medical education, and mentioning his illustrious professors affirmed the wisdom of his career choice. It sounded good that when he received his doctoral degree—for which his father had paid dearly—"the most famous physiologist in the world, Johannes Müller," personally conferred it.[143] But it is highly significant that in his medical essays of the late 1840s, Virchow continued to identify Müller as a physiologist. In "Scientific Method and Therapeutic Standpoints," an address delivered to the Berlin Society for Scientific Medicine in 1847, he declares, "It is a method which differentiates the Harveys, the Hallers, the Bells, the Magendies, and the Müllers from their lesser contemporaries. This method is the spirit of the natural sciences."[144] In this list, Müller becomes the culmination in a series of physiologists who experimented more and more actively to determine how blood circulated, muscles contracted, and nerves conducted signals. Virchow had studied anatomy and pathology, his chosen field, with Müller, but Müller's physiological experiments epitomized the method he wanted to promote.

Virchow turned much more critical when he discussed Müller's microscopic studies of tumors, investigations that served as a point of departure for his own work. In the first volume of the *Archive for Pathological Anatomy*, Virchow wrote that Müller's cancer studies did not have the impact they would have had if this "great observer" had drawn the right conclusions. Müller had identified pathological growth with embryonic growth, but instead of seeing tumors as intrinsic tissue developing wrong, he had continued to view them as extrinsic entities attacking the body. Despite his excellent microscopic technique, Müller had adopted the ontological view—the idea of diseases as independent entities—that Virchow wanted to avoid. He also criticized Müller's characterization of all tumors as "benign" or "malignant," writing that this was as "unscientific" as basing a botanical classification scheme on the categories "poisonous" and "non-poisonous."[145] He characterized one of Müller's drawings of a fatty growth as "very beautiful" but wrong.[146] In an 1855 article on fatty tumors, Virchow wrote that Müller's name for them, "cholesteatoma," was inaccurate, since only some of them contained cholesterol.[147] In the volumes of the *Archive for Pathological Anatomy* published during Müller's lifetime, Virchow's teacher appeared as a great observer who had drawn false conclusions from his observations.

In *Cellular Pathology*, a transcript of lectures Virchow held during the last three months of Müller's life, the pathologist presented his teacher's work as a crucial foundation for his own, but emphasized its inadequacy and incompleteness. He praised Müller for coining the term "connective tissue," a name he found "most appropriate" because the tissue occurred in so many different organs.[148] He questioned Müller's idea that pathological development resembled embryonic development, since at the time Müller proposed the notion, so little was known about embryonic growth. Virchow pointed out that tissue Müller had identified as a colloid (gelatinous) growth was actually badly swollen connective tissue.[149] In these cellular pathology lectures, Virchow aimed to unmask growths previously identified as extrinsic and invasive as normal bodily tissues that were developing wrong, and his treatment of Müller helped him to fulfill that goal. Müller had used a microscope competently and had faithfully recorded all that he saw, but his belief in a great plan of life—including that of tumors—had led him astray.

On the evening of 28 April 1858, Virchow wrote a long, emotional letter to his father, accusing him of financial mismanagement and unfairness to his wife.[150] Müller, who had died that morning, is nowhere mentioned. Since it was the last day of the semester break, Virchow must have learned of his teacher's death only at seven the next morning, when he returned to work at the Charité.

The Prussian Academy of Sciences chose the physiologist du Bois-Reymond, not Virchow, to deliver the memorial address for Müller, since du Bois-Reymond had worked with Müller for nineteen years and was probably going to inherit the physiological portion of Müller's chair. Virchow gave his own memorial address at the university's main auditorium sixteen days later, on 24 July 1858. The many resonances between du Bois-Reymond's and Virchow's elegies can be explained only through a complex feedback loop. Du Bois-Reymond spoke first, so Virchow may have adopted his phrasing and narrative tactics, but Virchow published first, and du Bois-Reymond studied Virchow's printed text in preparing his biography of Müller. In July 1858, Virchow had been working in Berlin for less than two years and was directing a pathological institute that was still growing. He had not yet been elected to political office. Although he enjoyed considerable prestige, he could not afford to antagonize university or government officials, since his work would require continued sources of funding. Like du Bois-Reymond, Virchow used his address to win approval for his own research and to spread his own ideas about how science should be done.[151]

Throughout the speech, Virchow shows his awareness that he is creating a historical narrative, calling Müller "a child of his time."[152] To call attention to key aspects of Müller's background, Virchow places them in short, sharp phrases at the beginnings of paragraphs. The line "Müller's father was a shoemaker" jolts the audience, many of whom would have come from more illustrious backgrounds. By calling Müller "the shoemaker's son," Virchow suggests that Müller may have dissected and drawn as well as he did because he came from a family where people worked with their hands.[153] Virchow introduces Müller as a figure of "powerful manhood" about to be honored by "the representatives of a great medical family."[154]

But Virchow's Müller is much more than a patriarch: he has magnetic powers. "By what secret means did he capture [*fesseln*] the beginner's heart," asks Virchow, "and keep it enchained [*gekettet*] to him for years? It is not without reason that some people said there was something demonic about him."[155] According to Virchow, Müller so terrified the auditors in his lectures that when fixed by one of his "linear looks," they would flee the room.[156] With dramatic, Byronic descriptions, Virchow depicts Müller's unrest: "Does he find peace in his research? . . . No, no! His countenance stays dark, two deep, angry folds lie between his brows, his dark look bores into the distance. He is being 'lashed by waves,' he says."[157] Virchow goes out of his way to depict Müller as a tormented man, a conduit of near supernatural forces.

Given Virchow's opposition to the Catholic Church, it is highly significant that he compares Müller to a priest. The abrupt sentence "Müller was raised in the faith of the Roman Catholic church" opens a paragraph, and like the shoemaker reference, it has been placed for maximum effect. Virchow draws attention to the facts that Müller wanted to be a priest when he was seven or eight years old, and considered studying theology when he entered the Bonn University.[158] Müller almost never mentions religion in his letters, and it does not appear to have been a major force in his life. Yet Virchow makes it a persistent theme in his address, quoting Müller's claim that "scientific research has something religious about it; I mean that it also has its cult."[159] In the extensive notes to his speech, Virchow reinforces this association of Müller with the priesthood. Even more than the main text, these notes must be explored as a consciously crafted history of science, since he wrote them not to impress a particular audience but to create a permanent record. Here Virchow writes:

> All I can say is that in his lectures and his solemn manner, Müller reminded me of a Catholic priest. . . . when he began his usual lectures with almost murmured words, or when he treated the central questions of physiology with almost religious zeal, every tone and expression, every gesture and look, seemed to betray the tradition of the Roman Catholic clergy.[160]

Late in his address, Virchow declares, "[Müller], too, . . . became an eternal priest of Nature; the cult that he served enchained [*fesselte*] his students as well, as though they were bound to him by a religious tie, and the earnest, priestly quality of his speech and gestures completed the impression of the awe with which everyone regarded him."[161] By reusing the verb *fesseln* (to bind or enchain), the same word he had used when asking what attached Müller's students to him, Virchow answers his own question. The magnetic priest Müller who subdued students with dark looks controlled them through irrational, emotional forces. By associating Müller with the priesthood, Virchow links him to the kind of thinking he most wanted to eliminate: submission to authority based on fear rather than rational knowledge.

But Virchow also stresses what he saw as Müller's finest trait: his commitment to unbiased observation and experiments. Like du Bois-Reymond's speech, Virchow's is a rescue story, and by showing what Müller might have been—a theorist who ruled by awe alone—he is better able to depict the torn, complex scientist that Müller became. Besides Roman Catholicism, the greatest danger to the youthful Müller was nature philosophy, which Virchow associates with Bonn. Probably to please his audience, Virchow credits the Berlin scientist Rudolphi with transforming Müller from a visionary dreamer into a microscopist and physiologist, showing that his teacher's real scientific work began when he came to Berlin. In dramatic tones, Virchow conveys Müller's drive and frenetic work pace: "He had to advance! . . . no peace, no rest!"[162]

Like du Bois-Reymond, Virchow makes Müller's first breakdown a turning point. After collapsing due to sleepless nights studying his own fantasy images, Müller embraced rigorous science and rejected fantasy for good. In reality, Müller had been doing physiological experiments for at least two years, but the sudden break made for a better story. In reviewing Müller's scientific achievements, Virchow emphasizes his physiology, giving his comparative anatomy little attention. He devotes considerable time to Müller's nerve root studies, the work he seems to have respected most. In his notes, Virchow offers a powerful example of Müller's desire to investigate living phenomena:

> In 1846, when I was prosector at the Charité Hospital, I kept encountering [a] pathological form of the spleen. . . . Since Müller had especially investigated the structure of the spleen, I wandered over to him with one of these, seeking an explanation for the follicular origin of the grains and some hint about the nature of the changes. Müller was not familiar with these changes; he was himself doubtful about whether they originated in the follicles. He said, "That's very strange, you have to investigate that!" When I told him that I already had, but that I was dissatisfied with the results, he said, "Then you must investigate further; that will certainly be very interesting!"[163]

This anecdote presents Müller as an inspirational figure, interested in everything and encouraging younger scientists to conduct research.

But when it comes to assessing Müller's own track record, Virchow proclaims, "Müller was no more of an experimenter than Haller was."[164] He denies absolutely that Müller founded experimental physiology, pointing out the earlier work of François Magendie and quoting Müller's critique of the French physiologist's animal experiments.

Instead, Virchow claims, Müller's strength as a scientist lay in his combination of philosophical theorizing and careful observation. He describes Müller's comparative anatomy briefly but accurately, noting that it occupied most of his time from 1834 onward and that he focused on "boundary" animals whose identification challenged classificatory schemes. Virchow recognizes all too well that Müller "oversaw a whole array of such different forms, among which a specific natural plan seemed to have been realized."[165] He also credits Müller with introducing the microscope as a key instrument for pathological studies, creating the impetus for his own work. Virchow concludes that:

> In the physiologist Müller, one admired not so much the genius of the discoverer, not so much the path-breaking flight of the visionary, but much more the methodological rigor of the investigator, the measured judgment, the secure calm, the perfection of knowledge. . . . Through him, the mystical and the fantastic were vanquished in the field of the organic. . . . Although he did not invent the "exact," the only scientific method, he certainly established it. Thus there is no Müller School in the sense of dogma, since he never professed any, only in the sense of methods. The scientific school that he inspired has no common theory, only common facts and much more, common methods.[166]

Since Virchow knew Müller as a teacher, not a collaborator, it makes sense that he depicted him as one who professed methods. This is the role that he played in Virchow's life, showing him the value of microscopic investigations but not forcing him to interpret his observations in a particular way. By this point in Virchow's speech, the audience must have seen Müller as a contradictory figure, a Catholic priest seeking the great plan of life and an experimenter who banished mysticism from the organic realm.

About Müller's politics Virchow is much less ambiguous, deferring this touchy subject until the end of his talk. Certainly the Berlin audience must have wanted to hear the former revolutionary assess the conservative rector's actions in 1848. Virchow leaves them in suspense until almost the last moment and then restricts himself to—for him—quite diplomatic statements: "Müller was no politician. . . . At that time, we stood in different camps."[167] Rather than antagonizing the listening officials, Virchow places his real thoughts about Müller's politics in the footnotes, for historians in the decades to come:

> Undoubtedly, the great political upheavals that filled Müller's earliest memories gave him the idea of a certain instability in state and political life. I still remember quite vividly what he said when I visited him to say goodbye on 19 February 1848, the night before I left for Silesia. He was astonished that I wanted to risk the dangers of hunger typhus, whereupon I replied that with a revolution threatening in France, one didn't know what was going to happen here at home. He was greatly shaken by this thought and said that this would be terrible, since socialism meant nothing but the general robbing of all property owners. He was by no means in agreement with the government, [but] when the revolution finally broke out in Berlin, he thought that the country was lost.[168]

Once the revolution erupted, Virchow recalls,

> The rector was seized by the greatest agitation. He trembled for the safety of the university, for whose treasures he believed he was personally responsible. Day and night it pulled him there, as though he himself had to stand watch. He ripped down inflammatory posters, he endangered himself by personally confronting the most violent students. . . . It was the unhappiest Rectorate since the University's founding. In a time in which everything was driven by the flow of politics, the man with perhaps the least political inclination was called upon to lead a body which because of its natural independence was least suited for unified leadership.[169]

In these notes, Virchow's Müller appears as well intentioned but old-fashioned, misinformed, fearful, and incompetent. As a character, he arouses more pity than respect. He displays personal courage and an admirable sense of responsibility, but no commitment to social reform. As in his science, he has the right impulses but misses the big picture.

The Müller of Virchow's memorial address never resolves his conflicting urges to dissect all of nature and to worship it unscathed. Most of Müller's students' memories change with time, but thirty-seven years later, Virchow still depicts his teacher in much the same way. In an 1895 essay, "One Hundred Years of General Pathology," Virchow praises and criticizes Müller for the same contrasting inclinations. By teaching physiology and comparative and pathological anatomy together, Müller was able to "present broad points of view, and to bring basic questions into the foreground of study."[170] Here again, Virchow recounts Müller's experimental study of spinal nerve roots and acknowledges his microscopic work. But he claims that Müller retarded the development of pathology in Germany by excluding clinical medicine from his research and teaching, and he implies that Froriep developed clinical pathology at the Charité largely in spite of Müller.[171]

In his memorial address, Virchow had said little about Müller's work on tumors, but in this 1895 essay, at the end of his career, he was willing to pass judgment. To his credit, Müller had begun investigating the structure of tumors as soon as Schwann's findings had suggested they were cellular. But in order to retain the concept of malignant tumors, Müller had sacrificed the histological principle he had used to classify benign tumors: the fact that they were composed of the body's own cells.[172] While Müller was willing to listen to Schwann and apply his results to pathology, he had feared the chaos that would ensue if he abandoned his old notions of order. For Müller, living matter and inorganic matter were fundamentally different, and cancerous tumors had to differ from healthy tissue. The Prussian king and Cultural Ministry granted scholarships, and angry socialists looted museums. Virchow did not see the world the same way.

Virchow's and Müller's Science

Ultimately, the best indication of Virchow's regard for Müller is not what the student said but what he did. To what degree did Müller's science influence Virchow's work? Virchow began his research with the topic that Froriep had suggested, investigating the relationships between venous inflammation and blood clotting. Through his rigorous microscopic and chemical studies of the blood, Virchow disproved Karl Rokitansky's dominant pathological theory that blood disorders underlay all diseases, which had been based on the assumption that blood produced the raw material that gave rise to sick cells. In the process, Virchow demonstrated that blood clots could be transported and that leukemia, "white blood," was a unique disease, not an imbalance of bodily humors.[173]

Virchow has become known today as the pathologist who made doctors think about disease at the cellular level. This legend gives him both too much and too little credit. Schwann's and Müller's microscopic studies, particularly Müller's tumor investigations, of the late 1830s had already motivated medical researchers to explore the cells of many bodily structures. But Schwann had believed that cells developed from a formless "cytoblastema," an idea that Rokitansky had embraced to show that diseases resulted from blood imbalances. Virchow's great contribution was his extensive microscopic studies indicating that all cells came from other cells.[174]

Once one accepted this idea, it became clear that pathological phenomena were not fundamentally different from normal ones; they were ordinary physiological processes occurring too fast, too slowly, or in the wrong place. Tumors, for instance, consisted of bodily cells that were growing too rapidly in places where they should not be growing. Virchow used this discovery to support his general theory that diseases were not mysterious or foreign, just ordinary bodily phenomena run amok. If this was right, then diseases could be controlled.

From 1860 onward, as Virchow's political career advanced, he acted on the consequences of his microscopic discoveries. Having revealed the life cycle of the *Trichina* worm, he published a popular book that led to a massive meat inspection program.[175] In the 1860s, he designed and raised money for a modern sewer system for Berlin. Of all of his public health projects, this one probably saved the most lives.[176] Because Virchow believed that people could best govern themselves if they were well educated, he often took part in educational reform. Thanks to Ernst Haeckel's outraged complaints, Virchow has become known for opposing the teaching of evolutionary theory. As an energetic popularizer of science, Virchow wanted children to learn as much modern science as possible, but he distrusted Darwin's theories and argued that young students should learn only established facts.[177] He never rejected evolutionary theory any more than he rejected bacteriology, once the evidence appeared that extrinsic organisms could cause diseases, but he feared that presenting unproved theories in schools would undermine people's confidence in science if these theories were someday overturned.

Müller's students took his science in different directions, and their individual uniqueness can be seen in how poorly they got along. Henle and Virchow offer a case in point. When Virchow began his microscopic studies of the blood and sought publishers for them, he turned, among others, to Henle, who by then was a full professor of anatomy in Heidelberg. Henle agreed to publish Virchow's work on fibers in his *Journal for Rational Medicine* but delayed printing the second part of it. "I must say that I was unpleasantly surprised when I found only half of it in this double volume," Virchow told Froriep. "Nowadays, when so much work is being done, it is hardly desirable for a paper to lie around for a year and a half; I find this rather unkind of Henle."[178] In the spring of 1849, when Virchow and Henle were engaged in a hot dispute, Henle considered going to Würzburg, possibly to talk the university out of hiring Virchow.[179] Virchow objected to Henle's entire project of rational medicine, which attributed pathological disorders to disturbances of the nerves. In his rational pathology, Henle had updated the eighteenth-century Scottish idea that since the nervous system provided the main link between a person and the outside world, diseases must affect the body through the nerves.[180] Virchow did not believe that diseases came "in" from the outside, and was trying to prove that they were just alterations of normal bodily processes. He attributed pathological phenomena to local disturbances of cells, arguing that inflammation could exist in uninnervated areas. In *Cellular Pathology*, he insisted that the nerves were not the only active elements of the body, warning readers that "besides nerves and nervous centers, other things exist, which are not a mere theater [*Substrat*] for the action of the nerves and blood, upon which these play their pranks."[181] He referred disparagingly to Henle's idea as "the aristocracy of the nerves."[182] Virchow also initially rejected Henle's idea that live, microscopic organisms could cause diseases. In this case Henle turned out to be right, but as Virchow pointed out, Henle had offered no evidence.[183] Both had learned the value of microscopy from Müller, but they disagreed radically about how to interpret what they saw.

Despite their differences, Virchow and Henle both belonged to the Müller school in Virchow's sense of the term. While they rejected some of his theories, they adopted his methods. Neither needed much encouragement to dissect cadavers or scrutinize living tissues, but Müller offered inspiring examples of physiological experiments, microscopic studies, and chemical analyses. Virchow's studies of tumors began with critiques of Müller's work, and his cellular pathology built on Schwann's, Henle's and Müller's observations of cells. In the late 1840s, Virchow almost fully embraced du Bois-Reymond's and Helmholtz's materialism, arguing that medical researchers should study the ways that physical and chemical laws governed the body. But as he developed his model of the cell as the fundamental living unit and pathological site, Virchow began arguing for a "new vitalism" that echoed Müller's older one.[184] In "Old and New Vitalism," the lead article in his *Archive* in 1856, he explained that while no one could defend the notion of a life force unifying each living organism, every cell possessed a vital quality that made it different from inert matter. Virchow never accepted Müller's idea of a life force realizing itself through a hierarchy of living forms, but he did believe that cells exerted forces that inorganic material did not.

In his cellular pathology and social reform programs, Virchow re-created another aspect of Müller's science: the nature philosopher's quest for an all-embracing system of knowledge.[185] Müller had sought a law that would explain all of life and had looked for it in every organism. Virchow never believed in a great plan of life, but he did think that basic biological knowledge could justify social theories and that observations of society could offer insights into the way the body worked.

Virchow shared Müller's passion for collecting, and if the pathologist wandered over to Müller with diseased spleens, he must have noticed the museum in which Müller was working. From the 1870s onward, Virchow's interest in anthropology drove him to collect skeletons from every known human race, and as director of the Pathological Institute, he continued to amass pathological and anatomical preparations. When Virchow died in 1902, the Institute's collection contained 20,000 objects, all labeled in his own writing.[186] And like Müller, Virchow believed in teaching the history of science and included it in his medical lectures. Among the scientists he most often mentioned was Johannes Müller.[187]

6

Banned from the Academy:

The Mentoring of Robert Remak

Robert Remak, who worked closely with Müller for twenty-five years, never delivered a memorial address. Neither the Berlin University nor the Academy of Sciences invited him to do so. Remak was never elected to the Academy of Sciences, despite Müller's efforts to make him a member. Although internationally known for his neurohistological and embryological studies, he never became a full professor at any Prussian university.

Like Virchow, Remak came from a poor family in rural Prussia (now Poland). Both scientists studied medicine in Berlin, took Müller's anatomy and physiology courses, became expert microscopists, and performed crucial studies of the ways that cells grow and reproduce. In 1848, both worked as democratic activists. Virchow married a woman named Mayer, the daughter of a prominent obstetrician; Remak, a woman named Meyer, the daughter of a wealthy banker. Both men became lecturers at the Berlin University in 1847, and both were candidates for Froriep's prosector position in 1846 and the Berlin pathology chair in 1856. Virchow, six years Remak's junior, got both jobs. Robert Remak, an Orthodox Jew, survived as a scientist only because of the ceaseless support of Müller and Müller's mentor, Alexander von Humboldt.

The Hopes of a Jewish Scientist

To appreciate Prussian scientists' attitudes toward Jews, one need only read some comments by Müller's other students, all written during Remak's career in Berlin. Jakob Henle wrote jokingly to his sister that she should be grateful to his father, for if he hadn't forced them to convert as teenagers, they would now be "filthy [*filzige*] Jews."[1] In April 1848, Emil du Bois-Reymond wrote to Carl Ludwig that the fighting on 18 March had been "nothing better than riots fomented by a nasty clique of Jewish litterateurs."[2] As a full professor in 1865, du Bois-Reymond told Hermann von Helmholtz that he could offer him three talented candidates for an assistantship in Heidelberg: Hermann, Bernstein, and Munk. According to du Bois-Reymond, Hermann was "probably the most distinguished of the three and possesses the greatest variety of knowledge . . . but he doesn't have a very pleasant outward appearance; he looks like a Jewish cantor." Helmholtz took Bernstein.[3] During the summer of 1840, Helmholtz complained to his parents about some Jewish passengers with whom he had sweated through a hot, cramped stagecoach ride:

> Across from us sat two Jewish women from Breslau and a Jewish high school junior who acted like a fifth grader, was unable to sit still for a moment and kept jabbering at us the whole way, while one of the Jewish women kept bickering with him the whole way, claiming he was ramming his arm into her side.[4]

When Rudolf Virchow knew that he was up for the Berlin pathology chair, he speculated to his father-in-law about how Alexander von Humboldt might be persuaded to support him. "Do you know a member of the Mendelssohn family?" he asked. "Since [Humboldt] sticks to the Jews, one must try to get at him through Jews."[5] While Ernst Haeckel was studying medicine in Würzburg, he wrote to his parents that two students he knew, "very nice people," had just fought duels with "the same shameless Jew-boy from Frankfurt, a fresh, repulsive person, and thoroughly hacked up his Jewish face, without suffering any damage themselves."[6] When one considers these scientists' careers in their entirety and judges them in the light of mid-century Prussian culture, none stands out as an anti-Semite. Their mean-spirited remarks about Jews were typical of educated Germans in the mid-nineteenth century. In the next decades, both Virchow and du Bois-Reymond would become known for their outspoken opposition to anti-Semitism. Hermann would not have been working for du Bois-Reymond if he had rejected Jewish scientists outright, nor would the many Jewish students he later mentored.[7] These comments by men who would become some of the most powerful scientists in the German-speaking world reflect the cultural climate in which they conducted their studies. These are the voices of Remak's colleagues.

The oldest son of Frederica Caro and Salomon Meier Remak, who managed a cigar shop and lottery office, Robert Remak was born in 1815, the year that Prussia annexed his region of Poland.[8] Posen (today Poznan), the city where he was raised, was about 20 percent Jewish, 13 percent German, and 67 percent Polish. The Prussian government encouraged its new Jewish subjects to align themselves with Prussia against the Poles, offering them the rights of citizens if they took German last names and spoke German. Although the Remaks' means were limited, they managed to send their gifted eldest son to the *Marien-Gymnasium* and then to the Polish Royal *Gymnasium*, where he received a good secondary education and was encouraged to support Polish nationalism. It is less clear how Remak's family afforded his medical training at the Berlin University. Since he was Jewish, the military institute that had opened the way for Helmholtz and Virchow was out of the question. A letter from Posen's magistrate in October 1833 requested that Remak be allowed to defer paying his lecture fees, since his father earned less than 500 taler per year, had three other children to support, and could contribute no more than 100 taler toward his son's medical studies. Local Polish aristocrats may have helped pay for his medical training.

Since the eighteenth century, many Jews had been coming to Prussia to study because they were banned from Poland's Catholic universities. Berlin's Collegium Medico-Chirurgicum had accepted Jewish students since 1730. The drive to make Prussian universities centers for science rather than theology helped open them to Jews, since it dissociated learning from religion.[9] Prussia's relative open-mindedness had its limits, however. Until 1842, the Academy of Sciences admitted no Jewish members; Frederick the Great twice rejected the candidacy of Moses Mendelssohn. In 1812, Prussia's Edict of Emancipation gave Jews full citizenship rights, including the ability to teach at universities, but the Cultural Ministry never honored it. Instead, Ministers Altenstein (1817–1840) and Eichhorn (1840–1848) did everything they could to prevent its enactment. David Ferdinand Koreff, the private physician of Prince Hardenberg, was appointed full professor at the Berlin medical faculty in 1816, but when the faculty told the king that Koreff was Jewish, he was quickly baptized.[10] Cultural Minister von Altenstein, who would so warmly support Müller's recruitment, viewed the Prussian universities as training grounds for loyal Christian servants of the state, and objected on principle to the concept of Jewish professors. In 1822, he convinced King Friedrich Wilhelm III to rescind the 1812 edict in part, preventing Jews from initiating academic careers. Minister Eichhorn shared Altenstein's views and was particularly determined to keep Jews from achieving positions of academic authority such as full professor or dean.[11] King Friedrich Wilhelm IV believed that Jews should be allowed to teach in nonreligious fields, so that limited opportunities began to open in the late 1840s. With rare exceptions, however, Jews were banned from academic careers in Prussia until the 1860s.[12]

Both Remak and Müller came to Berlin in 1833, and Remak began taking the new professor's courses as soon as he arrived. Besides Müller, Remak's most influential teacher was Christian Gottfried Ehrenberg (1795–1876), who taught zoology and aroused his interest in microscopic studies of the nervous system.[13] The writer of Remak's obituary in *The German Clinic* recalls that in 1836, the student from Posen was "a slender young man with prominent features, vivid eyes and quite closely cropped dark hair, . . . and with a great love of microscopic studies."[14] According to Remak's university diploma, he once received two weeks' detention for planning to fight a pistol duel, but he was never charged with belonging to a *Burschenschaft*.[15] By the mid-1830s, the Prussian government was trying to eliminate the political fraternities to which Müller and Henle had once belonged. *Burschenschaften* seem not to have appealed to Remak, who received his medical degree with honors in January 1838.[16]

Figure 6.1 Robert Remak around 1860. Source: "Gedenkblatt zum 100 jährigen Jubiläum der Medizinischen Fakultät Berlin," *Deutsche medizinische Wochenschrift* (10 October 1910): suppl. 58. Original photograph in the Bildarchiv des Instituts für Geschichte der Medizin, Freie Universität Berlin. Reproduced in Heinz-Peter Schmiedebach, *Robert Remak (1815–1865): Ein jüdischer Arzt im Spannungsfeld von Wissenschaft und Politik* (Stuttgart: Gustav Fischer, 1995).

Remak, however, proved to be a political dynamo. A radical democrat, he encouraged Virchow to get more politically involved, and he may well have been one of the "nasty Jewish litterateurs" to whom du Bois-Reymond was referring. The whole Remak family took a strong interest in politics; three younger brothers had to flee to America as a result of their student activism in 1848.[17] During the revolutionary period in Berlin, Remak belonged to the Democratic Club and played leading roles in the General Association of Berlin Doctors and the General Association of Associate Professors and Lecturers. In April 1848, he invited Virchow to join the Liberal Club, which became the Democratic Club when its members rejected the liberal plan for a constitutional monarchy. At the time, both Remak and Virchow stood left of the liberals, demanding a republic with free elections and the elimination of the standing army.[18] At the General Association of Berlin Doctors, Remak and Virchow argued about how medicine in Prussia could best be reformed, debating the kind of training that doctors should receive and the best ways that doctors could be made available to the poor. At the Berlin University, Remak and Virchow became the most politically active members of the medical faculty. In articulating the junior faculty's and lecturers' demands to the Senate, they worked closely together. A document drafted by Remak requested a voice in choosing deans and rectors and in assessing doctoral exams and awarding stipends. In November, when eighty professors signed an affirmation of loyalty to the king, both Remak and Virchow joined the eighteen rebels who issued an alternative statement of loyalty to the Constitutional Assembly.[19]

Remak's resistance to authoritarian governments may be linked to his feelings for Poland, which during his lifetime was held captive by Prussia, Austria, and Russia. He tried to promote communication between Polish- and German-speaking scientists, publishing important results in Polish and then citing them in his German papers. In 1841, he managed the translation into Polish of a manual for midwives.[20] But in 1850, when he was offered a full professorship of pathological anatomy in Krakow—the only such offer he ever received—he turned down the job. He wrote to the dean that he needed to concentrate on his embryological research and that he had an insufficient command of the language—an unlikely story, since his rejection letter was in Polish. Although Remak was studying cell division and embryological growth at the time, it is more likely that he resisted leaving Berlin for Poland. In response to an 1846 uprising, the Austrian government was attempting to remove Polish professors and replace them with German-speaking ones, and he may not have wanted to displace a Polish colleague.[21] Remak never thought of himself as a "Pole" or a "German." In his letters to his fiancée in 1848, he referred to "the Germans" and "the Poles" as though he belonged to neither group.[22]

In most respects, this was true. As an Orthodox Jew, Remak could not marry into a Christian family, which made the daughters of Berlin's professors and most of its doctors off limits. But like Virchow, he married well, selecting a compatible partner who would support his research and improve his social standing. In July 1848, he became the husband of Feodore Meyer (1828–1863), the daughter of the banker Eli Joachim Meyer. According to Remak's granddaughters, the banker allowed him to marry Feodore only if he promised to stay out of politics, but the banker must have approved of Remak's science.[23] Since the microscopist never received a salary for his teaching, and the income from his student lecture fees and private practice could not maintain a middle-class family, it is likely that Feodore's money supported Remak's family and research. Their son Ernst Julius (1849–1911) shared his father's interests but achieved what his father had not, working as assistant to the eminent Berlin neurologist Carl Westphal (1833–1890) and becoming a full professor in 1910.[24]

Remak's Scientific Achievements, 1835–1855

Considering the disadvantages that Remak faced, it is truly remarkable that he accomplished as much as he did. His observations of neuronal structure, cell reproduction, and embryological development underlie modern biology. Remak began his own research in 1835, during his second year of medical school, in response to Ehrenberg's and Müller's encouragement.[25] Initially he focused on the nervous system, making two discoveries crucial to understanding the way nerves work.

Remak showed first that myelinated nerves (those that appear white because they are insulated by layers of fatty membrane) are not hollow and filled with fluid, as most scientists thought in the 1830s, but are instead filled with a "primitive band," the smallest observable fiber in the nervous system.[26] Even more significantly, he demonstrated that sympathetic nerve fibers emerged from ganglia (grayish blobs of nerve cell bodies) and were attached to the cell bodies (then called "ganglionic cells").[27] This observation would allow neuroanatomists of the 1880s to define the neuron as a cell body and an axon, the modern name for Remak's "primitive band."

In an 1838 paper on the autonomic nervous system, Remak described sympathetic ganglia as "points of origin or boosters [*Verstärkungspunkte*] for nervous force [*Nervenkraft*]."[28] Müller's work sparked Remak's interest in the sympathetic nervous system, but he revealed much more about its organization than his teacher ever had. The student described unmyelinated fibers emerging from tiny ganglia in the kidneys, liver, spleen, lungs, larynx, tongue, and throat.[29] In his studies of the heart, Remak concluded that although that organ was largely independent of the brain—it continued to beat when removed from the body—it must be regulated by sympathetic ganglia, since the fibers that entered it emerged from those tiny gray masses.[30]

From 1843 onward, Remak became increasingly interested in the ways that cells grew and reproduced, unsurprising since he had worked with Müller while Schleiden and Schwann were studying cells. Like Virchow, Remak rejected Schwann's idea that cells grew like crystals out of a formless cytoblastema, although he reluctantly embraced Schleiden's notion that new cells might grow within existing cells like Russian dolls. "From the time that cell theory became known," wrote Remak, "the extracellular emergence of animal cells seemed as improbable to me as the spontaneous generation of organisms."[31] Over the next ten years, in exhaustive observations of blood corpuscles, embryos, and tumor cells, Remak demonstrated that new cells arise only through the division of extant cells. It is here that his work overlaps most closely with Virchow's, so much so that it is unclear who first said "omnis cellula e cellula [all cells (come) from cells]."[32] In 1854, a year before Virchow asserted that cells arise only through division, Remak wrote, "According to my observations, the cells of which the animal embryo consists multiply much more through successive divisions, starting with the fertilized egg."[33] In frog embryos, he could watch tadpoles develop from the first cleavage on.

Remak's studies of tumors reinforced the observations he had been making of embryological tissue. Müller had proposed that pathological growth closely resembled embryonic growth, an assertion that Remak took as a point of departure in his investigations of cells. In his claims about tumors, Remak stayed closer to Müller's teachings than did any of Müller's other students. Repeatedly, he referred to tumors as "parasitic," a notion that Henle and Virchow strongly opposed. Henle claimed that disease-causing agents (not diseases) parasitized the body, whereas Virchow rejected the whole notion that diseases were entities independent of the bodies in which they appeared. Remak, in contrast, wrote in 1841 that because of cancer's tendency to spread, "one certainly sees the cancerous tumor as something parasitic, endowed with a relatively independent life, and thus com-

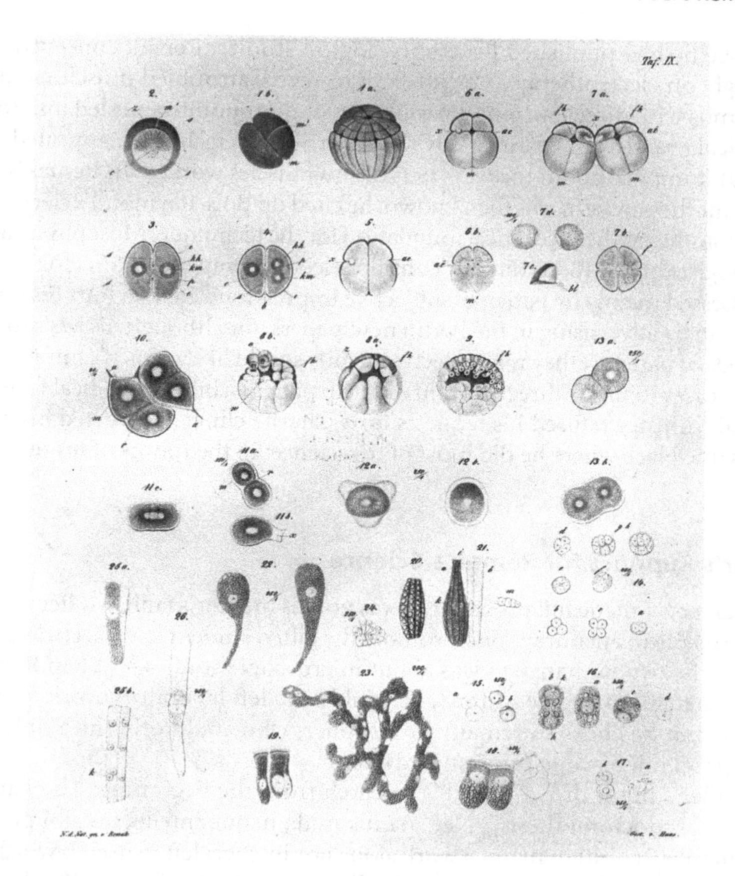

Figure 6.2 Cleavage and cell division of a fertilized frog (*Rana esculenta*) egg. Source: Robert Remak, *Untersuchungen über die Entwickelung der Wirbelthiere* (Berlin: G. Reimer, 1855), plate IX.

pares it with known material contagia, with syphilis and scabies."[34] He quickly qualified, though, that the agent that caused syphilis (unknown until 1905) had its origin outside of the body, whereas cancerous tumors were always intrinsic. He agreed with Virchow that pathological tissue emerged from normal tissue, offering evidence that Virchow used to prove his new version of cell theory.[35]

Remak is best known today for his embryological studies, published in 1850–1855.[36] Between 1843 and 1855, he identified the three embryonic germ layers and showed which organs develop from each.[37] From the outer layer come the brain and sensory organs, as well as the skin, hair, and nails. From the middle one arise the spinal column, the heart and blood vessels, the sex organs, and voluntary muscle. And from the inner one emerge the digestive tract, liver, gall bladder, pancreas, and lungs. Remak described the creation of all of these organs at the cellular level, leaving a wealth of drawings suggesting how they worked.

Once he had published his embryological studies, Remak concentrated increasingly on electrotherapy, the attempt to revive atrophied muscles and paralyzed limbs with direct current. Possibly career disappointments led him to focus on clinical practice, but more likely the controversial technique appealed to him for what it implied about the ways nerves and muscles worked. In Remak's papers on galvanotherapy, as it was then known, he cited du Bois-Reymond's electrophysiological studies as the theoretical foundation for the technique. Most physicians and physiologists of the 1850s did not consider electrotherapy scientific, for its results were assessed mainly by patients' subjective impressions. Remak hurt his scientific reputation by advertising in the Berlin newspapers, for although his ads brought in hundreds of patients, they made electrotherapy sound like a quack's miracle cure.[38] In some cases Remak's direct current did help patients, but the medical faculty and Cultural Ministry refused his requests for a Charité clinic. He treated his patients in the same place where he did most of his science: in the rooms of his own home.

Müller's Support for Remak's Science

If Remak never made full professor, it was not his mentor's fault. Müller did everything possible to encourage and promote the gifted student's research. From 1835 onward, the two exchanged ideas about microscopic studies.[39] When Remak received his medical degree in 1838, he could have left laboratory work for private practice, but he chose to remain with Müller, who could offer him only a little work space, microscope time, and advice.

Müller's input drove Remak's research from the beginning. The Latin epigraph to his doctoral thesis, "Nec manus nuda neque intellectus sibi permissus multum valet" (Neither manual work alone nor intellect left to itself is worth much [a saying of Francis Bacon]), recalls Müller's statements of the 1820s about how physiology should combine philosophy with close observation.[40] While studying invertebrate ganglia, Ehrenberg had become convinced that the sensory and motor roots of spinal nerves—which Müller had demonstrated were distinct—must have different structures. Together, Remak's two anatomy professors encouraged him to begin microscopic observations of nerves, but it was Müller who offered ongoing support.[41] References throughout Remak's publications indicate that he and his teacher shared their ideas regularly. In his first article, published in Müller's *Archive* in 1836, Remak cites the same scientists whose work Müller summarized in his yearly research report and criticizes their work the same way Müller did.[42] Müller listened closely to his student's findings, for he cited his results repeatedly in his *Handbook of Human Physiology.*

Müller didn't just accept Remak's findings on faith; the student's work interested him so much that he repeated it personally and confirmed his results. Remak ended his first article by thanking Ehrenberg and Müller "for kindly granting me the use of their microscopes, especially [Müller] . . . who had the kindness to test the main findings of my investigations through dissections."[43] Müller also "decisively confirmed" his student's "observations and interpretations" of sympathetic nerves in 1838.[44] Long after Remak had matured as a scientist, their common interests drove Müller to keep repeating the student's experiments, sometimes working with him directly. In his 1847 book on the intestinal nervous system, Remak thanked Müller for "conducting some follow-up investigations [*Nachsuchungen*] together with me."[45]

Müller also gave Remak full access to his precious Anatomical Museum so that his student could explore nerve structure, cell growth, and embryology from a comparative perspective. In the conclusion to his 1843 study of chick development, Remak thanked Müller for letting him use museum preparations.[46] By the late 1840s, this generosity must have become habitual, since Remak acknowledged Müller again in 1847 for "entrusting museum preparations" to him "with his usual willingness."[47]

Most significantly, Müller suggested new directions in which the student might take his work. Remak began his second paper, a microscopic study of nerve fibers, by recalling a conversation in which Müller had told him, "Certainly we are still barely acquainted with the real primitive fibers of the nervous system, and these so-called primitive fibers may still have much finer fibers within them."[48] In the introduction to his book on electrotherapy, Remak recalled that "since 1839, Müller had often encouraged me to produce a work on general pathology."[49] In his book on the intestinal nervous system, a work that he dedicated to Müller, Remak summed up his appreciation of his teacher:

> It is my privilege to thank Herr Geh. Müller for the lively interest that he has always taken in my investigations. By teaching and encouraging, this interest gave me the priceless opportunity to receive comments that flowed from the freest command of science, through which new observations were inspired.[50]

As every one of Müller's students testifies, the scientist always urged a young researcher to investigate further.

Besides guiding Remak's research, Müller offered him practical support with his career. Because Remak was Jewish, Müller could not appoint him prosector or museum helper in the 1830s—Minister von Altenstein would never have allowed it—but he backed him for any other available job. Müller included Remak's results in his yearly research reports in the *Archive* and published his articles even when he doubted the findings, such as the "primitive band." He assigned Remak articles on the nervous system and retina in the *Encyclopedic Dictionary of Medical Sciences*, an honor given only to his most trusted students.[51] When a chair of physiology became available in Vilnius (then in Russia, now in Lithuania) in the fall of 1839, Müller joined Alexander von Humboldt in recommending Remak for the position. Remak did not receive it, probably because the Russian government, like the Prussian one, did not relish hiring an unbaptized Jew. But it is likely that Müller's recommendation won Remak the valuable position as clinician Johann Schönlein's assistant in 1843.[52]

When Schönlein needed help at the Charité, he probably asked Müller for his best microscopist, hoping to put the student's skills to work in his clinic.

Because Remak was nonmilitary, Schönlein could give him no official position and could pay him only 200 taler, for which he had to reapply each year.[53] Grateful for his assistant's skill, Schönlein did everything he could to promote Remak's career, allowing him to study patients from his private practice. Schönlein also encouraged Remak to write reports on his work and send them to Cultural Minister Eichhorn. In 1845, Remak published these as *Diagnostic and Pathogenetic Studies*, in a single volume dedicated to Schönlein.[54] Despite all of Schönlein's efforts and government connections, however, he could not prevail against the military doctors and make Remak—not Virchow—the Charité prosector in 1846. Perhaps this defeat disillusioned Remak, who may have begun to see the clinical route as a hindrance, not a help, to his investigations. In 1849, he abandoned his job in Schönlein's clinic to make more time for his embryological research.[55]

That summer Müller recommended Remak (along with du Bois-Reymond, Helmholtz, and Ludwig) for the chair of physiology in Königsberg. "Given this opportunity," wrote Müller,

> I must also mention the achievements of the lecturer Dr. Remak with respectful recognition. His direction is not that of the aforementioned men but is oriented mainly toward anatomical-physiological microscopy with a regard for pathology and development, which constitutes a significant part of physiology.[56]

Suspecting that the Cultural Ministry might share du Bois-Reymond's disgust for Remak's physiology, Müller specified in his letter, "The direction of Dr. Remak in the field of physiology is as justified as the experimental one."[57]

When a position became available in the Academy of Sciences in the fall of 1853, Müller campaigned to get Remak elected. In his letter of recommendation, he wrote: "Remak's work is highly noteworthy because of the change in perspective brought about by its theoretical views on general development, because of its careful criticism of old and new observations, and because of an investigative rigor due equally to diligence and devotion." Suspecting that Remak's activity as a physician might work against him, Müller added that his student "had long since almost fully withdrawn from clinical practice and dedicated himself exclusively to the duties of science."[58] When Müller died in the spring of 1858, Remak lost his second strongest supporter.

Müller and Humboldt Help Remak Become a Lecturer, 1843–1847

Even with Müller's help, Remak could not have survived in the Prussian academic world. Since Jews were forbidden to teach, he needed not just an academic mentor but one whose connections went beyond the university and Cultural Ministry, directly to the Prussian king. He found one in energetic, science-loving Alexander von Humboldt, who had done so much for Müller and in the 1840s—when he was nearly eighty years old—began helping Müller's students.

Knowing that Humboldt liked to assist gifted young scientists, Remak sent him a copy of his doctoral thesis in 1838. In the dedication to his book on vertebrate development, Remak outlined the growth of their friendship. "Not only did I receive encouraging words," he recalled, "I won tireless advice and support against religious and political intolerance, which has made my life a struggle, has scattered my energies, and denied me a place for my [scientific] activities."[59] Humboldt read in detail everything that Remak sent him, often commenting in his letters on particular theories and drawings.

Like Müller, Humboldt wrote to the Russian cultural minister in 1839, recommending Remak for the chair of physiology in Vilnius:

> The works of Mr. Remak that I have been able to follow under the microscope together with my friend Mr. Ehrenberg show a great acuteness of observation, laborious diligence, and that comprehensive view that characterizes modern physiology. . . . The young man belongs to the Mosaic religion, and a delicacy of feeling for which you can hardly blame him prevents him from making himself a Christian out of a simple motive of self-interest. . . . I know that in our administration, which is very tolerant with regard to national beliefs, the difficulty [of a candidate's religion] can be overcome in certain cases, by granting a favor due to personal merit.[60]

Out of gratitude, Remak dedicated most of his major works to Humboldt. Although the two never experimented together, Remak regarded him as a fellow scientist, citing Humboldt's electrophysiological studies of the 1790s in the electrotherapy book he dedicated to Humboldt for his eighty-ninth birthday. Although Remak spent much more time with Müller, his friendship with Humboldt seems to have been closer.

Oddly, until 1847, Remak did not dedicate any of his writings to Müller. Henle, Schwann, du Bois-Reymond, and Helmholtz all offered their doctoral theses to their adviser, whereas Remak consigned his to a friend from Posen, the physician Bentkowsky. When Remak asked Humboldt's permission to dedicate the intestinal study to him, his aged mentor thanked him but suggested, "Couldn't you dedicate the book . . . jointly to me and Schönlein or Johannes Müller? Such an *accomplement* with learned friends would be a great pleasure to me."[61] Remak offered the monograph to Humboldt and Müller together, "with the deepest reverence, out of the most sincere gratitude."[62] More than once, Humboldt reminded Remak of the value of Müller's support, suggesting that when communicating with his more powerful mentor, the frustrated student may have sounded insufficiently appreciative of his teacher. In 1847 Humboldt told him, "Your beautiful works can have no higher blessing than the benevolent approval of a man who combines the noblest human feelings with his mental and scientific greatness."[63] Perhaps because Humboldt's assistance brought more results, Remak expressed more appreciation to him than to gloomy, hardworking Müller. With Alexander von Humboldt, Remak formed an emotional bond that he never experienced with his adviser.

From the time that Remak received his medical degree in 1838, he had hoped to complete his *Habilitation* and teach on the Berlin University's medical faculty. Consequently, as soon as he knew Humboldt well enough, he began asking him how to petition the king for permission to teach, which would be possible only if the Cultural Ministry allowed him to complete his *Habilitation.* In 1840, when Friedrich Wilhelm IV became king and Eichhorn replaced Altenstein as cultural minister, Remak thought that he had a chance, but Humboldt advised him not to apply. Between 1833–1834 and 1848–1849, the number of students enrolled at the Berlin medical faculty dropped from 407 to 194, while the number of instructors decreased only slightly, from forty (including fourteen lecturers) to thirty-six (including fourteen lecturers).[64] In the 1840s, Eichhorn was trying to discourage any scholar from completing his *Habilitation*, and the last thing he wanted was a Jewish lecturer.

When Remak insisted on trying anyway, Humboldt told him that he would help as long as Müller approved. Müller must have promised his support, since on 14 March, Remak wrote to the king, declaring that he had been preparing for a university career for eight years with physiological research and private instruction and asking the king to reverse the Cultural Ministry's refusal to let him teach. He then offered some reasons why unbaptized Jews should be admitted to a Prussian medical faculty:

> When all the adherents of the Mosaic religion, especially in Your Majesty's enlightened states, admit openly that they have developed their long rigid religious and moral status through a Christian education and hope to develop it still further, perhaps I may also be permitted to declare, respectfully and solemnly, that my education and character are intimately connected with the views rooted in Christianity which permeate the life of science and the state. But this ideal accordance seems to make a formal conversion to Christianity difficult rather than easy. . . . The well-known involuntary origin of such religious conversions often makes the convert's social status dubious. . . . And I need not mention the individual's duty to advance the community given to him by nature, rather than entering another community, even though it may give him inner satisfaction.[65]

Unmoved—or perhaps annoyed—by Remak's logic, the king rejected his petition.

But in 1847, as a liberal majority in the Prussian Assembly prepared to increase the rights of Jews, Remak vowed to try again.[66] At Müller's house, he, Humboldt, and his adviser carefully planned their strategy.[67] Humboldt told Remak how to write to the king and accompanied the young scientist's letter with one of his own in which he twice invoked Müller's and Schönlein's names. Humboldt described Remak as "born to Jewish, very orthodox and poor parents" and reminded the king of his expressed wish "to favor an outstanding Jewish scholar with a dispensation."[68] Remak's own letter, more assertive than that of 1843, presents teaching more as a right than a privilege:

> For eleven years I have been engaged in physiological and pathological investigations which would have opened an academic career for me six years ago, if I, as a Jew, had not been prevented by the highest cabinet order of 18 August 1822 from setting out on one. A conversion to Christianity would be impossible for me because of the orthodoxy of my parents, to whom I owe piety and love. Also, I have always had the greatest aversion to the thought of gaining external advantages through a change of religion.[69]

This time, the king said yes, instructing Eichhorn to give Remak a university teaching position for which he was qualified.

Asked to write to the Cultural Minister, Remak now turned cocky, wanting to know when he could become associate or full professor, even rector or dean, and lecturing Eichhorn on religious tolerance.[70] Horrified, Eichhorn did everything he could to stall Remak's appointment. In an angry letter, Humboldt scolded Remak, letting him know how close he had come to wrecking his career: "What an unfortunate thought, my dear Remak, to enter into religious discussions with a Christian Cultural Minister and make everything hard for me, when I have expended my warmest good will and most precious time."[71] Seeking support, Eichhorn wrote to all the professors in Prussia, asking them whether they thought Jews should be allowed to teach; 141 voted yes, and 102 voted no. The Berlin medical faculty sent a joint statement declaring that "in the interest of science," the Cultural Ministry should stick to "the principle of not considering religious differences." In his own letter, Müller wrote that the hiring of Jews did not violate any university statutes and that the universities of other German-speaking lands should also be opened to Jews. Schönlein backed him, but not all of the professors agreed. The chemist Mitscherlich warned against the "brazenness and impertinence" of the Jewish character, and the anatomist Schlemm suggested setting a quota of 8 percent.[72] Eichhorn managed to avoid making Remak a professor, telling the king that no positions were currently available, but when the monarch insisted, he had to make him a lecturer. On 14 October 1847, Remak gave his inaugural lecture as the first unbaptized Jew ever to teach on Berlin's medical faculty.[73]

Remak as a Teacher

As Remak emphasized in his letters to the king, he had already been teaching for six years. Since 1840, he had been giving private instruction in histology, and since he had entered Schönlein's clinic, he had been teaching the medical students microscopy.[74] Once Remak became a university lecturer, he offered two or three courses per semester, among them histology, microscopy, general anatomy, and vertebrate development. The anatomist Wilhelm Waldeyer recalls that in 1858–1859, when du Bois-Reymond took over Müller's physiology lectures and Reichert, his anatomy course, Remak was teaching histology and microscopic anatomy five hours a week and offering an embryology lab.[75] He attracted quite a respectable number of students, sixty-four total during his first two years, including eighteen in his winter 1848–1849 microscopy course. Even with all this teaching, however, Remak received no salary and could not support himself with his students' fees. Every year the medical faculty picked two lecturers, usually those who attracted the most students, to receive a stipend of 50 to 100 taler, but they never chose Remak as one of these.[76]

Remak did no better at seeking space in which to teach. In 1847, after he completed his *Habilitation*, the Charité administration let him use a corner of the autopsy room, but this must not have sufficed. From 1840 until 1851, he instructed students in microscopy in his own home, first at Leipzigerstrasse 18, then Unter den Linden.[77] At this time, Müller came to his rescue, offering him instruments and a corner of his museum. An announcement in the *General Medical Central News*, probably intended to attract doctors, proclaims:

> During the next semester [summer 1851], Dr. Remak will hold his histological and embryological lectures in the Anatomical Museum, which has been supplied with the necessary instruments for this purpose through the generosity of Prof. Johannes Müller. Until now Dr. Remak has had to hold his lectures in his own apartment, since a laboratory has been lacking.[78]

According to du Bois-Reymond, this new institutional space consisted of a window nook. In March he wrote to Helmholtz that Müller was setting up "a kind of physiological institute" at the museum and that, "to heighten the pleasure, a window nook of the pathological museum has been cleared out for Remak's microscopy courses."[79] With his anatomical collection growing, this was the only space that Müller could spare. In 1853, Remak spent his own money to rent two private rooms, in which he set up seven microscopes for himself and twenty-four students.[80] Like Henle, he believed that anyone interested in microscopy should be able to work with instruments firsthand.

These cramped quarters never bothered Remak's students, many of whom must have used microscopes in their own living space. Thinking back on his years as a Berlin medical student (1839–1841), the anatomist Albert Kölliker wrote:

> I remember with vivid satisfaction a private course at Remak's, in which this highly gifted investigator told a few eager listeners in his own apartment about his observations of the developing chicken embryo, explaining his findings with demonstrations of preparations and with discussions which remained unforgettable to me and which soon thereafter were presented to the whole scientific world in his famous great work, which founded a new era in embryology.[81]

Kölliker went on to explore nerve structure and cell reproduction in Würzburg, where he worked with Virchow and taught Ernst Haeckel. The anatomist Wilhelm

His (1831–1904), whose developmental studies in the 1880s helped to prove that neurons were individual cells, remembers that in 1851:

> Remak first lectured at the University but then he had us come to his apartment Unter den Linden, where he had set up his incubators and microscopes, and sacrificed many eggs to give us satisfactory demonstrations.[82]

Although Remak's teaching brought him little income and took time from his research, he must have put considerable energy into it, for his students loved him. In 1860, when the Berlin University celebrated its fiftieth year, the students chose the professors who would give speeches and make toasts. From the medical faculty, they picked du Bois-Reymond, Virchow, Mitscherlich, and Remak.[83]

Since Remak could not live on appreciation alone, he repeatedly applied to become an associate professor, his only hope of ever receiving a salary. The medical faculty did not always back him, and when then did, they had to petition the Cultural Minister six times before he was appointed.[84] In July 1847, just after he had been made a lecturer, Remak wrote to Cultural Minister Eichhorn, asking to be appointed associate professor. The microscopist argued that if he hadn't been Jewish, he would have been a lecturer since 1843. When Eichhorn submitted Remak's request to the medical faculty, however, they refused to support it, writing that such a speedy promotion for Remak would hurt other gifted young lecturers such as Ernst Brücke.[85] In 1851, with his Krakow offer in hand, Remak tried again, but the medical faculty argued that he (age thirty-six) was one of their youngest lecturers and would have to wait his turn. Remak made yet another failed attempt in 1853, when Müller was fighting to get him into the Academy of Sciences, but the timing was still not right.[86] Between 1848–1849 and 1854–1855, the number of lecturers on the medical faculty rose from fourteen to twenty-three, while the number of full professors sank from fourteen to eleven, and the number of associate professors, from eight to seven. Remak came up against "career traffic" as an increasing number of talented young scholars fought for faculty positions.[87]

In 1854, however, the medical faculty asked the cultural minister to promote Remak, writing that he numbered among "the most skillful observers" of the microscopic anatomy of tissues.[88] They pointed out that he had been publishing for over fifteen years and had an international reputation. But Remak also had a stubborn enemy on the faculty: the botanist Heinrich Schultz-Schultzenstein (1798–1871), who wrote that Remak's work was mere "hair-splitting" and that he was using the medical faculty to become a "Jewish martyr." The cultural minister may not have taken Schultz-Schultzenstein's objections too seriously, since the botanist opposed microscopy and animal experimentation in general and had also voted against promoting du Bois-Reymond. But in 1855, the cultural minister made the younger physiologist an associate professor and rejected Remak's plea, even though du Bois-Reymond had taught almost no university courses. Like Remak, the younger experimenter had Müller's and Humboldt's backing, but he had been a lecturer since 1846 and a member of the Academy of Sciences since 1851. When Remak learned of du Bois-Reymond's promotion, he wrote angrily to the cultural minister and the medical faculty that he must regard his deferral as a blow to his scientific honor, since the reforms of 1847 had left the Cultural Ministry with no legal basis for rejecting him on religious grounds.

In the end Remak became an associate professor for the same reason he had become a lecturer: eighty-nine-year-old Alexander von Humboldt lobbied the cultural minister and appealed directly to the king.[89] As had been the case in the 1840s, Remak and Humboldt tried to take advantage of a crucial regime change. In the fall of 1858, Wilhelm I, the conservative, militaristic brother of mentally ill Friedrich Wilhelm IV, effectively took over the Prussian government and appointed Moritz August von Bethmann-Hollweg (1795–1877) cultural minister. Humboldt immediately asked him to promote Remak, and the medical faculty, rallied by Schönlein, supported his plea, invoking the recently deceased Müller's name. In January 1859, Remak became an associate professor—but he never did receive a salary.[90]

No Support from Müller's Students

Although Müller and Humboldt gave Remak all possible help, the Jewish scientist received little aid from Müller's other students. Of the pupils considered in this study, none liked Remak, and several actively disliked him. As the three opponents for his thesis defense, Remak selected Bentkowsky, a Polish physician to whom he dedicated his dissertation; Josef Mianowski (1804–1879), director of the Vilnius Military Hospital; and Count Czapski, a law student. Possibly these men's families had supported his medical studies, but it is striking that he chose neither Henle nor Schwann, whom he must have known since all three had been doing microscopic work with Müller for the last four years.[91] Even though he was discussing his microsopic work with Müller during the late 1830s, Remak could not have socialized with Müller's assistants, since as an Orthodox Jew, he could eat only kosher food. He could not have joined Henle, Müller, and Schwann for second breakfast at the Anatomical Institute, where the three were almost certainly chomping on sausages.[92]

Du Bois-Reymond, who looked down on most anatomists, expressed annoyance when Müller recommended Remak for the Königsberg physiology job. In May 1849, he wrote to Carl Ludwig (whom he was hoping would get the position):

> Müller . . . went and spoiled everything. Not content to praise you, . . . he also . . . recommended Helmholtz and probably Remak, whom he declares also to be a physiologist. Remak is no dangerous rival because Schulze [of the Cultural Ministry] sees clearly about him, and the faculty did not propose him and will not have him forced upon them.[93]

Wanting only an experimental physiologist to receive the position, du Bois-Reymond viewed Müller's plug for Remak as cronyism or, at best, misguided charity. Unlike du Bois-Reymond and Helmholtz, Remak had never fully rejected Müller's vitalism, and his work remained more clinically oriented than theirs. While he had nothing against microscopy, du Bois-Reymond resented the potential misuse of a physiological job. Whatever Remak did, it was not physiology.

Virchow, whose research interests most closely coincided with Remak's, proved to be anything but a friend.[94] The two competed in 1846 for Froriep's prosector job, which Virchow won easily because of his microscopic and clinical skills and his military connections. During 1850, while both were studying cell growth and Remak was seeking a professorship, the two exchanged letters in which the younger but more powerful Virchow offered advice. Remak asked Virchow, who was by then a professor in Würzburg, for help obtaining a professorship of physiology in Erlangen, another Bavarian city. Virchow replied that this search had already been settled but suggested that Remak try for a chair of histology in Bern, Switzerland. He also told Remak quite openly that he suspected his religion was hurting him in his job hunt. Remak wrote back bitterly, "I was so naïve, in the year 1850, not to think of religious difficulties! I forgot that we were counting 1850 years *after Christ*."[95] Virchow replied with practical advice about whom to contact to win the Bern job. Unlike du Bois-Reymond, he respected and closely followed Remak's work, which was confirming his ideas about cellular reproduction.

But when Virchow published his manifesto, "Cellular Pathology," which opened the eighth volume of his *Archive* in 1855, he did not mention Remak, although he had often cited his work in the past.[96] Suspecting that Virchow was claiming priority for the idea that all cells arise through division, Remak wrote him an angry letter:

> You have set down the sentence "omnis cellula e cellula" as your own, without even mentioning my name. That you have made a fool of yourself to those knowledgeable in the field (since the necessary embryological discoveries were not at your disposal), neither I nor anyone else can remedy. But if you want to avoid a public debate on this matter, I would ask you to inform me at your earliest convenience how and when you would like to discuss it. It goes without saying that I reserve the right to take matters into my own hands, should your subsequent explanation fail to satisfy me either in form or in content.[97]

Remak probably never mailed this letter, since it now lies among his own papers. But his threatening tone reveals the outrage he must have felt. Virchow, the better writer and politician, became associated with the idea that cells come only from other cells, even though Remak—in more indirect ways—had expressed it first.

In February 1856, when the medical faculty asked the Cultural Ministry to fund a full professorship of pathology, it listed Virchow as its first candidate and Remak as its second. Although Virchow did write to his father-in-law, asking him to speak to his powerful friends, and he did suggest using the Mendelssohns to approach Humboldt, he never asked anyone to speak against Remak.[98] From a scientific standpoint, he took Remak seriously as a competitor, but politically, he knew that Remak had no chance. Müller, who was leading the campaign to establish the professorship, had listed him before Remak. Both candidates frightened the Cultural Ministry with their past political activism, but Remak was Jewish and had not yet been made an associate professor. Finally, Remak was not primarily a pathologist, whereas Virchow had helped to create the field. Virchow never worked against Remak, in part because he never had to.

Virchow did Remak the most harm in 1858, when the forty-three-year-old lecturer applied to become an associate professor on the Berlin medical faculty and Virchow was called on to assess Remak's electrotherapy book. The new pathology professor criticized it strongly and recommended denying Remak a Charité clinic. In the book, wrote Virchow, "a normally sharp-sighted and in many respects highly experienced observer in a relatively new field has assembled—without any particular selective criteria—clinical experiences which can have lasting value only after longer and steadier tracking."[99] Probably Virchow would have condemned the book even if Müller himself had written it. It was too empirical—the work of a clinician, not a professor of medical science. Like du Bois-Reymond, he did not feel that Remak was using his scientific talents to best advantage. At the end of his career, the Jewish scientist's work had taken a clinical turn. But in other respects, it was a lot like Müller's.

Remak's Science as a Continuation of Müller's

One can see the importance of Müller's science for Remak's by observing how often he cites Müller and noting which studies he chooses to cite. Remak refers to Müller's findings in almost every paper he writes. He appears to have known every detail of Müller's research, including studies under way while he was writing his own articles.

Like Virchow, Remak most often cites Müller's *On the Fine Structure of Pathological Tumors* (1840), which provided a point of departure for both scientists' work. In Remak's 1852 article denying extracellular cytogenesis, he mentions the evidence in Müller's tumor book that the creation of cells within cells is "a very widespread phenomenon."[100] Remak calls Müller's tumor studies "an indispensable foundation and preparation for further medical observations," for by carefully examining tumors, Müller had shown that they were made of cells and that many growths thought to be cancerous were actually benign.[101]

In the field of embryology, Remak relied on Müller's and Retzius's observations of developing frogs.[102] Müller's description of the "shaggy" yolk membrane in fish eggs offered Remak evidence that eggs were genuine cells.[103] In his 1852 article on cellular genesis, Remak cited the discovery that had given Müller so much trouble, "the snails (*Natica*) that are being generated in such a miraculous [*wunderbare*] way in the gut of a *Holothuria* (*Synapta digitata*)."[104] Müller had begun publishing this work the previous fall, and at the time Remak's paper appeared in the *Archive*, his teacher was deeply preoccupied with it. Given Remak's wording, which so closely follows Müller's, it is likely that Remak was one of the few scientists with whom Müller discussed this distressing research.

In his electrotherapy book, Remak declared that Müller's "dry and critical separation of physiology from other disciplines that had sunk into confusion brought about the transformation of [physiology]."[105] It was not Müller's physiology, however, but his microscopy, chemistry, and anatomy, that most inspired Remak's own work.

Remak's belief in the value of microscopy was his greatest inheritance from Müller. Remak used microscopes throughout his career, starting his studies with the Fraunhofer instrument that Rudolphi had given Müller in 1824 and an achromatic Pistor and Schiek microscope that belonged to Ehrenberg.[106] In Remak's scientific writing, he emphasized the importance of microscopic analysis, pointing out the differences between what the eye and the lens could resolve. In an 1840 article on the sympathetic nervous system, he wrote that the microscope offered a way to see "important, though not absolute, differences in those parts of the nervous system that differ so entirely from each other in their function but manifest no significant differences to the naked eye."[107] In the conclusion to his *Investigations of Vertebrate Development* (1855), Remak called his histological studies a "complement and support" to his unaided observations.[108] He could never have identified the three germ layers and their products without a microscope.

Besides embracing Müller's microscopy, Remak adopted his techniques of chemical analysis for characterizing tissues. Whereas Müller had identified tissue types, Remak used chemical analysis at the cellular level, attempting to characterize individual membranes. Treating his preparations with sulfuric, acetic, and chromic acid; alcohol; and mercuric chloride, he tried to establish or rule out affinities between cells. In an 1851 article on cells believed to contain blood corpuscles, he argued against Virchow's idea that pigmented particles observed in these cells consisted of degraded blood, since the pigment usually retained its color in sulfuric acid.[109] Remak's chemical analyses often focused on membranes because he was trying to demonstrate the mechanisms of cell cleavage. In a brief report in 1854, he announced that with a mixture of copper sulfate, wood vinegar, and alcohol, he had managed to isolate a membrane in the frog egg that remained after the first cleavage and then formed a cover surrounding the dividing cells.[110] In one of his final articles on cell theory in 1862, he warned that until embryologists had the kinds of differential stains that botanists did, they would never fully understand the membrane's role in cell division.[111]

Remak shows his indebtedness to Müller most clearly in his determination to view growth and development from a broad, comparative perspective. In his 1843 study of chick development, Remak thanked Müller for their discussions "of comparative anatomical questions, of which . . . my work is an extension."[112] Like Müller, whenever Remak explored any particular structure, he studied it in as many animals as possible. In his first published research in 1836, he compared innervation to the skin and muscles in humans, calves, pigeons, fish, and frogs.[113] A cautious observer, Remak viewed comparative anatomy as an incentive for research rather than an excuse to theorize. "Above all," he warned in 1844, "one must be careful not to impose conclusions drawn in the frog on the entire animal kingdom."[114] He also considered the dimension of time in comparative anatomy, as Müller did in the late 1840s, trying to compare structures in animals and embryos of as many different ages as possible. In Remak's 1854 study of particulate liver cells, he observed cells in fifty embryos of different ages. He concluded his 1858 investigation of intestinal ganglia by citing Müller's work on fish and amphibians, and writing, "Only a comparative study of the remaining vertebrates will yield a full explanation."[115] Remak's writings never suggest that he viewed life as a great plan realized by a unique organic force, but he shared Müller's view that one could understand a structure or function only by observing it in many different animals.

Accordingly, two of Remak's favorite words in his scientific writing are "analogy" and "analogical." Like Müller, he tried to build scientific knowledge by comparing structures and processes in different animals. He opened an 1841 study of the brain and spinal cord by saying that "the analogy to birds" should long have suggested that mammalian brains also had a thin white outer layer. Later in the same article, he described tiny, club-shaped extensions coming out of some spinal nerves and wrote, "I can recall an analogy to these processes only in the hearts of mammals."[116] For Remak, mapping connections in the animal kingdom never took first priority, as it did with Müller, but he did not believe that cell structure and embryological development could be understood without it.

Remak also followed Müller in his quest to link anatomy (the study of structure) with physiology (the study of function). Both scientists understood physiology in the broadest possible sense, which is why, to du Bois-Reymond's annoyance, Müller recommended Remak for the Königsberg physiology chair. In the introduction to his electrotherapy book, Remak declared that "as an independent science, [physiology] encompasses all of organic nature."[117] Remak began his career by exploring the possibility that sensory and motor nerve roots—whose different functions Müller had shown—might have different microscopic structures.[118] Working mainly as an anatomist, Remak referred to physiology most often in his conclusions when discussing future experiments and the broader significance of his work. At the end of his intestinal nerve study, he speculated that "there could be several physiologically distinct types of nerve fibers, just as there are several nervous systems."[119] He concluded an 1853 paper by declaring that ganglious nerve fibers "open a new, unpredictable path and new problems for neurophysiological research."[120] Although du Bois-Reymond saw Remak as hopelessly anatomical, Remak respected experimental physiology and viewed it as indispensable to scientific knowledge. Responding to critics in 1844, he wrote, "I cannot comprehend how someone can prove that the gray fibers are not nerve fibers without conducting physiological experiments."[121] Just as anatomical results could inspire physiological studies, unexpected physiological findings could engender new anatomical and clinical techniques. Remak asserted in his electrotherapy book that "the discovery of new medical treatments builds almost exclusively on physiological foundations."[122]

As an investigator, Remak felt himself pulled toward both the clinic and pure research. Among Müller's students considered in this book, he showed a commitment to practical medicine second only to Virchow's. Remak thus differed significantly from Müller in his motivation to do science. Like his teacher, he loved microscopy and dissections, but he never shared Müller's fascination with animal forms. Instead, Remak loved cell structure and embryology, and his writing leaves it unclear whether he was studying the egg to understand the chicken or the chicken to understand the egg. In the introduction to his embryology book, Remak called his comparative studies of chicks and batrachians "time-robbing" and said that his clinical work had "delayed" his embryological studies, implying that his real interest was development.[123] As an embryologist, Remak described his goal as learning the "genetic meaning" of particular structures, which in the 1850s implied their embryonic origin or the tissues into which they developed.[124] As early as 1841, Remak stated that the brain's development revealed its cellular organization.[125] By the early 1850s, he confessed that he was studying frog embryos to learn about cell division, "to follow the developmental history of the tissues back to the first cleavage."[126] He called his observations of tadpoles the "keystone" in the "reform of cell theory."[127] At the same time, he admitted studying cells so that he could learn more about the ways that animals grew. Granular liver cells, he wrote, were of interest "for the development of tissues" because of the "clarity" with which one could watch their grains multiplying and their membranes dividing.[128] While Müller had compared finished forms, Remak wanted to study forms in the making, equally fascinated by their primary components and the ways that these components combined.

As an anatomist intrigued by cell structure, Remak called clinical research "hovering between heaven and earth."[129] While working for Schönlein, he felt torn between his loyalties to pure science and to his patients. Remak had long valued the practical benefits of research, writing in 1841 that the real reason to study tumors was to learn how to diagnose and remove them.[130] In the introduction to his electrotherapy book, he explained that his collaboration with Schönlein had aroused his interest in clinical studies and that he now wanted to use his physiological knowledge to cure patients. Remak often published his findings in medical journals such as *The German Clinic* and *Medical News*, whose readers were principally doctors. Since his work appeared in Müller's *Archive* almost yearly, Remak probably chose these clinical journals not because the best scientific ones rejected him, but because he wanted doctors to learn about the latest developments in microscopy, anatomy, and embryology.

In almost all of his articles, Remak emphasized how much there was left to know about how the human body worked, and he tried to inspire new researchers to join the effort. He closed the first segment of his 1840 article on the sympathetic nervous system by writing, "I have called [this work to] the attention of the medical public, whose interest in physiological research is now more active than ever."[131] Remak especially tried to draw physicians' attention to Müller's findings. He pointed out that "Müller's newly discovered techniques for the investigation of tumors have simplified the solution of the most important [diagnostic] problems from a practical perspective."[132] But he also argued that Müller had much more to offer clinicians than his pathological studies. His microscopic and physiological investigations were crucial for their work.

Remak as a Scientific Writer

Remak's research reports often seem to be written for beginners, giving readers methodological tips about details that can make or break experiments. In his 1837 article on nerve fibers, he carefully tells readers how to lay dissected nerves out on a microscope slide, as though he were in the room talking to them:

> If one takes a very thin . . . strand of a fresh cerebrospinal nerve from any vertebrate . . . and frees it from any loosely associated cellular tissue, . . . and then, with a sharp knife, cuts a piece out of this strand, half a line to one line (1.09–2.18 mm) in length, covers it with a rather large and heavy glass slip without exerting any pressure, and studies it under an appropriate magnification (150 to 200 times), one will see the parallel, densely packed nerve fibers.[133]

From the encouraging tone and detailed instructions—particularly the tip not to crush the nerve with the cover slip—one can see that he wants readers to repeat his observations and see the nerve fibers for themselves. It is likely that having squashed a number of nerves, Remak did not want investigators to deny the nerve fibers' existence because of faulty preparations.

Remak's articles abound with these helpful hints, for he knew that the way microscopists prepared their specimens would shape their theories about the structures they contained. He told his readers that the best way to moisten a crawfish's nerve was to apply a drop of the crustacean's own blood, a tip that he shared with his microscopy students.[134] In the simplest possible language, he explained how to obtain blood from a chick embryo for microscopic observation:

> One need only take an egg between the third and sixth day of incubation, open it on one side and then lay it down on the unopened side until the embryo has risen to the top of the yolk fluid, cut open a blood vessel, and catch the drops that emerge on a dry glass plate.[135]

He advised readers to keep the egg as warm as possible and to warm the slide before using it. It is highly significant that Remak offered his microscopy students and the readers of Müller's *Archive* (in which this article appeared) the same kind of methodological help. It suggests that his goal, as a teacher and writer, was to inspire future experimenters who could confirm his results when some authorities did not.

Often, Remak invoked methodological differences to explain discrepancies between what he and other prominent microscopists saw. He pointed out, for instance, that gray nerve fibers were much easier to see in freshly slaughtered animals than in older carcasses, suggesting that people who were having trouble seeing them should try looking at fresher preparations.[136] When arguing against the existence of blood corpuscles within cells, he explained that pigmented particles and small blood clots looked like blood cells and could be mistaken for them.[137] By attributing interpretations and theories to differences in method, Remak gave his critics a graceful way out. As with his students and general readers, he invited them to look again, using his techniques.

Perhaps because of this rational, diplomatic approach, Remak's findings never brought him scientific fame. Other investigators, particularly Virchow, became known for ideas for which Remak had provided key evidence, not because they stole his findings but because they wrote about them more forcefully.[138] In his 1837 article on microscopy, Remak claimed that "dryness of expression and the simplest and most exact possible narration of findings do the greatest service to communications of microscopic observations."[139] In later publications, he lived up to his word. As a writer, he could not compete with the rousing orators Virchow and du Bois-Reymond, or with Müller, who had such a gift for making readers see what he saw.

Remak chose his words with judicious care, since he often had to coin new terms to describe unknown structures. In 1838, he proposed calling the slender gray nerve fibers he was observing "organic" rather than "sympathetic," a "barbaric name."[140] "Sympathetic"—the name that stuck in the end—carried eighteenth-century baggage, conjuring images of nerves that created mysterious affinities. "Organic" better described their function, since many went to bodily organs, and Remak wanted simple, accurate terminology. When naming the inner fibers of nerves, he yielded to Jan Purkinje, admitting that his own term "primitive band" might be misleading. "Axis cylinder," the Czech microscopist's term, better represented the tiny threads that they were both observing.[141] Purkinje's term has since been shortened to "axon."

As a microscopist describing unfamiliar structures to his readers, Remak offered them some metaphors that conveyed vivid images. He referred to the "mosaic-like appearance" of nerve cross sections and told of "biscuit-shaped" and "dumbbell-shaped" blood corpuscles. Dividing blood cells looked like *Semmelpaare*, two rolls baked together with a furrow between them.[142] In one of his most intriguing metaphors, he compared the "wave-like concentric furrows" of a ray's electric organ to the acoustician Ernst von Chladni's (1756–1827) geometric sound figures representing sonic waves.[143]

Finding words to represent visual phenomena challenged Remak as it still does every anatomist and physiologist. In his scientific writing, Remak showed great sensitivity to his readers, sometimes leading them straight into the microscopic scene. He opened an 1854 article on liver cells with a cinematic zoom-in:

> Two years ago, in rabbit embryos, between the cylindrical areas of the liver lobes, I found round, colorless, transparent bodies 1/100 to 1/60 of a line in length, in which . . . I could distinguish a smooth surrounding membrane, a thick wall of delicate concentric layers, and a sharply bounded cavity full of grains.[144]

Such an opening could not fail to interest readers, but unlike Virchow and du Bois-Reymond, Remak was never a popularizer, and his writing never reached as many people.

When he appealed to a broader public, Remak never met with the success of these two virtuoso speakers. His electrotherapy attracted hundreds of patients, but his book on the subject did not win European doctors' respect. To assess the differences between his writing and Virchow's, it is worth examining one of Remak's case histories, written to show the value of electrotherapy:

> [The case] concerns a forty-nine-year-old weaver's wife, Henriette Paul. She is the mother of seven children, very thin and old-looking. For the past seventeen years she has suffered from "cramps" in her feet and arms, with a swelling of the joints that is particularly pronounced in the fingers, wasting and trembling of the limbs, weakness in the lower back, etc. . . . She cannot dress herself without help and can bring a spoon to her mouth only with difficulty. . . . On the day after [a six-minute application of direct current to her arm and chest muscles] the woman reported that for the first time in ten years she could fix her hair with her right hand and bring a spoon to her mouth without trembling. . . . [After four weeks] the treatment had to be suspended, since the gouty swellings of the joints did not favor another application of galvanic current.[145]

When one compares this report with Virchow's story of the hemorrhaging engineer, one detects a major difference in style. Virchow's patient died, and Remak's remained uncured, so that neither could boast of a successful case. Because Virchow's case appeared in an article, and Remak's in a book, one might expect some differences in format, and the difference in the patients' genders is also significant. Still, the contrast is striking. Virchow offers readers an adventure story, emphasizing action—both the engineer's and the doctor's—whereas Remak presents his patient as being, not doing. Remak also seems to distrust the woman's descriptions of her own symptoms, and he does not discuss—as Virchow would have—the living conditions and expected activities of a working-class mother of seven. As a writer, Remak fails to convey the excitement of medicine, even when describing a new treatment.

If he avoided popularization, Remak also avoided historicism. He does not seem to have wanted to shape the historical record of Müller. For this reason, du Bois-Reymond and Virchow have left the most widely read accounts of their teacher, although Remak worked with him longer than either of these other students did. In 1844 Remak attacked Henle, who was crediting Purkinje with discovering the axis cylinder. "The need to express hypotheses," he wrote, "seems to arise more in Henle than in other observers; whereas out of prudence and justice, when it concerns the rights of others, a writer, particularly a historical-critical one, cannot go too far in the other direction."[146] Remak resented Henle for taking control of scientific history, for telling his own version—in which Remak barely figured—of how the axis cylinder was discovered. In this case, Remak fought narrative with narrative, describing how he had shown Purkinje his "primitive band" in the fall of 1837. In a footnote, he offered a telling detail. Struggling to see the boundaries of the "primitive band," someone had joked that only pure being had no contours. True to form, Henle had quipped, "We generally deal here with impure being."[147] By introducing Henle's personality, Remak for once told a better story and retained his priority, something he very rarely did. If he had dared to write history, as Müller's other students did, we would have to weigh their accounts against that of a scientist who worked with him longer and may have known him much better.

7

Ernst Haeckel's Evolving Narratives

Anyone who is beginning to suspect that scientists define themselves by rejecting their mentors' ideas should look hard at the career of Ernst Haeckel. Haeckel's best-selling books belie the notion that each scientific generation establishes its identity by rebelling against the previous one.[1] Unlike du Bois-Reymond, Haeckel built his scientific reputation by showing his *likeness* to Müller—but he had to re-create Müller in order to do it.

When Haeckel depicts his teachers in his letters, scientific books, and popular lectures, he consciously shapes his readers' perceptions of them. Simultaneously, he presents himself as their loyal successor. Like du Bois-Reymond, he is fully aware that he is writing history, controlling future generations' understandings of his mentors by creating them as characters. At times, Haeckel grants himself importance as a witness. In 1856, when his teacher Virchow told him that he was leaving Würzburg, he began a letter to his parents: "'*Alea jacta est*—the Rubicon is crossed—I have just signed [the papers]!' with these words, . . . my honored teacher and now supervisor Virchow surprised me on Saturday at the railway station in order to announce his definite acceptance of the call to Berlin."[2] In this description, Haeckel writes himself into the history of science, suggesting that he is the first to hear of Virchow's decision. In an autobiographical sketch composed late in life, Haeckel boasted of his devotion to Darwin: "Among German scientists, no one has spoken out so early, so openly, and so unconditionally in favor of the Darwinian theory of evolution, and none has striven so hard through his own comprehensive studies to promote it, further extend it, and draw it out to its final consequences, as Haeckel."[3] Here, too, he associates himself with a better-known figure, in this case depicting himself as the loyal disciple who will "promote" and "extend" the master's work. When writing about Müller, he uses similar tactics, defining himself first as an extender, then as a brilliant successor who solved problems his teacher was just beginning to articulate.[4] As a scientific writer, he creates authority both by witnessing the glories of respected figures and by remolding them so that they bear witness to his own. That Haeckel's depiction of Müller is a literary narrative, designed to promote Haeckel's own evolving theories, can be seen from the way that it changes with time.

A Subtle Rebel

Born in 1834, Haeckel came from a middle-class family that valued culture and education. His father, a high-ranking Prussian bureaucrat, oversaw school and church matters in the district of Merseburg. Haeckel's parents approved of his early interest in botany and the other natural sciences, but like Helmholtz's, they urged him to channel his scientific energies in a direction that would allow him to earn a living: general medicine.[5] Despite their insistence that he study medicine, Haeckel's parents supported his early research, as had du Bois-Reymond's. In medical school, Haeckel examined preparations under his own Schiek microscope, and his father paid for his research trips to Helgoland (1854), Nice (1856), and southern Italy (1859–1860). Haeckel regarded his life from age twenty to twenty-five as his *Entwicklungsjahre* (years of personal growth). In his letters to his parents, he mentions having read Goethe's autobiography, *Poetry and Truth*, a narrative of artistic development.[6] Haeckel found the book perfectly applicable to an emerging scientist, and his own notion of his "years of development" probably has its origins in Goethe. He believed that he had the right to spend his early twenties learning about science and culture rather than settling right into a paying job. By August 1859, concerned about Haeckel's swaying inclinations toward science and art, his father decided that his son had grown enough and insisted that he focus on his medical career. Haeckel stayed in Italy for another eight months, but in this time he completed his research on radiolaria, the microscopic organisms that would bring him his first book, an academic job, and the ability to start a family.[7]

Haeckel's most revealing portrait of Müller comes from his letters to his parents, written between 1852 and 1856, while Haeckel was studying medicine in Würzburg. Between the ages of eighteen and twenty-four, Haeckel moved back and forth several times between the Berlin University, where he could live at his parents' house and hear Müller's lectures, and Würzburg, where he could learn cutting-edge science from Virchow and Kölliker.[8] Haeckel began his studies in Berlin in the spring of 1852 but stayed only one semester, during which he heard about Müller but did not enroll in his courses. For the next three semesters (from October 1852 until April 1854) he remained in Würzburg, returning to Berlin for the summer semester of 1854 and the winter one of 1854–1855, during which time he had considerable contact with Müller. He then went back to Würzburg until the fall of 1856, returning to Berlin in 1856–1857 for his doctoral research.[9]

Figure 7.1 Ernst Haeckel with his parents in 1853. Source: Ernst Haeckel, *Briefe an die Eltern 1852–1856* (Leipzig: K. F. Koehler, 1921), frontispiece.

Haeckel wrote to his parents only when he was away from Berlin, so that his comments on Müller are either anticipatory or retrospective, but they are far closer in time to the experiences they describe than any of his subsequent portraits. Edited and published by Haeckel's student Heinrich Schmidt in 1921, the medical student's letters home offer the least processed remarks on his teachers. At the time that he wrote them, he had relatively little at stake. Encountering scientific ideas for the first time, he did not need to tailor his depictions of his mentors to suit theories in which he was investing his reputation.

Still, Haeckel's youthful letters have a distinct narrative purpose. He tells the stories in them to convey a compound message. Simultaneously, the young medical student is trying to convince his parents that (1) he is a hardworking, responsible son; (2) medicine is horrible: and (3) science is wonderful and he should devote himself to that instead. A perceptive, articulate writer like du Bois-Reymond (though without his arrogance and wit), Haeckel negotiates these conflicting goals with the greatest diplomatic skill, so that his rebellion against his parents' plans is quite subtle. One begins to believe that he is such a good medical student, he should never have to become a doctor.

Haeckel's letters reveal a compulsion to narrate that continued throughout his career. In his science and in his personal writing, he is a champion storyteller. When Haeckel writes, he does his best to convey the wonder of natural forms and to awaken his readers' interest in science. In 1853, he told his parents that he was dissecting polyps, jellyfish, and coral from 5 A.M. until 10 P.M. and that his "zoological passion" had prepared him for the "much higher pleasure" of anatomy, which allowed him to study "the wonderful inner structure of animals."[10] When his microscope revealed "a forest full of the loveliest, tiniest forms on a rotten sausage rind full of mold," he thought to himself, "oh, if you could only show this to your dear parents, so that they could enjoy it with you!"[11] This early desire to share what he was discovering developed into a lifelong commitment to popularize science; to convey not just discoveries but the process of discovering. In the first chapter of his best-selling *The History of Creation* (1868), Haeckel wrote, "The highest triumph of the human mind, the true knowledge of the most general laws of nature, ought not to remain the private possession of a privileged class of savants, but ought to become the common property of all mankind."[12] Like du Bois-Reymond and Virchow, he saw the popularization of science as part of his job, and his comments on Müller should be read in the context of his other story-telling.

As a student, Haeckel felt obligated to describe his scientific activities for more personal reasons. His parents were supporting him, and as a dutiful son, he rendered "accounts" of everything he experienced. Haeckel seems to have been genuinely fond of his parents, respecting their judgments in all but his career choice and missing their company immensely. Like Müller in his letters to Nanny, the young Haeckel wrote to convey detailed visual pictures of his surroundings so that his readers could be there with him. When he arrived in Würzburg, he devoted half a page to a verbal map of his room, giving its dimensions, the locations of the windows and all pieces of furniture, and the positions of his books. When he moved a year later, he sent his parents an actual map.[13] To complement this spatial information, he sent schedules listing what he did in each hour of the day. For instance, in the winter semester of 1853–1854, a day looked like this:

> 8–10 dissection exercises (12), 10–11 materia medica (5), 11–1 practical (!) chemical work in the laboratory (8), 1–2 lunch at the Harmonie . . . 2–3 physiological chemistry (2), 3–4 general pathology and therapy with special emphasis on pathological anatomy, with Virchow (5), 4–5 theoretical obstetrics with Scanzoni . . . (5), from 5–6 is the only free hour in the day; from 6–8 on Fridays and Saturdays I also have a microscopy course on the study of normal animal tissue with Kölliker.[14]

Haeckel's enumeration suggests the avalanche of information that landed on him each day and the work expected of a nineteen-year-old student. As soon as he finished his classes, Haeckel hurried home and studied until late at night, a fact of which he carefully apprised his parents.

During his years in Würzburg, Haeckel's letters home portrayed him as the most diligent student imaginable. As a Protestant in Catholic Bavaria, he often commented on the strangeness of local celebrations, remarking that the wild Carnival festivities "have not affected me in the least."[15] Near the end of his first year, he wrote that he had begun avoiding his friends because "when I go out to bars with them, all they ever do is urge me to drink beer and fall in love, which they view as the only remedy that will turn me into a human being, but which to me seem equally horrible and superfluous."[16] During the spring of 1856, Haeckel expressed his attitude toward his work in terms similar to Müller's: "When I know clearly and distinctly what I have to do and then, all day long without interruption, can hurry from one completed task to the next without pause or rest, then in the evening I'm quite happy and think to myself with pleasure that the day was not lost."[17] If one can believe what he writes to his parents, no student could have been better behaved. Haeckel depicts himself as a near martyr to science.

The next step in his narrative logic is to convey the horrors of clinical medicine. Whereas Helmholtz had quickly accepted medicine as the most sensible career choice, Haeckel never stopped resisting. Near the start of his first few semesters, his emotions broke out as his weakly internalized exhortations collapsed before the reality of diseased patients and fetid cadavers. Haeckel's vows to continue were as genuine as his expressions of despair. He wanted to please his parents, but he could not see himself as a doctor. Early in his first semester, he wrote:

> I am now firmly convinced, like other, much brighter men before me, that I can never be a practical physician, that I can never study medicine. Don't think, dear parents, that I have had this insight because of the first disgust [*Ekel*] with dissections, or the "mephitic air of the dissecting hall and the *cadaverum sordes.*" For the most part, I've already overcome that kind of unpleasantness and would be willing and able to surmount it. But it's a very different thing to study a healthy body, even a diseased body, and to study the disease itself. For this I have an insurmountable disgust [*Abscheu*] (for which weak nerves and hypochondria may well be partly responsible) to which I can never reconcile myself.[18]

In this passage, one can hear the eighteen-year-old student struggling with himself as he realizes that medicine appeals to him neither practically nor intellectually. Haeckel soon got used to dissecting, but he wanted to study healthy animals.

In the 1850s, it took a strong stomach to learn medicine. The anatomy lecture hall where Haeckel took his classes "reeked barbarically of rotting flesh" and was so "extraordinarily small and narrow" that the students sat like packed fish and Kölliker could barely get behind his desk.[19] When Haeckel assisted Virchow with the pathology lab during the hot summer of 1856, the stench was "somewhat unpleasant even for Virchow."[20] In contrast to the medical dissecting hall, Virchow's private workroom seemed almost cozy, "a small one-windowed chamber that looked so mystically topsy-turvy and ingeniously sloppy that a witch's kitchen, or better, the laboratory of a medieval alchemist could give only a feeble idea of it."[21] Virchow was examining diseased tissues under the microscope, and his workroom could not have smelled much better than the dissecting hall, but

Haeckel romanticized the space where research was conducted. He reserved the words *Schmutz* (filth) and *Gestank* (stench) for medicine, along with *greulich* (horrible), the same word that he used for love and beer.[22]

When Haeckel began his fourth semester, he wrote that although his father's arguments over the summer had convinced him to complete his studies, he now had new reasons to believe he could never become a doctor. He could overcome his disgust for rotting bodies but not his contempt for "the monstrous imperfection, unreliability, and uncertainty of the whole art of healing." Medical knowledge struck Haeckel as oxymoronic, and he complained that he was receiving instruction "without systematic rules or order."[23] When Haeckel offers his parents a parody of "that quack, charlatan, and jumping-jack" Rinecker teaching pharmacology, he begins to sound like du Bois-Reymond:

> Gentlemen! Today we come to the constitutional application of mercury! Here, too, as everywhere in the teaching of remedies, we find an utter lack of determined rules and reliable experiences specifying the application and use of the same. It is much more the case that every doctor makes his own rules and then determines from his patients how much so-and-so he can give them without aggravating the disease until it causes their death. Yes, gentlemen, that is precisely the beautiful and attractive thing in the art of medicine, that it so lacks a fixed and general basis, rule and order, so that every doctor can treat or ruin his sick patients as he wills. . . . Is the art of medicine not a beautiful science?[24]

Haeckel followed this fictional harangue with a line of exclamation points and question marks. He could hardly believe what he was hearing. For him, anecdotal lectures communicating no demonstrable rules or theoretical system conveyed no real knowledge. There was nothing beautiful about disorder.

To fulfill his third narrative goal, Haeckel contrasted scenarios like this one with descriptions of the wonders he was learning in Kölliker's comparative anatomy and Virchow's pathology courses, with their emphasis on microscopy and detailed structural analysis. In February 1854, Haeckel suggested to his father that he might become a ship's doctor and thus realize his boyhood dream of travel, but his plan seems to have received a lukewarm response.[25] In December 1855, he let his parents know what terrible sacrifices his medical courses were forcing him to make. Virchow had asked him "in his room, under his guidance, to conduct a fine, microscopic-anatomical investigation of 'cyst formation in the *plexus chorioidei* of the brain,'" but he had had to turn him down to save time for his coursework.[26] Just after his twenty-second birthday in 1856 (a point in each year during which he underwent deep reflections), Haeckel thanked his father for forcing him to study medicine against his will. Writing that his years of effort, particularly Virchow's cellular and anthropological approach to medicine, had taught him to appreciate nature in new ways, he acknowledged that it had also given him a useful way to earn a living.[27] Here he appears to have capitulated rather than rebelled, subtly or otherwise, but by the end of his life, when he composed his autobiography as a successful scientist, he had changed the story. According to that account, when he passed his state medical exam in March 1858, he threw his clinical manuals out the window.[28]

Haeckel's Affinity for Müller's Science

Haeckel had found himself drawn to the natural sciences since childhood because of some inborn inclinations. Strong and athletic, he liked working outdoors, especially wandering through the woods to collect botanical specimens. During his research trip to Helgoland in 1854, he rose early each morning to swim. Like Müller and most of his students, Haeckel also had a talent for drawing. Considering the richness and beauty of his illustrations, it is not surprising that he sought a field in which he could put his artistic energy to work. He called drawing "my only real joy, which at any time can drive away bad thoughts."[29] All of Haeckel's talents drove him toward comparative anatomy: his love of the outdoors, his gift for drawing, and his ability to turn images into stories.

In medical school, Haeckel found that he learned better from visual than from verbal representations. He wrote to his parents that "there is in me some real, sensory element that makes it much easier for me to grasp and retain thoughts and facts, so that they make a much deeper impression, when they're taken in through images than when they're simply put down, naked and dry, in words."[30] To learn anatomy, Haeckel kept illustrated notebooks in which he would draw everything he had been learning in class. He called this mental processing *nachzeichnen* (after-drawing), the best way that he knew to memorize complex structures. Near the end of his first semester in Würzburg, his teachers increased their pace so that they could cram in all the material they had not yet covered. Kölliker doubled his lecture time and taught all the blood vessels in fourteen days. Haeckel complained that his fingers were going lame and that he could no longer keep up with his notebook, "which is really turning into a splendid illustrated work."[31] When Kölliker then rushed through the peripheral nerves, Haeckel wrote that he could "hardly think and draw [*nachzeichnen*]" the material, "let alone write it down."[32] Throughout his career, drawing would remain a way to make sense of the images confronting him. *Nachzeichnen* became a form of narrative through which he could tell stories on his own terms. Haeckel judged himself and his readers well, for his 1862 book on radiolaria caught the attention of naturalists throughout Europe with its extraordinary illustrations.[33]

When drawing what he saw under the microscope, Haeckel discovered that he could look through the lens with his left eye while using his right eye to oversee what he was sketching. Franz Leydig (1821–1908), who taught him microscopic anatomy, said that he had never seen anyone do this before.[34] In microscopy, which Haeckel loved from the start, the need to convey what one is seeing through images and words becomes all-important, since the microscopist must describe things that only s/he can see. Since Haeckel's parents were well-off and supported his scientific endeavors, they ordered him a Schiek microscope after his first year

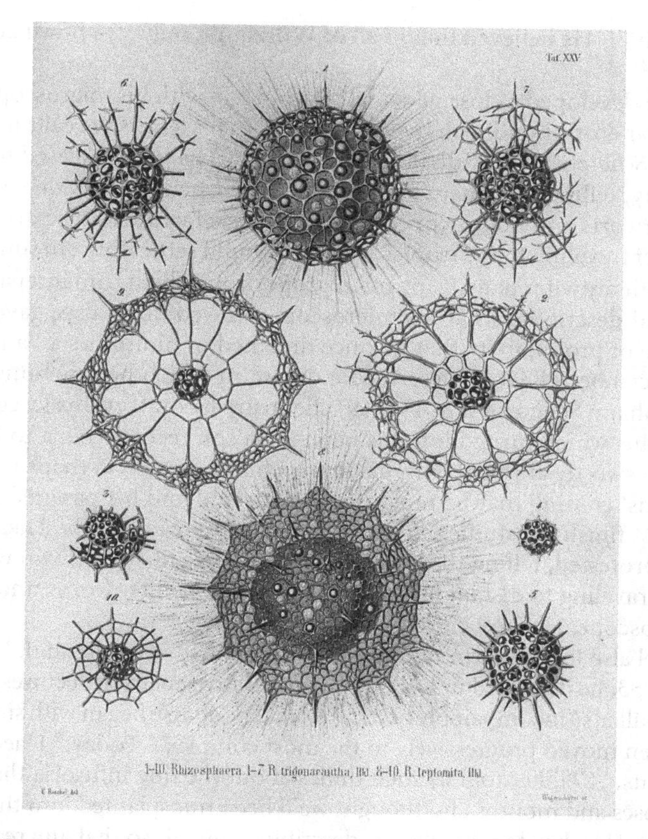

Figure 7.2 Ernst Haeckel combined his scientific and artistic talents in preparing his drawings of radiolaria, the microscopic marine organisms that helped to make his scientific reputation. Source: Ernst Haeckel, *Die Radiolarien* (Berlin: Georg Reimer, 1862), plate XXV.

in medical school. Excited, Haeckel asked his mother to stop by Schiek's workshop and make sure that his instrument would be ready in time for Virchow's pathology course, advising her to compliment Schiek's collection of copper-plate etchings.[35] Schiek delivered the microscope by the August due date, and by the end of August 1853, Haeckel reported that he was dissecting frogs, mice, snails, and locusts, and using the instrument to study their tissues. "The bliss of the microscope need not be described any further," he wrote. "It is really the highest that I know."[36] When the other students arrived, they all wanted to see Haeckel's instrument. "My microscope has been a huge hit here," he told his parents, "and I have to take it out and show it to everyone. Naturally I'm thoroughly envied because of it. But most people think: if they had had to live a dog's life as I did for a whole semester to save for a microscope of this caliber, they would much rather do without it!"[37] He believed that in all of Würzburg, only Virchow had a better instrument.[38]

Haeckel developed a close personal relationship with his microscope, laughing at his near erotic bond with the instrument. In his letters, he calls it "my darling" [*mein Schatz*] and "my divine microscope."[39] His friends teased him about his misogyny, telling him that God hadn't wanted man to be alone, so he had created liverworts and microscopes.[40] Twice Haeckel wrote to his parents about his fantasy of traveling to the tropics, "where I would settle down in some primeval forest with my wife (namely my microscope, from which I am inseparable)."[41] Drawing and describing what his microscope allowed him to see gave Haeckel such a sense of productivity that he once dreamed of himself as a Faustian creator. In December 1853, he described a dream in which he saw himself in his professor Johann Scherer's (1814–1869) chemistry laboratory, busily cooking up a homunculus which turned into his nephew (a son recently born to his sister) and "smiled sweetly at [him] with his lovely child's eyes."[42] Perhaps responding to his friends' chiding that he must find a girlfriend and his parents' insistence that he must finish his studies so that he could support a family, Haeckel's unconscious protested, telling them that he could create life in his own way. In his dreams of traveling to distant lands and observing tiny life- forms under his beloved microscope, Haeckel closely resembled Müller.

Haeckel also loved marine animals the same way that Müller did. During the spring of 1853, he delighted in Kölliker's comparative anatomy lectures, although he found Müller's "incomparably better." Kölliker's course began with single-celled animals, then moved progressively to the most complex. "Today," Haeckel wrote to his parents, "Kölliker told us such things about the tiny infusoria that we covered our noses and mouths and thought we'd been transplanted into the realm of fairy tales."[43] Haeckel, too, excelled at describing animals so that any reader could appreciate their wealth of forms. On his research trip to Helgoland, he wrote:

> Invigorated by an early morning swim, I began to gather on the beach the first, best seaweed, along with a whole host of algae, parasites, crustaceans, worms, mollusks, etc. lying on and among it, so as to follow my first acquaintance with the sea as a bathing place with a second one with its wonderful, diverse [*mannigfaltig*], magnificent inhabitants.[44]

In this letter, as in his early scientific works, Haeckel described marine organisms with Müller's language, having adopted the vocabulary of his mentor's anatomy lectures. In the foreword to his book on radiolaria, he referred to "the zoological-systematic interest which was vigorously stimulated by the wonderful multiplicity [*Mannigfaltigkeit*] of the new . . . extraordinarily beautiful forms" discovered by Müller.[45] Haeckel shared Müller's awe at life's diverse forms and his determination to discover their affinities.

Once Haeckel began building his own classificatory system based on evolutionary theory, however, he abandoned Müller's words. By the mid-1860s, he had developed his own phylogenetic approach to comparative anatomy in which organisms were related through common lines of descent rather than the action of life force.

In *General Morphology*, he used the term *Mannigfaltigkeit* to mock the pre-Darwinian approach to comparative anatomy: "It seeks out the endlessly diverse [*mannigfaltigen*] forms, the outer and inner relationships among the forms of animal and plant bodies and delights in their beauty, admires their multiplicity [*Mannigfaltigkeit*], and is astounded by their purposefulness [*Zweckmässigkeit*]."[46] Yet Haeckel's phylogenetic trees, which bring animals' relationships to life before the reader's eyes, serve much the same purpose as Müller's museum.

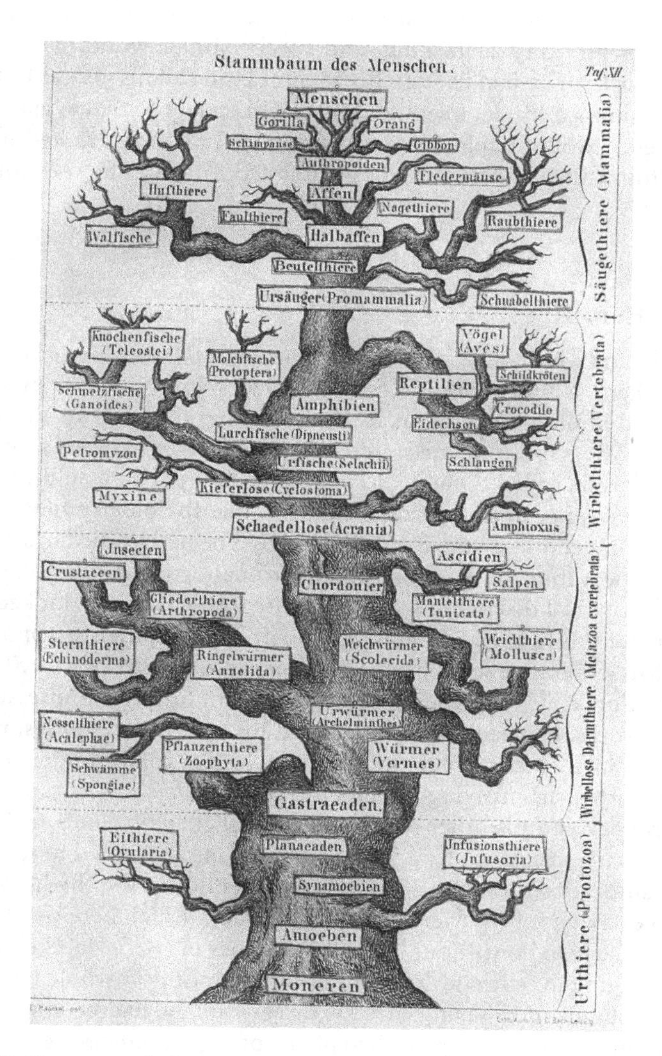

Figure 7.3 Ernst Haeckel's phylogenetic tree, showing human beings' relations to other animals, suggests how Müller should have arranged his museum. Source: Ernst Haeckel, *Anthropogenie* (Leipzig: Wilhelm Engelmann, 1874), plate XII.

Whether drawing or writing, Haeckel needed to put things in order. Like Müller, he loved seeking patterns in vast numbers of different forms and then presenting those forms so that the trends became obvious even to the most skeptical viewers. His best-selling books became portable museums in which any reader could see the solution to the mystery of life.

Even after rejecting "diversity" and "purposefulness," Haeckel retained Müller as his mentor. For the narrative of his scientific development, he wanted an authoritative precursor, not the foil du Bois-Reymond had sought. Rather than setting up Müller's science as opposed to his own, he presented it as the next best thing. Only evolution was lacking, and surely Müller would have embraced Darwin's theories if he had lived to hear them! Using all his gifts as a storyteller, Haeckel refashioned Müller over the course of his career so that as Haeckel's own ideas changed, Müller could remain his honored mentor. He had found him only after four other scientific heroes had failed him, and he was not about to give him up.

Haeckel's Heroes

As a scientist and writer, Haeckel looked for heroes, and when he didn't find them, he created them. Unlike du Bois-Reymond, Haeckel respected, even sought out, authority and idolized his famous teachers, writing that next to Virchow's "gigantic mind," he felt like a "miserable worm."[47] During Haeckel's "years of development," as he called them, he longed for a figure who would provide both intellectual guidance and emotional support, like the older philosophers who taught the younger ones in the Platonic dialogues he had studied in school. From his teachers, he wanted not just instruction on anatomy and microscopy but personal, philosophical discussions on the nature of life. Haeckel looked to senior researchers for this kind of relationship, and one by one, they disappointed him. Probably Müller became so special to Haeckel because he died before he could let him down. Consequently, when Haeckel built him up as an impressive mentor in his later works, he needed only to expand his positive memories, not to alter any negative ones. Haeckel's relationship with Müller can be understood only in light of his earlier dissatisfying interactions.

Like Müller and his other students, Haeckel craved friends with whom he could discuss his scientific ideas. Listing the reasons he wanted to return to Berlin for the summer semester of 1854, Haeckel mentioned the "highly stimulating and useful exchanges with friends who do research, which here [in Würzburg] I pretty much have to do without."[48] In the summer of 1855, Haeckel wrote pages about his longing for a friend "to whom I can pour out my whole heart and who really understands me," saying that for three years, he had been vainly seeking this kind of intimacy. Although he had plenty of acquaintances, he wanted "a real friend . . . whose heart and soul you can press right up against yours."[49] The relationship Haeckel desired would involve shared scientific curiosity, philosophical assumptions, and emotional closeness, an intimate scientific bond like that of

Müller and Henle in the late 1820s. The loneliness that Haeckel expressed must not be mistaken for the stereotypical isolation of the "lonely scientist," however. Haeckel wanted and actively sought intimacy—particularly with his scientific mentors.

Haeckel's first hero died two years before he was born, so that he remained an untarnished idol throughout his career. Like Müller and most other nineteenth-century German scientists, Haeckel revered Johann Wolfgang von Goethe, quoting him throughout his scientific and popular works. Since so many nineteenth-century naturalists made use of Goethe, it is important to consider which works Haeckel quoted and how he used them. Whereas Müller quoted Goethe mainly in his philosophical works of the 1820s, Haeckel relied on him throughout his life.

Goethe's fascination with diverse life-forms and his quest for a general form that could produce all of the others struck Haeckel as the right approach to the life sciences.[50] Haeckel opened *General Morphology*, a book written for anatomists and zoologists, with an epigraph by Goethe asserting that "nature eternally creates new forms." Several chapters begin with Goethe's insights into comparative anatomy. Haeckel does not identify the texts from which they come (possibly because they would have been familiar to his readers), labeling them simply "Goethe." He does not analyze or criticize the statements, treating them more as the utterances of a prophet. Goethe's belief that the discovery of a generalized form would explain the patterns in animals' bodies was not an evolutionary theory. He did not think that descent from a common ancestor gave rise to the trends in animals' appearances, although Darwin respected Goethe's comparative anatomy and invoked it when offering his own explanation of the relationships among living things.[51] Haeckel, however, treated Goethe as a seer who had grasped intuitively that organisms evolve and that animals should be classified according to their hereditary relationships.

Haeckel introduced *The Evolution of Man* (1874), a collection of his open lectures on Darwinian theory, with Goethe's "Prometheus," a poem admired by Romantic writers for its expression of defiance. "I know nothing poorer under the sun than you gods," Prometheus proclaims. "I honor you? What for?" In this book, Haeckel opens each lecture with a statement by a respected biologist, such as Carl Ernst von Baer or Thomas Henry Huxley (1825–1895), leaving it to readers to relate Goethe's poem to the book. Probably he hopes that the rebellious hero of a past generation will make an equally good spokesperson for Darwinian scientists, and that the latter will refuse to honor zoologists who merely delight in natural forms rather than classifying them systematically. Biologically, Haeckel's alignment of Goethe and Darwin is problematic, for the two had very different understandings of life-forms. What for Goethe represented the realization of an ideal principle was for Darwin the result of random chance.[52]

Unlike Goethe, who served as a lifelong inspiration, the first living scientist whom Haeckel admired quickly disappointed him. When Haeckel was fourteen, his parents gave him Matthias Schleiden's *The Plants and Their Life* (1848). Schleiden, who in the late 1830s had discovered that all plants were made of cells and caused Schwann to propose the same for animals, emphasized the cellular structure and life functions of plants, doing his best to make readers respect botany as an exact science. His ideas offered a system that might be used to characterize all living things. So impressed was Haeckel by Schleiden's work that when he passed his *Abitur* in March 1852, he planned to begin studying botany with Schleiden at the University of Jena, but a rheumatic inflammation of his right knee forced him to move to his parents' house and begin his medical studies in Berlin.[53]

Haeckel's first semester in Berlin may have offered him a new perspective on Schleiden's botany, since when he recovered, he enrolled at Würzburg, not Jena, and soon after his transfer, his opinion of Schleiden declined. In February 1853, Haeckel wrote that "although Herr Schleiden is a true genius and utterly unique, I am not so blindly enthusiastic about him as I was earlier."[54] By the spring of 1854, when he was deciding to return to Berlin and study with Müller, Haeckel's misgivings about Schleiden had turned to utter disillusionment:

> With respect to Schleiden, I've reached the same point of view as all other German botanists, namely that in Schleiden, as he is now, there is nothing, not even the smallest thing, to respect or to make him stand out. The outstanding, unsurpassed works through which Schleiden so quickly and excitingly made such a great, immortal name for himself as an unknown young lecturer will remain unforgotten for all time; but his independence and originality have turned into inflated self-interest and conceited contempt for all others, often much more thorough and exact, in any case more modest and considerate investigators, and now Schleiden, whose originality sinks further from day to day, is good only for . . . negating and protesting everything.[55]

Haeckel does not cite any recent works by Schleiden and is probably just repeating criticism he has heard from his teachers and fellow students. Haeckel never met Schleiden, so that his change of heart was not based on any personal experience. Still, this first disappointment left a deep impression and may have inspired his later conviction that scientists fester with age.

The zoologist and anatomist Carl Gegenbaur (1826–1903), a mentor only eight years older than Haeckel, did the most to advance his scientific career. If Gegenbaur reciprocated the emotions Haeckel expressed for him in the dedication to *General Morphology*, then Haeckel did find in Gegenbaur the intimate friend he had been seeking. Like Müller and Schwann, Haeckel and Gegenbaur met on a walk in the woods. During a botanical excursion near Würzburg in 1853, they discovered that they had a lot in common.[56] Both came from families of well-educated civil servants, and both had decided reluctantly to study medicine since their real love, natural science, seemed to offer no way to earn a living.[57]

Gegenbaur had begun studying medicine at the University of Würzburg in 1845 and had completed his medical degree there in 1851. Soon afterward, he had met Müller in Berlin and had studied marine organisms on Helgoland, as Müller had suggested. Having caught Müller's passion for marine invertebrates, Gegenbaur had then traveled to Messina with his Würzburg professor Albert Kölliker, and conducted research there for over a year.[58] In late 1852, Haeckel described a "beautiful, clear, interesting" lecture in which Kölliker reported his and Gegenbaur's observations of simple marine organisms, including "an important discovery that he had made simultaneously with our Johannes Müller, his teacher, who at the same time . . . was conducting similar studies near Trieste, on the Adriatic Sea."[59] When Haeckel met Gegenbaur in the summer of 1853, Gegenbaur had just returned from Sicily and was talking excitedly about Müller and marine biology. It was probably Gegenbaur who convinced Haeckel to leave Würzburg in the spring of 1854 and study comparative anatomy with Müller in Berlin.[60]

In retrospect, Haeckel played up Müller's influence on Gegenbaur. In a 1904 lecture on the history of biology in Jena, Haeckel proclaimed that although Gegenbaur had been taught medicine by Virchow and Kölliker:

> He was influenced in an excellent way by Johannes Müller in Berlin [Virchow and Kölliker's teacher]. Because he combined their methods of comparative anatomy, because he connected the most careful empirical observation of individual morphological appearances with comprehensive philosophical judgments about their general relationships, he successfully completed that great morphological work equal to the famous creations of J. Müller.[61]

Gegenbaur appealed to Haeckel because of the way he combined theory with close observation, just as Müller had advocated. Haeckel learned to compare anatomical forms partly from Gegenbaur, but he preferred attributing the influence to Gegenbaur's more famous teacher.[62] According to Haeckel, Gegenbaur drew his value from his relation to Müller.

Haeckel's early writings show some anxiety about competition with Gegenbaur. When the older scientist completed his *Habilitation* in Würzburg in early 1854, Haeckel wrote to his parents:

> On 28 January a young lecturer completed his *Habilitation* [Karl Gegenbaur], indeed, once again in comparative anatomy and histology! If only so many people wouldn't lean on this beautiful subject! There's hardly any room left for other people, and in the end, what's going to become of all of these lecturers?[63]

Gegenbaur, however, thought that there was plenty of room in the field for them both, and worked actively to advance Haeckel's career.

When Haeckel passed his state medical exams in March 1858, he was planning to follow in Gegenbaur's footsteps and perform the research for his *Habilitation* with Müller in Berlin. Devastated by Müller's unexpected death, Haeckel might never have regained his momentum as a scholar if Gegenbaur had not urged him to travel to Messina and conduct research there in 1859. Gegenbaur also got Haeckel his first academic job, nagging him to complete his *Habilitation* so that he could qualify for a lectureship at Jena, where Gegenbaur had been the zoology professor since 1856.[64]

In the 1860s in Jena, Haeckel and Gegenbaur functioned as a team to promote an evolutionary approach to morphology. Possibly Gegenbaur's decision to study vertebrate anatomy after 1860, and Haeckel's choice to focus on the invertebrates, minimized competition and jealousy between the two. Haeckel may have preferred marine invertebrates because his senior colleague had the vertebrates covered.[65] With Gegenbaur's support, Haeckel rose to associate professor of zoology and director of Jena's Zoological Museum in 1862.[66] The increased salary allowed him to marry Anna Sethe, to whom he had been engaged since September 1858. Sadly, both Sethe and Gegenbaur's wife died in 1864, an additional bond between the scientists that Haeckel perceived as a stroke of fate.[67]

In his publications, Haeckel never expressed anything but gratitude to Gegenbaur, and they remained on good terms for most of their lives. Haeckel dedicated *General Morphology*, his first comprehensive anatomical work, to Gegenbaur with an effusive, six-page declaration of scientific love. In this dedication, Haeckel wrote that he needed to tell the story of their "brotherly friendship bond" and recalled the day they had met in the Gutenberger Forest. "Since that day," Haeckel proclaimed, "a strange parallelism in our fates has drawn the indissoluble bond between us tighter and tighter. . . . Here, with the happiest possible division of labor, we have built up our common scientific field . . . so that much in the following work which appears to be my achievement was really awakened and nourished by you." Haeckel called Gegenbaur "my brother in fate, my academic colleague, and my hiking partner."[68] Haeckel became engaged to Anna Sethe soon after Müller's death, recovering from his scientific loss through his relationship with her. When she died, he found solace in his scientific friendship with Gegenbaur. The emotion sustaining his personal and scientific relationships seems to have flowed from a common source.

Although Haeckel found intimate friendship with Gegenbaur, his early mentor never acquired star status. He may have been a loyal supporter, but he could not be revered as a scientific master. Despite Haeckel's bond with Gegenbaur in the early 1850s, he regretted that he could not get closer to his two famous professors, Albert Kölliker and Rudolf Virchow. He hoped to win their approval, perform work under their guidance, and engage in intimate discussions about the philosophical meaning of their research.

A gifted microscopist with a range of interests as broad as Müller's, Kölliker had come to Würzburg in 1847 as a professor of physiology and comparative anatomy.[69] After beginning his medical studies at the University of Zurich, Kölliker had learned comparative anatomy from Müller and microscopy from Henle and Remak during 1839–1841 in Berlin. From 1842 to 1844 he had worked as Henle's assistant and prosector in Zurich, and had risen to associate professor of physiology and comparative anatomy there in 1844 when Henle had left for Heidelberg. By the time Kölliker came to Würzburg, he had published influential studies on invertebrate development, sympathetic nerves, red blood cell formation, and muscle structure. As an anatomist, he had acquired an international reputation and was known for his aversion to theorizing and his determination to check personally every fact that he published.

Figure 7.4 Rudolf Virchow (left) and Albert Kölliker (right) in Würzburg in 1850, a few years before Haeckel took their classes and tried to win their favor. Seated are their colleagues Johann Joseph von Scherer, Franz von Kiwisch, and Franz von Rinecker. Source: Bildarchiv Preussicher Kulturbesitz, Berlin. Reproduced in Constantin Goschler, *Rudolf Virchow: Mediziner, Anthropologe, Politiker* (Cologne: Böhlau, 2002), illus. 2, printed with kind permission of the Bildarchiv Preussischer Kulturbesitz.

Together with Virchow, Kölliker attracted medical students to Würzburg. Between 1849 and 1855, the enrollment increased from ninety-eight to 388.[70] Haeckel was just one of those who had come to learn from the star anatomist, and his letters home suggest that he expected a genius. His comments on Kölliker began as soon as his classes did. He wrote that Kölliker possessed "a perfect masculine beauty, the like of which I have seldom seen; his black eyes, especially, are quite splendid."[71] From the beginning, Haeckel hoped to get Kölliker's attention, but less than a month into his lectures, he reported regretfully that he had not gotten to know Kölliker any better.[72] By the following July, however, Kölliker had invited him as his personal guest to a faculty picnic, an honor extended to only a few students. Unfortunately, Kölliker's wife had not brought enough to eat, so that the favored few all went hungry.[73]

Days later, in July 1853, Kölliker opened his workroom to the young student. Proudly, Haeckel wrote to his parents:

> I've now gotten to know Kölliker somewhat better. Last week I brought him some eggs from a mole cricket (*Gryllotalpa*) that I had dug up in the botanical garden. He invited me to study them and the development of the young animals that came out of the eggs and offered me a microscope that I could use any time in his room in the Anatomical Institute.[74]

Because of his demanding coursework, Haeckel waited until the following December to take real advantage of Kölliker's offer. At the time, the anatomist's workroom was filled with Englishmen who had come to study anatomy full-time under Kölliker's supervision, and one Sunday, while sketching a microscopic preparation of the optic nerve, Haeckel "dared to tell" Kölliker how much he envied the foreign visitors. Kölliker replied that he "had certainly offered [Haeckel] the same opportunity."[75] If Haeckel is accurately representing the exchange, Kölliker sounds almost disappointed that the first-year student did not join him sooner as a collaborator.

Yet even when Kölliker respected him as an anatomist and microscopist, Haeckel wanted more from his teacher. In March 1854, when he thought of leaving Würzburg for Berlin, he admitted that "despite all the admiration I still have for Kölliker's extraordinary anatomical talents and capabilities, my wish to come in closer contact with him seems to have diminished." [76] In the same paragraph, he described his keen disappointment with Schleiden. When Haeckel returned from Berlin in the summer of 1855, Kölliker gave him free run of the Anatomical Museum—something that Müller had never done. Haeckel bragged:

> Kölliker has opened the Comparative Anatomy Museum to me in the most obliging (and for him almost incomprehensibly liberal) way. I can pick up the key from him at any time and look at all the preparations as the spirit moves me. Even the cabinets with the very interesting, informative preparations in alcohol are open to me, and I'm allowed to study them at least from the outside (which I could not do even once in Berlin with Müller, who merely put the skeletons, etc. at my disposal.)[77]

In the fall of 1855, Haeckel reported that he was spending a great deal of time at the Anatomical Museum, for Kölliker had now given him his own key.[78] Perhaps Kölliker's kindest gesture to his student was to invite him along on a research trip to Nice in the late summer of 1856. Considering that Kölliker taught Haeckel anatomy much longer than Müller did, personally supervising his dissections and microscopic preparations, and repeatedly inviting him to work with his own instruments and collections in his private space, why did Müller become the heroic mentor of Haeckel's narratives, and not Kölliker? This question can best be answered after a consideration of Haeckel's comments on Virchow, for Virchow—equally important for Haeckel's scientific development—failed to qualify for the same reasons.

When Virchow left Berlin for Würzburg in 1849, he became the first professor of pathological anatomy in the German territories.[79] The Bavarian university was delighted to have him and the new students he attracted, and organized festivals to show its appreciation. When Virchow turned down a prestigious offer from Zurich, the students held a torchlight parade in his honor. Haeckel, however, reported that the students' sympathies for Virchow were not so strong, and he wished that they were marching for Kölliker instead.[80] But Haeckel, with his keen interest in microscopy and his love of celebrity, did everything possible to win Virchow's respect. He reported proudly that at the faculty picnic to which Kölliker had invited him, he had gotten to know Virchow personally.

In the winter of 1853–1854, Haeckel found himself awed and frustrated by Virchow's pathology course. "For the most part," he wrote to his parents, "this lecture course covers things that are still unpublished and that have been recently discovered by Virchow himself." The hundred-seat lecture hall was invariably full; no one ever missed class "since here [you can] hear things that you would otherwise never read or experience." Haeckel assessed Virchow's lectures as "difficult, but extraordinarily beautiful," reflecting that "I have never before seen such concise language, penetrating force, tightly drawn conclusions, sharp logic, and at the same time such extremely graphic descriptions and compelling stimulation in lectures, as are combined here."[81] Haeckel was particularly excited by Virchow's descriptions of disease at the cellular level. He called the idea that all living things are made of cells "the wonder of all wonders" and vowed "to devote all my powers to investigating the cell."[82] When Haeckel returned from Berlin in the spring of 1855, he was equally pleased by Virchow's private course on pathological anatomy and microscopy that linked close analyses of diseased tissues to clinical observations. Having just come from Müller's classes, Haeckel called Virchow's lectures "among the best and most informative that I've ever heard." Virchow wanted each student to see for himself what diseases could do to bodies, since for him, narrative was never enough. Haeckel reports that:

> We sit in groups of thirty to forty at two long tables, in the middle of which runs a little train track where the microscopes roll on wheels and are shoved from one person to the next. There, in an hour, one often gets to see great numbers of the strangest and oddest pathological preparations while Virchow gives really outstanding lectures (naturally in keeping with the material that has just come into his hands from the clinic).... It's exactly this connection between the clinical-pathological, anatomical and microscopic findings, the way one obtains it in the clearest, most comfortable fashion as a total, unified image of disease, that is so extraordinarily interesting, informative and important. And one looks in vain for something like this in Berlin, where pathological anatomy is absolutely unthinkable![83]

Haeckel admired Virchow's tendency to present the "big picture," drawing connections between findings on the wards and under the microscope. In the theoretical course that Virchow offered in the afternoons, Haeckel found that the lectures were "certainly not fluent and smooth, but fresh, compact, and outstanding because of the characteristic high and general standpoint from which all things are regarded, so that the driest, most singular detail becomes attractive."[84] Just after Haeckel's birthday in February 1856, when he thanked his father for forcing him to study medicine, he credited Virchow—not Müller—with teaching him medicine in a way that inspired him, showing the connections between research and clinical practice and the need for reliable knowledge of each.[85]

It took Haeckel longer to get Virchow's attention than Kölliker's, but he won the pathologist's respect as well. Early in the 1855–1856 winter semester, Virchow asked Haeckel to write up a few lectures from his practical pathological anatomy course and submit the reports to the *Viennese Medical Weekly*, whose editor had been asking Virchow for a contribution of this kind. Perhaps Haeckel had shown Virchow his illustrated notebooks of his lectures, which contained drawings better than the ones Virchow published in *Cellular Pathology* (1858).[86] Haeckel sent the journal a short paper on the connections between typhus and tuberculosis, which received a negative response from scientists opposed to Virchow's cellular approach to diseases. Identifying with Virchow, Haeckel wrote that "we" followed up with a fourteen-page explanation. With Virchow's approval, Haeckel submitted two more such articles on disorders of the uterus and the ovaries, commenting to his parents, "How sweet to be attacked on Virchow's behalf!"[87]

In the summer of 1856, during the last months before Virchow returned to Berlin, the pathologist made Haeckel his assistant. This meant that the Pathological Museum was in his charge and that he could work most of the day right next to his famous teacher. Despite receiving a position coveted by most of his class, Haeckel remained unsatisfied, for Virchow's personality and style in the tiny workroom differed dramatically from his own. "The most ticklish and difficult thing about my position now is my relationship with my boss," he confessed to his parents:

> On the whole, Virchow has been very nice and friendly to me up to this point. But he's much too reserved and cautious for me to conclude that he's satisfied with me. At first this was clearly not the case. My whole essence, my whole way of handling things is too different from his to meet with his approval. Unfortunately, nature has not granted me even a trace of the divine calm, coolness, and constancy with which he invariably regards all things, with the greatest objectivity and clarity, and my haste, heat, and unrest are thus none too pleasant to him.[88]

While Haeckel admired Virchow's concentration and perceptiveness, he grew increasingly frustrated with his lack of emotional expressiveness. He wanted a teacher who would communicate more than the best way to prepare tissues. About two weeks later, he complained:

> If only Virchow weren't so extremely reserved, so that he never even hints at what he really wants and means. For instance, he hasn't even once let me hear a word of praise or reproach, although he's had ample opportunity, especially for the latter. . . . I've gotten to the point where I think about every sentence I say to him for fifteen minutes beforehand and then turn it around in my mouth ten times before saying it. . . . The most rewarding response that I have so far been able to extract from him was once when I presented an idea to him that I'd had about a microscopic preparation which seemed to have wonderful potential. "Yes,"

said Virchow with his usual calm, after he had heard me out, "I also had this idea once, during a certain period of my life."[89]

In mid-June 1856, Haeckel reported that his "personal relationship" with Virchow continued "cold and objective," although "highly profitable," since he believed that he needed to temper his own "terrible subjectivity." At the same time, he envied his friend Johannes Lachmann, who was working with Müller in Berlin: "How much happier is Lachmann in this respect with his divine Johannes Müller, with whom it must be a true joy to relieve him of the most boring, unprofitable tasks!"[90] When Virchow left for Berlin that fall, Haeckel went, too, but he did not "follow" Virchow or work in Berlin as his assistant. Virchow's prestigious pathology chair included a salary for his own prosector, who would choose his own assistants, and Haeckel needed to prepare for his state medical exams.[91] But by then, he had also dismissed Virchow as the kind of mentor he was seeking.

Haeckel wanted a scientific hero who would discuss the secrets of life, a respected scientist ready to form a close emotional bond with a younger one and talk about philosophy, love, and religion as well as anatomical discoveries. Gegenbaur almost fulfilled this role, but Haeckel's colleague at Jena was too close in age, and his work had not yet won him international acclaim. While Kölliker and Virchow respected Haeckel's abilities, he worked for them during their busiest, most productive scientific periods. Neither had time to philosophize with a young assistant, although they gladly accepted his help dissecting and drawing. More significantly, Kölliker and Virchow objected to speculative theorizing on principle, discouraging their students from expressing ideas that were not rooted in extensive observations. But Müller, depressed and declining in 1854–1855, might well have taken time to discuss the organization of life with a sympathetic young student. As a senior anatomist, he might have felt more comfortable confiding his dreams and doubts to an adoring pupil than to a critical junior colleague like du Bois-Reymond. Finally, Haeckel's experiences with Virchow and Kölliker were those of everyday reality. He received most of his medical training in Würzburg, so that Müller was never his "boss" and was less often the subject of student gossip. While Haeckel struggled in Virchow's workroom, the grass looked greener in Müller's. Haeckel's wording in the passage comparing his plight with Lachmann's suggests that he imagined Müller's assistants as martyrs who took worldly cares upon themselves in order to liberate a great mind. For Haeckel, Müller retained his divine status—particularly after his death—whereas Virchow stayed alive to bother him.

In 1877, more than two decades after their collaboration, Haeckel and Virchow became involved in a sharp debate about whether evolutionary theory should be taught in public schools. At the annual meeting of German scientists and physicians in Munich, Haeckel proposed that schoolchildren should learn about evolution, but in a lecture held four days later, "The Freedom of Science in the Modern State," Virchow contradicted him. Virchow did not oppose Darwin's views, but he argued that evolution was still a theory, and science at the secondary school level should limit itself to demonstrable facts.[92] At the core of their debate was a fundamental disagreement about what science was. Virchow, who had used his training under Müller to make pathology a rigorous, empirical science, feared that Haeckel was bringing back the worst aspects of nature philosophy.[93] Haeckel, who valued Müller's philosophical side as much as his empirical one, suspected that Virchow opposed Goethe's and Müller's comprehensive approach to science.[94] In his 1905 lecture "The Struggle over Ideas of Development," Haeckel claimed that after he met Virchow in 1852, he "soon entered into the closest friendly relations with him as a special student and admiring assistant."[95] This statement is an exaggeration at best. As scientists, Haeckel and Virchow failed to connect.[96] They differed not just in their styles but in their values and goals. Taught by Müller during different periods of his life, they responded to different aspects of his science, developing them in ways that suited their own personalities and understandings of scientific knowledge.

Haeckel Rewrites Müller

Because Haeckel's evaluation of Müller evolved over six decades, one learns more about their relationship by examining his comments in the order in which he wrote them than by assembling a chronological record of their contact through sources written at different times. Considering the huge mentoring role that Haeckel eventually assigned to Müller, this contact was limited: (1) he may have encountered Müller during his first semester of medical school in 1852; (2) he attended Müller's comparative anatomy and physiology lectures in 1854; (3) he talked to Müller in 1854–1855 while studying specimens at the Anatomical Museum; (4) he worked with Müller briefly during his research trip to Helgoland in 1854 and met him in Nice in 1856; (5) he may have discussed his research with Müller in 1856–1857 while preparing his doctoral thesis; and (6) he may have talked to Müller while studying for his Prussian state medical exams in 1857–1858.[97]

In the spring of 1853, when Kölliker's comparative anatomy lectures were starting, Haeckel wrote to his parents:

> Exactly these lectures that I'm hearing now, I could have heard nowhere better than in Berlin, . . . and since there I could have heard them from one of the greatest, most noble men, from Johannes Müller, who made such a captivating impression on me, it hurts two or three times as much. Physiology and comparative anatomy (the two most interesting subjects there are) are exactly what he delivers incomparably better than Kölliker.[98]

Haeckel does not say when or how Müller made this "captivating impression." In the summer of 1852, he was not formally enrolled in Müller's courses. He may have met Müller or audited his lectures, but he is probably incorporating the claims of fellow students.

Figure 7.5 Johannes Müller in 1854, the summer he and Haeckel fished together on Helgoland, as painted by Frau Bauman-Jerichow. Source: Wilhelm Haberling, *Johannes Müller: Das Leben des rheinischen Naturforschers* (Leipzig: Akademische Verlagsgesellschaft, 1924), plate VII, 417.

At the time, Haeckel saw Müller as the founder of comparative anatomy.[99] On two occasions he called Müller's lectures "classical"—lectures that he may not yet have heard.[100] He regarded Müller as a hero before he knew him personally, so that his expectations shaped his narrative of their actual meeting. In a diary entry in 1855, Haeckel wrote about his first contact with Müller:

> Here, for the first time, I got to know an authority respected by all and whom I set up as a scientific ideal, for my close relations with him during this time (in the museum, etc.) made comparative anatomy my favorite science from then onward.[101]

Written at least eight months, perhaps as long as three years, after his first meeting with Müller, Haeckel's account is constructed with future readers in mind. Like du Bois-Reymond, he seems aware that Müller will someday be known through his descriptions of him, and even in this early diary entry, Müller becomes part of a narrative defining him in relation to Haeckel's career.

Haeckel's joyous accounts of his days with Müller on Helgoland are much more spontaneous. His letters home make it clear that while he hoped Müller would come to the remote North Sea island, he by no means expected it:

> On Tuesday afternoon . . . [when we looked over] the passengers who had just arrived on the steamer, we saw among them, with no little surprise and delight, Johannes Müller, our great and noble authority, whose presence we had so desired but still hardly hoped for.[102]

Müller had come to study the development of starfish and sea urchin embryos, and Haeckel soon altered his work habits so that he could stay close to his teacher. The student gave up sorting, drying, and dissecting the seaweed and mollusks he had been collecting, and spent hours fishing with Müller and his son Max. Müller taught Haeckel how to skim the waves for plankton with his specially designed net, and he communicated his fever to gather as much material as possible in the time available:

> Since Müller and his son have been here, our own daily work has begun with comparative anatomy. At 8 A.M., in the company of this guiding star, we go out to sea for one to two hours and skim the surface with a butterfly net, so that we obtain the richest selection of the most exciting creatures to view under the microscope. These excursions, by the way, are not just very instructive, but also amusing, since old Müller almost always entertains us in a very funny and witty way. But since the zoological treasures we find this way keep us busy all day with the microscope, we have too little time left for other things, and our occupation with seaweed and larger sea animals has not amounted to much since then.[103]

Müller was never happier than when collecting and examining marine life with his young students, and one can see why he would have liked the tractable, admiring Haeckel. But this collaboration which so inspired the student lasted only two weeks.[104] It proved crucial, since Haeckel would use Müller's technique of pelagic fishery for the radiolaria studies that launched his career, but those who read his later references to it are likely to be surprised how little time the mentor and student actually spent together.

Despite his successes in Würzburg, Haeckel passed the spring of 1856 moping and longing for Müller. Placed in charge of Virchow's Pathological Museum, he thought what a "divine pleasure" it would be to work with Müller that summer, and told his parents that "gloomy and miserable, I wandered through the cold rooms of the Pathological-Anatomical Museum and thought longingly of the comparative anatomical treasures in the other wing of the anatomy building."[105] Haeckel missed the excitement of comparing unfamiliar life-forms, but he also seems to have established an emotional connection with Müller that he never had with any other teacher. When Haeckel learned that he would be accompanying Kölliker to Nice and that Müller might be there as well, he raved, "That would really be wonderful, if we ran into the most god-like of all scientists there!"[106]

Yet Müller did not serve formally as Haeckel's adviser. In the fall of 1856, Haeckel returned to Berlin to do the research for his doctoral dissertation on the crayfish. Between October 1856 and March 1857, when he received his degree, he may have discussed his work with Müller, but there is no record of their interaction. Haeckel's doctorate was conferred by Christian Gottfried Ehrenberg, the microscopist who had encouraged Remak and who in that year was serving as dean of the medical faculty. Since Müller published Haeckel's study of the crayfish in his *Archive*, the two almost certainly exchanged ideas at some point.[107] Their discussions may have continued between October 1857 and March 1858, when Haeckel was in Berlin studying for his state medical exams.

Probably Haeckel was talking to Müller, for when his hero died in April 1858, Haeckel had arranged to do research in his workrooms.[108] In Müller, he believed he had found the master whose teaching included intimate, personal discussions of the nature of life. While mourning the loss, Haeckel had to rethink his own career in practical as well as intellectual terms. After some deliberation, he took Gegenbaur's advice to study marine life in Sicily. Haeckel's new vision of himself and Müller emerges in his detailed study of radiolaria, the microscopic marine organisms Müller had loved. In this, his first book, Haeckel presents his life and work as a continuation of his mentor's. Dedicated to Müller, the radiolaria monograph opens with a foreword discussing how Haeckel's findings relate to those presented in Müller's last paper. Haeckel states that he learned pelagic fishery from Müller and that it was very fortunate that he took Müller's article along to Italy. "I regard it as the solid foundation," he writes, "on which it was possible to carry out the extensive construction of my monograph."[109] Müller's paper was more descriptive than theoretical, not likely to inspire a young scientist as his *Handbook of Human Physiology* might. But Haeckel loved Müller's last work because of the way it demonstrated affinities between life-forms not previously seen as related. It was the kind of work he wanted to do.

When Haeckel began observing the intricate, single-celled organisms in Messina, fifty-eight species were known, of which Müller had described forty. Haeckel discovered an additional 144.[110] While he dared to criticize Müller's classification scheme in light of these new findings, he represented his work as an extension of his teacher's, conducted with his assumptions and methods. He concluded his foreword by stating, "I gratefully honor in Johannes Müller the teacher who influenced my scientific development more than all others, determining and guiding, from whom I obtained the most powerful encouragement to investigate the animal world, as he also personally introduced me to the most delightful study of pelagic fauna."[111] When Haeckel created himself as a scientist, he chose and modeled his mentor.

Soon after Haeckel returned from Sicily, however, he read a book that changed his relationship to Müller's anatomy: Heinrich Georg Bronn's (1800–1862) translation of *The Origin of Species*.[112] Haeckel at once embraced the idea that living things had evolved from common ancestors, for he saw in Darwin's theory the solution to the puzzle of how all life-forms were related. To Haeckel, common descent explained the resemblances between plants and animals much more satisfactorily than the action of life force. Philosophically, Haeckel would take the notion of evolution much farther, using it as the scientific underpinning of monism, his belief that matter and mind are inseparable. From the early 1860s onward, Haeckel began editing his descriptions of Müller so that they supported his new understanding of comparative anatomy. His references to his mentor in *General Morphology* (1866) are thus strikingly different from those in his radiolaria study.

In the foreword to *General Morphology*, Haeckel again credits Müller with having introduced him to comparative anatomy, but he presents Goethe and Jean Baptiste Lamarck (1744–1829), not Müller, as Darwin's precursors. In Haeckel's history of the field, Müller represents the best of the old, dualistic school, whose comparative anatomists believed that there was a life force independent of the physical forces acting in animals' bodies. "When I later had the priceless good fortune to become acquainted with the empirical foundations and commanding perspectives of dualistic morphology," he writes, "with their full dimensions and content, through immediate contact with my unforgettable teacher Johannes Müller, that monistic opposition was already quietly beginning to develop."[113] Here Haeckel presents Müller's comprehensive vision of the animal kingdom as one that needed to be transcended. His own work is now no longer a direct continuation of Müller's, but a new kind of comparative anatomy that developed in response to it. In his first chapter, he writes that "since the all-too-early and insufficiently mourned death of Johannes Müller (1858) whose powerful authority during his lifetime still managed to uphold a kind of strict order over the wide domain of organic morphology, an increasing wildness and anarchy have been tearing up this [field]."[114] According to Haeckel, Müller failed to perceive the secret linking all life-forms, but his comprehensive vision kept them in order. In the eight years since his death, that order had begun to erode. By depicting Müller as a fallen emperor, Haeckel presented himself as the vigorous leader who would restore order with his monistic, evolutionary morphology.[115]

In his successful popular work *The History of Creation* (1868), Haeckel assigned Müller only a minor role. In this book and in his next one, *The Evolution of Man* (1874), he called Müller a zoologist, identifying him only by that aspect of his work closest to his own.[116] In this first passionate popularization of Darwin's theory, Haeckel structures the story of comparative anatomy in terms of heroes and villains, so that Müller, being neither, is marginalized. Goethe and Lamarck function as Darwin's unappreciated precursors; Cuvier (Müller's role model), as the "formidable opponent" of evolutionary theory. Like Cuvier, Müller did not believe that species had developed over time, but he died before being forced to embrace or reject Darwin's theory. Because of the way Haeckel tells the story, there is little to praise in Müller's work. Haeckel gives him credit, however, for his "attempt to establish a connected anatomical developmental series."[117]

Lay readers loved Haeckel's writing, and in Germany, more people learned about Darwin's evolutionary theory from Haeckel than from any other author. His greatest popular success, *The Riddle of the Universe* (1899), sold 400,000 copies and was translated into more than thirty languages, prompting Vladimir Ilych Lenin (1870–1924) to speculate that popular science books might become a weapon in the class struggle.[118] In this best-selling work, which attacked organized religion and explained how human beings were related to other animals, Müller played a much more prominent part. This time Haeckel presented his teacher as a physiologist, and in his third chapter, "Our Life," included a four-page elegy to Müller, calling him the "creator" of "'comparative physiology,' which embraces the whole sphere of life phenomena."[119] Here Haeckel depicted Müller as an initial splash from whom ripples spread, presenting later scientists (Schwann, Remak, Kölliker) as Müller's students:

> Nearly every one of the great biologists who have taught and worked in Germany for the last sixty years was, directly or indirectly, a pupil of Johannes Müller. . . . Müller was originally a vitalist, like all the physiologists of his time. Nevertheless, the current idea of a vital force took a novel form in his speculations, and gradually transformed itself into the very opposite. For he attempted to explain the phenomena of life mechanically in every department of physiology. His "transfigured" vital force was not above the physical and chemical laws of the rest of nature, but entirely bound up with them.[120]

Here, for the first time in his works, Haeckel presents Müller as a rigorous experimentalist who reformed anatomy and physiology. "This distinguished biologist," he wrote, "having a comprehensive knowledge of the whole field of organic nature, of morphology, and of physiology, introduced the 'exact methods' of observation and experiment into the whole province of physiology, and, with consummate skill, combined them with the comparative methods."[121] These portraits differ radically from Haeckel's depictions of the 1860s and 1870s, in which Müller appears as a gifted anatomist of the old, vitalistic school. In the intervening decades, Haeckel had developed a need for a new kind of precursor, and his portrait of Müller changed accordingly.

Once again, the driving force behind Haeckel's narrative was his need to perceive his own work as the continuation of a legendary scientist's. In *The Riddle of the Universe*, he was arguing that the scientifically determined laws of physics, not the revealed laws of God, sufficed to explain how the universe worked. No comparative anatomist could function as his predecessor, so he played up Müller's physiological achievements. In Haeckel's references to Müller in *The Riddle of the Universe*, one can see the striking personality differences between Haeckel and du Bois-Reymond, who had also praised Müller's experimental physiology. Haeckel explains away Müller's vitalism because he himself is not a vitalist, and his scientific identity depends on demonstrating continuity with his mentor. Du Bois-Reymond attacks Müller's vitalism because his scientific identity depends upon showing their differences.

Late in his career, Haeckel's international reputation as a comparative anatomist and embryologist led him to prepare a short (twenty-three-page) autobiographical sketch.[122] In this narrative, written perhaps five or six decades after Haeckel and Müller's trip to Helgoland, Haeckel has altered the details. By this time, Haeckel regarded himself as an expert analyst of life-forms, and Müller the anatomist again became a suitable precursor. Writing about himself in the third person, Haeckel explains:

> His eager private studies at the Berlin Comparative Anatomical Museum soon brought him into closer personal contact with Johannes Müller, and during the next fall vacation, the great teacher allowed [*gestattete*] the enthusiastic pupil to accompany him on an excursion to Helgoland and occupy himself each day for a whole month [*einen Monat hindurch*] with catching and observing lower marine animals and their immature forms.[123]

On Helgoland, the student and teacher had worked together for two weeks at the most, and if Müller had "allowed Haeckel to accompany him," why was Haeckel so surprised to see him when he arrived? Haeckel's references to his own enthusiasm and Müller's response to it ring true, but in order to emphasize Müller's role as his mentor, Haeckel's memory more than doubled the time of their interaction.

Haeckel's choice of words in his autobiographical sketch reveals the central importance that he assigns Müller in his life narrative. Haeckel calls Müller's death *jäh* (sudden), a strong, violent word suggesting a stroke of fate or an unexpected precipice after a stretch of level ground:

> Suddenly, in April 1858, the highly celebrated master was torn from the scientific world through his sudden [*jähen*] death. Disconsolate over this loss, Haeckel found himself among the six closest students who with burning tears bore the coffin of their most deeply loved and honored teacher upon their own shoulders.[124]

Two sentences later, Haeckel tells readers how his engagement to his cousin, Anna Sethe, helped him to recover from this pain. Interestingly, he uses the same word, *jäh*, to describe his young wife's death from an abdominal infection in 1864. Certainly both deaths were unexpected, and then, as now, *jäh* was a common term for a sudden loss. When he applies the same term to both situations, however, he links his two greatest emotional losses as major turning points in his life: 1858, the year in which he lost his mentor and was forced to become an independent scientist (except for Gegenbaur's help), and 1864, the year in which he lost his wife and his old religion and philosophy. Haeckel uses the same word for each because in each case, he felt the same emotions.

In Haeckel's last public lectures, he created his most vivid, emotional portraits of Müller. Written fifty years or more after his and Müller's trip to Helgoland, they show the relationship he would have liked to have with all of his heroes. Addressing a general audience at the Berlin Singakademie, Haeckel confided:

> In the summer of 1854, I had several remarkable conversations with Johannes Müller, whom I rank highest among all my famous teachers. His lectures on comparative anatomy and physiology—the most ingenious and stimulating that I have ever heard—had so captivated me that I asked and obtained his permission to study and sketch the skeletons and other preparations in his wonderful Comparative Anatomy Museum. . . . Müller (then fifty-four years old) had a habit of

> spending every Sunday afternoon alone in the museum. There he would walk up and down for hours in the spacious rooms, his hands folded behind his back, busy with thoughts about the secret relationships among vertebrates, whose "holy mystery [*Rätsel*]" was preached in such a penetrating way by the aligned skeletons. But every so often my great master would turn his attention sideways to the little table on which I (as a twenty-year-old student) was sitting in a window nook, conscientiously drawing the skulls of mammals, reptiles, amphibians, and fish. Then I was permitted to ask for explanations of especially difficult anatomical relationships, and once I dared to ask him shyly: "Couldn't all these vertebrates, whose inner skeletal construction is the same despite all the outer differences, originally have descended from a common ancestral form?" Slowly the great master shook his thought-filled head and said, "Yes, if only we knew! If you could ever solve this riddle [*Rätsel*], you would reach the highest heights!" A couple of months later, in September 1854, I was permitted to accompany Müller to Helgoland and learned from him the magnificent wonders of the sea world. As we were fishing together in the boat and catching some beautiful medusae, I asked how their remarkable alternation of generations could be explained. Couldn't the medusae, from whose eggs polyps still develop today, also originally have arisen from the more simply organized polyps? To this forward question I again heard the resigned answer, "Yes, there we face great mysteries [*laute Rätseln*]. About the origin of species [*Ursprung der Arten*] we know almost nothing![125]

By re-creating these "remarkable conversations," Haeckel portrays himself as the scientist who solved the "riddles" Müller could not. Twice he puts the title of his best-seller (*Die Welträtsel*, or *The Riddle of the Universe*) in Müller's mouth, and their questions and answers are expressed in evolutionary terms (*Ursprung der Arten*) that the two would probably not have used in 1854. In the next paragraph, Haeckel declares that if Müller had lived to read *The Origin of Species*, he "didn't doubt in the least that this surprising solution to the dark mystery of creation [*Schöpfungsrätsel*] would have moved Müller deeply and would have driven him, after careful deliberation, to endorse it fully."[126] It is extremely unlikely that a fifty-eight-year-old Müller would have embraced Darwin's ideas, since few comparative anatomists of his generation did.[127] But as a character in Haeckel's story, he remains under the writer's control. Haeckel makes him the previous step in a progressive array of scientists not unlike the skeletons in Müller's museum, and his edited memory of Müller advertises his own work.

When Haeckel praised Müller's exact methods in *The Riddle of the Universe*, he contrasted Müller's "comprehensive point of view" with "the one-sided and narrow methods of those modern physiologists who think to discover the nature of the vital phenomena by the exclusive aid of chemical and physical experiments."[128] The quintessential "modern physiologist" was Emil du Bois-Reymond, with whom Haeckel had never been on good terms, since neither respected the other's way of building scientific knowledge. Haeckel particularly objected to du Bois-Reymond's 1872 lecture "On the Limits of Our Knowledge of Nature." In this speech at the annual meeting of German scientists and physicians, du Bois-Reymond had identified seven problems that would probably never be solved by science, among them the origins of life and human consciousness.[129] Du Bois-Reymond's point had been that science could create reliable, useful knowledge only if investigators focused on soluble problems and avoided speculation. Haeckel, who wanted to combine science with philosophy, had protested as early as 1874, and by 1899, even though du Bois-Reymond had been dead for three years, his criticism remained harsh.[130] In *The Riddle of the Universe*, Haeckel rarely called du Bois-Reymond a scientist. Instead, with sarcasm worthy of his opponent, he praised him as "the famous orator" and "the distinguished orator of the Berlin Academy."[131] The conflict between Haeckel and du Bois-Reymond suggests the complexity of Müller's science. Both students admired their teacher, but they respected different aspects of his work and took it in very different directions.

As a grant applicant, Haeckel had reasons to resent du Bois-Reymond that were far more practical than epistemological. In April 1881, he asked the Academy of Sciences for 12,000 marks for a research trip to Ceylon to study lower marine animals. The Academy turned him down. Nathanael Pringsheim (1823–1894) responded to the application, writing, "Haeckel's general scientific position and the dogmatic-speculative direction of his thought, which, as is well known, he expresses in such a pronounced way in the treatment of general biological problems, do not permit the Academy to support his efforts."[132] Although du Bois-Reymond did not write the report, Pringsheim expresses his position exactly, and as a secretary of the Academy, the senior physiologist may have persuaded many to deny the grant. In early May 1881, du Bois-Reymond wrote to Helmholtz (who had apparently voted in favor):

> Since I can understand your position in this Haeckel business only on the assumption that you know his writings only from hearsay, allow me to offer you, for your information, the enclosed exercise in idiocy [Haeckel's *The Perigenesis of the Plastidules*]. This is no youthful work, but is only five years old. This has nothing to do with personalities; you should simply judge for yourself whether a person who writes such stuff is worthy of the Academy's support.[133]

Haeckel made the trip anyway, perhaps with his own funds, but he used his power as a writer to take revenge. In *The Riddle of the Universe*, he classed du Bois Reymond with Virchow, Wilhelm Wundt (1832–1920), Karl Ernst von Baer, and Immanuel Kant as thinkers who had undergone an "entire change of philosophical principles" as they aged. He attributed the change to "a gradual decay of the brain."[134]

Because Müller died at a younger age, he never underwent this decay. Haeckel's accusation that du Bois-Reymond and Virchow saw the body and mind as separate is scientifically unjustified; their research demonstrated the interdependence of matter and mind much more definitively than Haeckel's ever did.

Haeckel's resentment of du Bois-Reymond is better explained as the resistance of a slightly rebellious student to one who had rebelled more entirely. Haeckel rejected Müller's notion of life force, but he shared his belief that close observation of animals' structures would yield the secret of life's organization. Du Bois-Reymond thought the origin of life was unknowable, but that physiological experiments could yield valuable information about how living things worked. Sixteen years older than Haeckel, a successor to Müller and a leader of the Academy of Sciences, du Bois-Reymond represented the establishment. The senior physiologist had perceived the call for rigorous experimentalism as liberating, but Haeckel thought it denied his power of expression.

Orderly Narratives: Müller's Impact on Haeckel's Science

In Haeckel's early writing, his favorite word is "order." Over the course of his career, his vision of life changed considerably, but he always shared Müller's passion for comparing, sorting, and arranging. Again and again, in a vast number of different contexts, he upheld *etwas Ordentliches* as his ideal and a lack of *Ordnung* as a serious defect.[135] He believed that August Schenk (1815–1891), his botany professor during his first semester at Würzburg, was an excellent researcher but had remained unknown because he was "too lazy to write something orderly, to put together a great work."[136] To Haeckel, an "orderly" book meant a comprehensive, systematic study, which he regarded as the greatest scientific achievement. In his fantasies about traveling to the tropics with his microscope, the goal was "to achieve something orderly."[137] During his trip to Helgoland, despite his excitement about fishing with Müller, it bothered him that he had to gather material as fast as he could without taking time to make sense of it. After reporting that Müller was leaving soon, Haeckel wrote that he had decided not to go to Zurich, mainly "because I want to put the treasures I've collected here in order as soon as possible."[138] Haeckel's main objection to the teaching in his clinically oriented courses was their lack of "systematic rules and order." In Virchow's classes, he took notes as fast as he could, then copied them out and made drawings at night, trying "to establish some order, to digest and learn the material."[139] While Haeckel made these references to "order" as a very young student, he retained his belief that the best scientific work was systematic and comprehensive. His quest for order was the heart of his science, just as it was the driving force of his writing. In this respect, Haeckel's science resembled Müller's, even though the organizing principle was evolution rather than life force.

In Virchow's lectures, Haeckel found it hard to follow the "red thread."[140] Like Müller, he used Goethe's—more particularly, his character Ottilie's—metaphor for continuity of meaning. With all the facts and observations, it was hard for the young Haeckel to determine the main argument Virchow wanted to make about disease. In *The Evolution of Man*, Haeckel offered readers a "red thread" of his own design: his "biogenetic law" proclaiming that each developing organism recapitulated the stages of its ancestral development. Anticipating his description of Müller in the museum, Haeckel told readers that his biogenetic law was "Ariadne's thread," the only way "for understanding to find its path through this entangled labyrinth of forms."[141] Any good science, he believed, needed a red thread, just as good writing did. Darwin's idea that similar-looking animals had descended from common ancestors offered comparative anatomists the chance to organize their museums as narratives, and in his later works, this is what Haeckel tried to do.

While Haeckel admired Kölliker's and Virchow's perceptiveness, he preferred Müller's science to theirs, for Müller combined science with philosophy in ways that the younger anatomist and pathologist refused to do.[142] When Kölliker described the scientific viewpoint in the introduction to his physiology course, Haeckel had some misgivings:

> The scientific investigator must proceed purely empirically and critically; he may employ only objective research, observations, and experiments and at most draw up and derive general laws from the results he has found. He must never turn teleological, idealistic or dynamic, or, in a word, philosophical [*naturphilosophisch*]. Although I really must respect this realistic, empirical method of research, with its absolute objectivity, as the right one, I just don't like it. A general, philosophical [*naturphilosophische*] perspective and overview of the whole through investigation of the particular has a special appeal to me, and it's something I need.[143]

For Haeckel, science needed philosophy as much as it needed experiments. A hard-thinking scientist required theory to make comparisons and draw connections.

For the same reasons, Haeckel found Virchow's science less appealing than Müller's. The frustrated student wrote to his parents that "Virchow is a man of reason through and through, a rationalist and a materialist. He regards life as the sum of the functions of the different individual, material, chemical and anatomical organs."[144] Although Haeckel won Virchow's respect as a scientific observer, Haeckel never understood science as objective. In his 1878 reply to Virchow's claim that high school students should learn only established facts in their science courses, Haeckel asked, "Where do we find a border between objective and subjective knowledge? *Is* there such a thing as objective science?"[145]

Although all of Haeckel's drawings and classifications are based on the most careful observations, fact-finding was not the part of science he most valued. Haeckel got most excited when he saw something that made all of his observations fall into a pattern. In the summer of 1856, he observed a cyst in the plexus chorioideus of a mental patient that suddenly offered "the key," clarifying everything he had seen in cysts in this area. "All the observations that have been so puzzling [*rätselhaft*] up until now became clear once and for all," he wrote.[146] As an established scientist, Haeckel continued to seek laws and formulas that would cause everything to fall into place. In *General Morphology*, he deplored the lack of "cohesiveness [*Zusammenhang*] and unity" in anatomy and embryology, expressing his desire for comprehensive knowledge in architectural terms. He called for "a frame [*Gerüst*] erected on a solid foundation, according to a determined plan, a framework [*Fachwerk*] of beams which instead of enclosed walls and habitable rooms offers continuous space [*Zimmerwerk*] and empty rooms."[147] Haeckel's vision of the best epistemological space seems based on his fond memories of the "spacious rooms" in Müller's museum.

The life-changing insight that allowed Haeckel to reorder all of his knowledge was Darwin's theory that modern species had evolved from ancestral ones. From the time that Haeckel encountered this idea in 1860, he referred to it throughout his works as the "solution" to a "puzzle." "Here, all at once, in the simplest way, the great problem of the origin of species, of the kinship of organic forms was solved," he proclaimed in his autobiographical sketch. Darwin's theory also offered "the solution of the most difficult philosophical problems," permitting a new "unified perspective."[148] By 1899, when Haeckel published *The Riddle of the Universe*, he described evolution as an alchemical philosopher's stone. "All the partial questions of creation are indivisibly connected," he asserted, and "they represent one single, comprehensive 'cosmic problem,' and . . . the key to this problem is found in the one magic word—evolution."[149] Haeckel organizes his narratives—both the story of his own life and his story of how life developed—the same way that he organizes his science: by showing how sudden realizations permit the creation of an all-encompassing structure.

Haeckel's science therefore has much more in common with Müller's than with Virchow's or du Bois-Reymond's. Müller's older students had good reason to suspect that his youngest was bringing back nature philosophy. In *The Riddle of the Universe*, Haeckel wrote that "the whole drama of nature apparently consists in an alternation of movement and repose," a statement strikingly similar to Müller's 1822 claim that "flexion and extension are the two poles and marks of life in motion."[150] In Haeckel's 1904 lecture on biology in Jena, he defended the work of the nature philosopher Lorenz Oken, who had published Müller's first article in his journal, *Isis*:

> This nature philosophy later acquired the worst possible reputation because many of its fundamental ideas, which were correct in and of themselves, were exaggerated and distorted through many fantastic representations. But it would be wrong simply to reject it because of this. For the best that biology has ever achieved in the knowledge of general relationships and laws has been philosophical feats [*Taten*], the products of thought, not thoughtless observation.[151]

The distinguishing feature of nature philosophy was its quest for general laws, which dominated Haeckel's work as well. He opened *The Evolution of Man* with Goethe's claim that "the higher up you go, the more generalized your view becomes."

Haeckel's narratives of the history of science share his science's quest for "generalized views." To tell the story of biology's development, he relies on several sudden, sharp breaks, precipices as *jäh* as Müller's death. Haeckel introduces these in his later works through his near mystical use of dates, emphasizing a few crucial years as the end of one phase and the beginning of another. He stresses, for instance, that Darwin was born in 1809, the year in which Lamarck published *Zoological Philosophy*, and that Darwin reached Cambridge in 1828, the year that the embryologist Karl Ernst von Baer published *The Evolution of Animals.*[152] But the year that Haeckel endows with the greatest meaning is 1858, that of Müller's death. "By a very remarkable coincidence," Haeckel writes:

> Johannes Müller died in the year 1858, which saw the publication of Darwin's first communication concerning his famous theory. The theory of selection solved the great problem that had mastered Müller—the question of the origin of orderly arrangements from purely mechanical causes.[153]

Rather than emphasizing continuity of thought, Haeckel tells his story in terms of a dramatic revelation that changed all scientists' ways of thinking. The old school's finest representative had to die in order for the new school to be born.

When describing his own life, Haeckel depends just as heavily on narrative breaks. His belief in God must have waned during his early scientific years, but according to his autobiographical sketch, it disintegrated suddenly in 1864, the year of his wife's death. As a young student, Haeckel found that his scientific studies reinforced his religious faith. Disturbed by Karl Vogt's (1817–1895) materialism in 1853, he wrote to his parents, "I can't comprehend how the same people who work with these magnificent wonders and study their individual characteristics can doubt and simply deny the wise, creative power of God."[154] By 1868, however, he was writing that "where faith commences, science ends."[155] Over the course of fifteen years, his anatomical studies and his reading of Darwin and Vogt must have persuaded him that material causes sufficed to explain the diverse forms of animals, but that is not the way that Haeckel tells the story. According to him, the terrible "stroke of fate" in 1864 "completed his total break with church beliefs and drove him into the arms of the most radical realistic philosophy."[156] Müller's death had driven him into Anna Sethe's arms; her death drove him into Vogt's.

The dramatic reconstructions that made Haeckel's stories best-sellers did not satisfy all of his fellow scientists. In 1868, the Swiss anatomist Ludwig Rütimeyer (1825–1895) accused Haeckel of "arbitrarily modeling and generalizing" some of the drawings in *The History of Creation* so that at particular stages, the embryos of different animals looked as similar as possible. Rütimeyer even claimed that Haeckel had used the same image for the dog, the chicken, and the turtle.[157] Haeckel's widely circulated drawings showing that ontogeny recapitulates phylogeny have now become infamous, today called "one of the most famous fakes in biology."[158] But given Haeckel's passionate love of nature and his reverence for comparative anatomy, it is extremely unlikely that he ever consciously falsified his drawings of animals' development. It is much more probable that, viewing evolution as a magic formula for ordering life's diverse forms, he saw concordances and drew what he saw.

In 1876, when the zoologist Carl Semper (1832–1893) accused Haeckel of falsifying the drawings in *The Evolution of Man*, Haeckel replied in the foreword to the third edition:

> If one claims that my schematic illustrations are "invented" and uses that as evidence to accuse me of "falsifying science," then the same must be true of all other diagrams used in teaching thousands of times each day. All schematic illustrations are "invented!"[159]

Haeckel recognized that his drawings presented development from a certain perspective, but he challenged his critics to show him illustrations that did not serve a particular theory. His science, like his history of science, follows a "red thread" to which he actively calls attention.

As histories, Haeckel's narratives were not so much orderings as reorderings. Each time his scientific beliefs shifted, he altered his representation of Müller so that his master prophesied his current ideas. While Haeckel's portrait of Müller is one of the most self-serving, Haeckel is also aware that there can be no objective depiction of his hero. He therefore emphasizes his relation to his teacher, highlighting his qualifications as his chronicler. Haeckel's relationship to Müller may have been close, in terms of exchanging scientific confidences, but it was also very brief. In retrospect, Haeckel exaggerated the degree of their intimacy and Müller's impact on his science. The Berlin anatomist did influence Haeckel's work by teaching him how to skim the waves for plankton and by encouraging his already active tendency to compare animals' forms. He also reinforced Haeckel's belief that system-building and central guiding principles were crucial to science. But Haeckel's depiction of himself as one of Müller's closest students fulfilled his own need to define himself scientifically. The loving, open-hearted adviser Müller was not one whom he experienced, but one whom he created.

Afterword

Remembering and Dismembering a Scientist

On 28 April 1858, Johannes Müller lost the power to tell his own story. From the moment of his death, it fell into the hands of other narrators, each with his own agenda and sense of order.

On the day of Müller's death, du Bois-Reymond wrote to Helmholtz, "This morning at seven o'clock, Johannes Müller was found dead in his bed, felled by a stroke [*vom Schlage getroffen*]. He had been suffering for a long time from dizziness, exhaustion, and anxiety. But no one had imagined such an end. At five o'clock he still felt fine." Helmholtz answered, "[J. Müller's death] has surprised and shocked me, as it has everyone who knew him."[1] In these first responses, one hears mainly amazement. At fifty-six, Müller was far from well, but no one had expected him to die.

By the time du Bois-Reymond published his memorial address for Müller, he had already altered the story. In the final section of his biography ("The End"), he described Müller's failing health from 1856 onward so that his teacher's death seemed almost expected. Müller had suffered from insomnia for many years, he wrote, and had been fighting it with "large doses of that treacherous narcotic," opium. He had also experienced heart palpitations and pains near his liver. In the winter of 1856–1857, Müller had come down with a fever so bad that he had had to cancel his classes for the first time since 1827. Fearing that he would die, he had put his affairs in order, summoned his son from Cologne, and prohibited an autopsy in case of his death, but by spring he was back in the museum. During the winter of 1857–1858, for the first time ever, Müller had begun complaining that he had too much work. His insomnia had grown worse, and he had suffered from dizzy spells so frequent that he no longer dared to climb his library ladders.

The last known portrait of Johannes Müller, a photograph by S. Friedlander taken in the summer of 1857, after he recovered from his severe illness. Source: Wilhelm Haberling, *Johannes Müller: Das Leben des rheinischen Naturforschers* (Leipzig: Akademische Verlagsgesellschaft, 1924), plate IX, 449.

According to du Bois-Reymond, "dark omens came over him." A few days before his death, he again sent for his son—a doctor—and made an appointment with his own physician, Dr. Böhm. But on the morning of 28 April, "he was found dead in his bed, just two hours after speaking with Nanny, cheerful and apparently well." Du Bois-Reymond attributed his death to "the rupture of a great vessel."[2]

When Virchow gave his memorial address on 24 July, he offered a condensed, dramatic version very close to du Bois-Reymond's account. According to Virchow:

> [Müller's] health began to suffer, his spirits grew changeable and moody, his irritability increased, he complained of pains in his head and sleepless nights. The foreboding of death came over him. He put all his affairs in order, public and private, he telegraphed for his son to come from Cologne, he made an appointment for the following day to discuss his health, and—when the morning (28 April) came, his wife found a corpse.[3]

Though less detailed, Virchow's story follows du Bois-Reymond's so closely that it may have been based more on his colleague's speech than on any firsthand knowledge. But then, du Bois-Reymond carefully read Virchow's memorial address while expanding his own into a publishable biography, so it is impossible to say who copied from whom. The two narratives shaped one another, just as they have been shaping readers' understandings of Müller ever since.

But where did du Bois-Reymond get his information between 28 April and 8 July? Who knew the body and habits of the physiological specimen Müller? An unpublished letter from the physicist Gustav Magnus to Jakob Henle offers a slightly different version of his colleague's death:

> Early last Wednesday Joh. Müller was found dead in his bed. For some time Müller had been noticeably quiet, even more withdrawn than usual. All winter long he had declined every invitation and avoided everyone. I confess that when I heard the news yesterday, I let this slip, suggesting [*mich plaudern liess*] that his death was no natural one. But this is not true; he had a stroke [*er ist vom Schlage getroffen*], and indeed quite an obvious auditory stroke [*Gehörschlage*]. At least that's how his doctor Böhm explained it. . . . Joh. Schulze, to whom I spoke yesterday, has . . . seen the body. He said that [Nanny] was in [Müller's] room at five that morning and had spoken with him. It seemed to her as though he wanted to sleep a little longer. When she came back at seven, he was dead.[4]

Apparently, the best sources of information were Nanny, the last person to see Müller alive, and his physician, who diagnosed a stroke. In his letter to Helmholtz, du Bois-Reymond used Böhm's—or Magnus's—exact words, opting for the more graphic "rupture of a great vessel" in his published memorial address. What strikes one in Magnus's account—totally absent from the students' speeches—is the rumor of an "unnatural" death. Magnus, the senior physicist in Berlin, knew Müller socially as well as academically. He had remained friends with Henle because of their common passion for music, and with du Bois-Reymond because of their commitment to physical experiments. Having noticed Müller's depression and withdrawal, Magnus felt guilty about suggesting suicide but acknowledged that it was a possibility. In noncommittal terms, he told Henle that he was more inclined to believe the doctor.

Unlike Magnus, young Ernst Haeckel had no compunction about spreading the suicide tale. In a letter to his family, he wrote:

> The rumor spread quickly (which we students closest to him regarded as true) that in a fit of despair, our deeply suffering master had put a rash end to his life with a dose of morphine, saving himself after a long period of illness. Like the other students closest to him, . . . I myself was convinced that under the circumstances, this sad outcome was the best one; the life of an invalid, without any mental work, would have been unbearable for this great, restless, creative man.[5]

By 1900, Haeckel had accepted this rumor as fact, writing to his friend Franziska von Altenhausen, "The great question of suicide . . . has occupied me since my youth—my great, esteemed master and teacher ended his nervous suffering with morphine; with bitter tears, I carried his coffin to the grave."[6]

In 1858, Haeckel could not have known whether Müller took his own life. He had fished with him on Helgoland and was planning to collaborate with him on his *Habilitation* project, but he was never close to Müller in the combined intellectual and emotional sense that Henle had been. Probably Haeckel was chosen as a pallbearer as much for his athletic build as for his fascination with Müller's work. Those organizing the funeral may have wanted him to represent a new generation of Müller pupils. His repeated use of the phrase "the students closest to [Müller]" suggests a competition and a determination to promote his own story over the official one. Since du Bois-Reymond and Virchow ignored the suicide rumor, they were excluded from this imagined group of "closest" students.

To be fair to Haeckel, Müller did confide in his youngest students in ways that he never did with his junior colleagues. In 1853, the medical student Albert Gunther wrote to his brother:

> Whatever he may be to others, so far as I am concerned, Müller is the nicest of men. He has never been known to give anyone so much of his time, and we talk for hours. As I speak frankly to him, so he does to me. . . . He told me lately that he was subject to fits of depression which he cannot shake off, and quite bowl him over. This embitters many hours of his life, and can render traveling unendurable. . . . Last week he said good-bye as he was about to leave for Sicily. He gave no one else any idea of where he was going. Anyone who asked him, he just said: Potsdam.[7]

Gunther's account, written at the time rather than in retrospect, supports Haeckel's claims that Müller was severely depressed and was emotionally closer to his young students than to his colleagues du Bois-Reymond and Virchow. Gunther and Haeckel, who adored Müller's comparative anatomy, would have been far less critical, making much more pleasant companions.

Yet even Gunther's description of Müller must be read with a wary eye. As a narrative, it bears a close resemblance to Bidder's, du Bois-Reymond's, and Haeckel's proud chronicles of how a gifted student won an eminent scientist's respect. In May 1853 Gunther wrote to his brother, "Müller put everything [in the museum] at my disposal . . . I was soon able to convince him that I could improve on many of the descriptions. So he gave instructions that whatever I wanted should be opened and given to me. He then conducted me to his laboratory and put a microscope at my disposal."[8] When Gunther left for Bonn a year later, Müller told him to offer Professor F. H. Troschel a copy of Gunther's article on Neckar fishes, along with a business card on which he had written, "Dr. Gunther, ichthyologist of consequence." In his diary, Gunther reported that Müller then told him, "in a voice charged with unexpected emotion, 'It is but seldom that we have a student it is a pleasure to teach, and for such we have every confidence in the future. *You* are one of those: God bless you!'"[9] Appearing in Gunther's diary and letters to his brother, these comments were recorded as a young scientist tried to make sense of his own life and work. While not intended for the public, his portrait of Müller had a narrative purpose: like Bidder, du Bois-Reymond, and Haeckel, Gunther was trying to identify his scientific strengths and was seeking affirmation of his own talent.

From the available accounts, it is not possible to determine what killed Müller. He had forbidden an autopsy, not just because he was a Roman Catholic but probably also because for twenty-five years, he had watched Berlin medical students dissect cadavers and would personally have known the man—very likely Virchow—wielding the knife. With so many of his systems failing, Müller might have died of heart or liver failure as well as a stroke, and he was a known user of opium. In the 1850s, laudanum (opium dissolved in alcohol) was widely consumed and readily available in most pharmacies. In his last years, Müller seems to have recommended it just as Freud would encourage cocaine use in the 1880s. In May 1853 du Bois-Reymond wrote to Helmholtz, "Müller thinks you should use opium for your colic and not *Marienbader.*"[10] There is no reason to doubt Nanny's report that Müller was talkative at 5 A.M. If he were using the drug to help him sleep, why would he be taking it in the early morning? And if he took an intentional overdose, why would he do so after an ordinary conversation with his wife, on a day when he had scheduled a doctor's appointment to learn how he could improve his health? No common denominator of truth emerges, no matter how one adjusts these fractions.

The students' versions of Müller's death yield no common truth because they were designed to promote particular perspectives. They were created to define the younger investigators' identities and to shape the future of science. As the new Berlin professors of physiology and pathology, du Bois-Reymond and Virchow needed to portray themselves as Müller's ideal successors, and they did so by adjusting their mentor to fit their own scientific ideals. Even if they had suspected suicide, they could not have mentioned it to their eminent audiences. To say that a senior professor had intentionally ended his life would have insulted the Berlin University and the government that funded their research. But Haeckel, who at twenty-four seemed better suited to carry a body than to give a public address, had nothing to lose by insisting that Müller had committed suicide. As a writer, he gained authority by presenting himself as an insider who knew the real truth. Müller's last student created his own role by portraying a lonely, suffering master whose work he would continue in his future career. His characterization, which contradicted those of the senior students, was driven by the same narrative needs.

These varying accounts of Müller's death epitomize the different ways in which the students depicted their teacher. Like the descriptions in their letters, the narratives show some common ground: Müller was depressed; he had been sick; he took opium; yet no one expected him to die. While they contradict each other on some points, they overlap considerably. If we were detectives reconstructing the "real" Müller, the obvious move would be to seek the truth in these common elements. The overlap in their accounts, however, may reflect their mutual influence and their borrowings from each other's tales. I would thus urge readers to consider everything, even if mentioned by only one writer.

In writing about his teacher, each student created a different Müller, emphasizing those parts of his science and character that resonated with his own. The pupils dismembered Müller in order to remember him, dissecting his complex science and personality and pointing out the structures they most valued—or hated—at the time. Müller's students portrayed him during different periods of his life, and their descriptions of him altered as they aged. Seeking truth by comparing their stories is a daunting task, since it demands profound knowledge of the writers. When seven young artists paint a model, a comparison of their works says as much about the painters as about the body represented. In weighing the students' accounts, readers should bear in mind the three issues I raised in the introduction: (1) the kind of relationship each pupil had with Müller, (2) the kind of science he practiced, and Müller's influence on it, and (3) the ways that relationship and influence might have affected the stories he told.

Jakob Henle, who worked with Müller from 1827 to 1840, is the only student with whom he had a close, emotional relationship. Henle respected his teacher and at first seems genuinely to have enjoyed their scientific intimacy. For Henle, Müller's "square" personality was at first a source of amusement, but later became oppressive. Unlike Müller's later students, Henle appears to have broken with his mentor not because of a fundamental difference in scientific values but because he thought that Müller was jealous and was crippling his career. As an anatomist, he shared Müller's fascination with living forms and was more of an observer than an experimenter. Henle knew that Müller had gotten him started in anatomy, and perhaps saved him from prison, but he believed that Müller was thwarting his efforts to teach microscopy.

Theodore Schwann collaborated with Müller from 1831 to 1839, drawn by the physiologist's innovative spinal nerve root studies. While the two were not intimate, their friendship lasted, and Schwann made considerable sacrifices to keep working with Müller in Berlin. As a scientist, he owed Müller for urging him to use microscopes and to conduct experiments. Müller disappointed Schwann, however, when he shifted his focus from experimental physiology to comparative anatomy. In his quiet way, Schwann seems to have rejected Müller's notion of life force as utterly as the more outspoken du Bois-Reymond. Twenty years after his collaboration with Müller, Schwann claimed that their scientific thinking had always gone in different directions.

Du Bois-Reymond, the wittiest writer of the seven, made himself the keeper of Müller's memory. The physiologist worked with Müller from 1840 to 1858, giving him nineteen years to study his mentor's scientific character. During that time, the two learned to respect each other, but du Bois-Reymond's consciousness of his dependency seems to have kept any true friendship from developing. Scientifically, Müller's vitalism and compulsive collecting exasperated him no end. The strategies of these two investigators could not have been more opposed: Müller accumulated, scrutinized, and arranged specimens to learn how a unique life force realized itself in the animal kingdom, whereas du Bois-Reymond designed experiments to show that the same forces acting in inorganic nature made animals' bodies work. Still, he knew that Müller had given him his lifelong project and was trying to boost his career. The university structure heightened the tension between them, since du Bois-Reymond could not become the Berlin physiology professor until Müller retired, and he needed to stay in the Prussian capital, where skilled machinists helped to design his instruments. In his memorial address, now the best-known account of Müller, he asserted his own scientific identity by showing that he would continue the experimental physiology that Müller had begun but had then abandoned.[11]

Helmholtz, who discussed his science with Müller only in the early 1840s, seems never to have known him intimately. As a physiologist, he looked to Müller as a role model in experimental design. His mentor may have influenced him most on a pragmatic level, however: shortening his military service, offering him temporary employment, and recommending him for his first academic job. Occasional comments in Helmholtz's letters show his lack of interest in Müller's animals, but since he soon had his own professorship, he regarded comparative anatomy as foreign but not insidious. For Helmholtz, it was simply a different field, not an expensive hobby that prevented his supervisor from buying the equipment he needed. When he wrote about Müller thirty-five years later, he emphasized his mentor's close observation and rigorous analysis.

Virchow studied in the same military medical school as Helmholtz, yet he saw a different side of Müller. Clinically oriented, he spent much less time with his teacher than the physiologically inclined students did. As the Charité prosector, Virchow experienced Müller as a rival who plundered the hospital's collection of pathological specimens. Scientifically, however, he admired Müller's microscopic studies of tissues and used his teacher's work on tumors as a point of departure. On at least some occasions, he sought Müller out, talking not just about diseased spleens but also about Prussian politics. Although Müller never shared Virchow's liberal views, he allowed the pathologist to combine his scientific and political careers by supporting him for the Berlin pathology chair. Like du Bois-Reymond's memorial address, Virchow's elegy for Müller emerged from a choppy sea of reverence, distrust, and gratitude.

Robert Remak, who worked with Müller longer than any of the other students considered here (1835–1858), delivered no memorial address, and his references to Müller in the personal letters known to us are respectful but minimal. Since his faith forbade him from sharing meals with Müller and his students, he probably didn't know his mentor as well as Henle did, maintaining a looser collaborative relationship. Müller shaped Remak's science by encouraging him to do microscopic studies and working hard to win him an academic appointment. In his scientific articles, Remak often cited Müller's microscopic observations, which he regarded as fundamental to neuroanatomy and embryology. It is useless to speculate what Remak might have said, had he been invited to talk about Müller, but if his personal letters are someday examined, his voice should be heeded as carefully as Virchow's and du Bois-Reymond's.

Ernst Haeckel, a full twenty-five years younger than Henle, knew Müller only in the 1850s. He belongs to a different generation than the other students considered here, but I have included Haeckel's voice for that very reason. Like du Bois-Reymond, Haeckel became a self-proclaimed guardian of Müller's memory, and he was the only student to have written that his teacher committed suicide. Haeckel seems to have had some intimate scientific conversations with Müller, but he exaggerated the extent of their sporadic interactions between 1852 and 1858.

Müller died before he could foster Haeckel's career, but he inspired Haeckel with his systematic approach to life. Even more than Henle, Haeckel loved Müller the comparative anatomist, making his viewpoint strikingly different from those of the other students. His frequent invocations of Müller make it clear how greatly one scientist's perception of another can change over fifty years. In Haeckel's case, the Müller he described ranged from a painstaking anatomist to an evolutionist manqué to an experimental physiologist who offered evidence for materialism.

If we superimpose these accounts, as some late nineteenth-century scientists superimposed photographs, a few common features emerge. Müller was a shoemaker's son. He worked too hard. He went out of his way to help bright students, pushing them to do experiments and use microscopes. He performed crucial physiological studies in his youth, but from the 1830s onward devoted himself increasingly to comparative anatomy, driven by his passion for the Berlin Anatomical Museum. Politically he was conservative, fearing poverty and social change. He died unexpectedly, opening the way for students who felt that he was holding them back.

What Müller's students seem to have missed was the degree to which he supported them even when he opposed their philosophical assumptions. He helped make du Bois-Reymond an Academy member and associate professor in the 1850s, when the physiologist was complaining piteously about his mentor's obsession with rococo plankton. He campaigned to bring Virchow back to Berlin after the pathologist had fought to destroy the Prussian monarchy, and he recommended Helmholtz and du Bois-Reymond—strong opponents of life force—for the Königsberg physiology job. It is discrepancies like these that make it valuable to hear Müller's voice with those of all seven students. Even the most perceptive learners can overlook crucial aspects of their teacher.

I therefore hope that readers seeking the truth about Müller's science will look not just at the overlap between these stories but also at each element of every perspective. Müller plugged his students' work in letters to the cultural minister but tried to keep Henle from teaching microscopy. He was silent and gloomy in the dissecting room and jovial while skimming the North Sea waves. He took the Charité's best pathological specimens but urged his rival collector to study diseased spleens. He believed in a force unique to living organisms but expended tremendous energy helping young scientists who regarded nature in fundamentally different ways. In novels, literary writers can be accused of inconsistency in their character development, but the same standard cannot be applied to scientific writers. Perhaps it should not be applied to novelists either. Human consciousness is notoriously inconsistent, and while the students' biases must be considered, they may all be right. They simply stressed different aspects of their teacher.

In exploring the bonds among Müller and his students, I have tried to shed some light on what motivates scientists and on how they get their ideas. Hopefully, these insights about eight gifted researchers will have value that transcends this case history. As my own experience attests, the driving forces of science are mostly internal. Of course, it is no small matter to define what is "internal" to science, but as a former scientist, I can say that when working in a lab, one decides what to do next mainly on the basis of what one is finding and on what other scientists are doing. The social milieu always affects laboratory life, however, in the form of funding and academic structures.

Certainly culture shapes science, especially through its responses to new technological developments. Müller's students were well aware of cultural influences. In 1851 du Bois-Reymond wrote to Ludwig, "How the age is saturated with certain ideas shows itself strangely in the fact that in several passages Helmholtz and I have met, even to the point of using the same expressions, without either one of us knowing about the other."[12] In his lecture on animal movement and in Helmholtz's talk on measuring small time intervals, both scientists had shown how muscle contractions could be monitored through electrical circuitry. Invoking culture to account for a scientific idea can be a useful strategy, but its success depends on how one defines culture. If one considers that both du Bois-Reymond and Helmholtz studied medicine in Berlin during 1839–1842 and took Müller's physiology course, it is not surprising that they developed some of the same ideas. But if one attributes their thoughts to some "spirit of the times," one falls back on a non-explanation. References to a zeitgeist explain nineteenth-century physiology about as well as magic explains gravity.

The case of Müller and his students illustrates two social forces that can affect scientific thinking: those of income and social class. For Müller, who worked frenetically in part because he feared poverty, and Virchow, who saw medicine as a path to social reform, keeping these material factors in mind makes good sense. Helmholtz might not have made the contributions to sensory physiology that he did if he had not been forced to enter science through medicine, on a military scholarship. Awareness of their social and economic positions did influence some of Müller's students' choices about what kinds of experiments to do. Remak might not have turned to electrotherapy—hurting his scientific reputation—if he could have supported himself as a professor and embryologist.

But like art, science is never fully determined by social conditions. If it were, all scientists from the same income brackets would be doing the same experiments. A more tempting explanation for new approaches to scientific problems is the force of personality. Gabriel Finkelstein has written that "the similarity and difference in [Müller's and du Bois-Reymond's] characters explain much of the similarity and difference in their science."[13] The problem here is the enormity of the term "character." Finkelstein's claim is a bold one, and he is probably right, but "character" encompasses constitution, upbringing, experiences, and personal style. The psychohistorian Peter Loewenberg calls character "one of the most useful of all psychodynamic clinical categories for the historian" and suggests that historical figures can be classified according to personality.[14] It is highly doubtful whether one can learn anything about Müller or his students by classifying them in twentieth-century psychological terms, but personality differences could explain why scientists such as Henle, Virchow, du Bois-Reymond, and Haeckel took such different approaches to living organisms, something for which cultural and social studies of science cannot account.

The complex relationships among Müller and his students support another explanation of why scientists perform the experiments and develop the ideas they do, one that complements studies of culture, society, and character. In a group of scientists sharing money, space, and equipment, *the interaction of personalities* is as crucial as the character of any one individual.[15] Members of scientific groups form their identities by distinguishing themselves from others. In particular, students often define themselves in contrast to the principal investigator. The experiments they conduct may develop in any number of directions, but initially their contributions are often responses to what they perceive as their adviser's errors. Their work is therefore driven not just by their own personalities but by those of the scientists they are reacting against.

In the case of Müller and his students, interactions did shape the course of science. Schwann realized that animals were composed of cells not during his long hours of studying tissues but during a conversation with Schleiden, who had claimed the same for plants. Du Bois-Reymond might never have undertaken electrophysiological experiments if Müller hadn't shown him Matteucci's findings. More interested in animal forms than in physics, Müller thought that animal electricity was worth studying but didn't want to do it himself. His cocky, persistent student seemed just the one for the job.

The cramped quarters in which Müller and his students worked made it impossible for them *not* to interact. Several shared work space and equipment, and the small rooms in which they investigated kept them from being lonely scientists. At the Berlin Anatomical Institute and Anatomical Museum, they did their dissections and microscopy in the same small rooms. Even when they worked in their own apartments, they often visited each other, demonstrating their findings, sharing ideas, and joking around. Neither they nor their universities could have afforded the space for them to work alone, and their letters indicate that even if they had had it, they wouldn't have wanted it. One trait common to the comparative anatomist Müller and his most hard-nosed experimental students was the need for scientific intimacy. In their personal letters, he and du Bois-Reymond expressed the same powerful longing for daily contact with scientific friends who shared their philosophical assumptions and dealt with the same everyday problems.

Müller and his students practiced a science of closeness, a science grounded in intimacies of two kinds. First, it involved decades-long relationships with particular animals (the sea urchin, the frog) and instruments (the microscope, the galvanometer), so that they could build factual knowledge only by knowing every quirk of their animals and apparatus. They could create scientific knowledge (*science*) only through intimate acquaintance (*connaissance*). Second, their work depended on the close relationships they had with each other. As their letters reveal, they helped find each other jobs and apartments; they traveled, ate, and occasionally slept together; they inspired, criticized, and praised each other's work; and sometimes they scolded each other for writing it up badly. Donna Haraway has proposed that the loving relationship underlies all intellectual work, and this thought can be applied to Müller's students in two senses.[16] Their science grew out of their close relationships to particular instruments, animals, and projects, and out of their complex bonds to each other and Müller.

Despite their closeness, it would be wrong to think of Müller's students as a scientific group in the modern sense. The students considered in this book worked together in twos and threes but never as a unified "lab." Schwann and Henle performed some of their research in the same places during 1833–1839, and du Bois-Reymond, Helmholtz, Virchow, and Remak all had some contact with Müller during the years 1839–1845. Haeckel worked with Müller much later (1852–1858), and knew Virchow and du Bois-Reymond only as established scientists. As Virchow pointed out, there was no Müller "school," except in the sense of methods, so that these students bore no common theoretical stamp.

Perhaps the most surprising finding of this study is how little Müller did to bring his students together. He, Henle, and Schwann worked in the same room as a scientific group might do today, but why did Müller not introduce du Bois-Reymond, Helmholtz, and Remak in 1841–1842, given their common interest in the nervous system? Virchow and Helmholtz were military medical students, not university students, so that their course requirements and clinical duties would have taken them in different directions from du Bois-Reymond. Remak was three years older than du Bois-Reymond and six years older than Helmholtz, and had begun studying medicine much earlier. Even if he had known these younger researchers, he would have had difficulty socializing with them, since he could eat only kosher food. The fact remains, however, that all of these students were communicating with Müller in 1841–1842, sharing their findings with him, and benefiting from his advice. It is puzzling, even disturbing, that he never called a joint meeting. As a mentor, he seems to have acted more as a prime mover than a creator of networks, trying to help each student but making no special effort to bring them together.

This reluctance to form a cooperative band of researchers becomes less surprising when one considers Müller's research strategies and his own development as a scientist. To the end of his life, he stuck to pelagic fishery. Rather than raising embryos in the lab and watching them develop, he skimmed the waves, building his science on the best plankton he could scoop up. As a teacher, he operated the same way, scanning the dissecting hall for talent and investing his energy in the students he "found" or who had the self-confidence to approach him on their own. Arleen Tuchman's claim that Müller's science was "an activity for the chosen few" relies heavily on Friedrich Bidder's self-promoting memoirs, but the other students' accounts confirm it.[17] Müller found his students; he did not raise them, and he did not believe that every medical student could be turned into a researcher. Since he himself had been fished out of a Koblenz high school by the Prussian reformer Johannes Schulze, Müller believed in a system based on government support of inborn talent. Müller's science was not elitist in the sense of religion or social class; he fostered the sons of Jewish shopkeepers, Catholic publishers, and Protestant diplomats with equal zeal. If the student was bright, good with his hands, and eager to learn how life worked, Müller could forgive him almost anything, including denials that life involved a unique force manifesting itself in a grand plan.

When one considers the diversity of Müller's students' ideas, his laissez-faire teaching looks more successful. It is just as well that he created no "school" of Müllerian researchers, preferring instead to find projects for scientific inclinations already present.[18] I opened this study by claiming that Müller never had a lab, and I will close it by claiming that he didn't want one. While he never stopped longing for research space, he had no wish to lead a unified group. Müller craved scientific intimacy, gladly sharing his thoughts with younger researchers, but he viewed science as a cooperation among friends doing related but independent projects. He did not see it as his job to tell others how to conduct their research or to introduce young scholars investigating related problems. Perhaps he feared what they might do if they combined forces.

Both Müller's own letters and those of his students testify to his insecurity, his growing fear that his approach to nature would stop yielding knowledge. Still, Freud's myth of oppressed sons joining forces to kill their father distorts rather than explains these scientists' relationships. Müller was never authoritarian, avoiding confrontations with Schwann and du Bois-Reymond even when their books blatantly contradicted his assumptions about nature. When his students felt restricted by his rule, they left to take new jobs, professorships he had helped them obtain. Most significantly, they never banded together. They complained about him but, scattered in hospital wards, cramped apartments, and distant cities, they never mutinied. His way of advising them ensured that they never had to. If there was any killing, it was done in retrospect, as they dismembered and remembered him through their writing.

Instead of a Freudian drama, what emerges from this study of scientists' relationships is the realization that one cannot separate life and work. In the 1970s, the French philosopher Jacques Derrida criticized people's tendency to think in pairs of opposites such as black/white and male/female, arguing that these culturally approved dichotomies promote prejudices by privileging one term over the other. I have found that even the most ardent dichotomy-busters cling to the division between personal life and work. We need it for survival, since we have to be able to criticize each other's thinking without resorting to ad hominem attacks. But aren't thoughts the most personal things of all?

The stories of Müller and his students show the degree to which science is driven by personal passion. The ideal of objective analysis underlies Western science, yet scientists' motivations for working sixty to eighty hours per week have their roots in individual emotions. However necessary the distinction may be for civilized debates, it is simply not true that work is one thing, and personal life is another. Life drives science just as—in some people—science drives life. The relationships among Müller and his students show to what degree the interplay of personalities can affect science. As creators of knowledge, we live our work, and we are what we do.

Abbreviations

ABBAW: Archiv der Berlin-Brandenburgischen Akademie der Wissenschaften

GSPK: Geheimes Staatsarchiv Preussischer Kulturbesitz

***Müllers Archiv*:** *Archiv für Anatomie und Physiologie und für wissenschaftliche Medizin*

HP: Johannes Müller, *Handbuch der Physiologie des Menschen*, 3rd ed., 2 vols. (Coblenz: J. Hölscher, 1838, 1840).

MKAWB: Monatsberichte der königlichen Akademie der Wissenschaften zu Berlin

NL: Nachlass

SBPK, HA, SD: Staatsbibliothek zu Berlin-Preussischer Kulturbesitz, Handschriften Abteilung, Sammlung Darmstaedter

UH: Universitätsarchiv Heidelberg

***Virchows Archiv*:** *Archiv für pathologische Anatomie und Physiologie und für klinische Medizin*

Notes

Introduction

1. L. J. Rather has written that "Schwann and Henle were hard at work in Müller's laboratory"; Timothy Lenoir, that "[things] were beginning to heat up right within the walls of Müller's own lab!" Both are seasoned historians of science and probably used the word "lab" only for convenience, but statements like these have encouraged readers to see Müller's "lab" as a unified space. See L. J. Rather, Patricia Rather, and John B. Frerichs, *Johannes Müller and the Nineteenth-Century Origins of Tumor Cell Theory* (Canton, MA: Science History Publications, 1986), 14; and Timothy Lenoir, *The Strategy of Life: Teleology and Mechanics in Nineteenth- Century German Biology* (Chicago: University of Chicago Press, 1989), 149.

2. Among the historians who have pointed out the students' biases in describing Müller are Gabriel Finkelstein, Frederic L. Holmes, Nicholas Jardine, and Lynn Nyhart. Finkelstein has written that "the true function of [du Bois-Reymond's] 'Memorial [Address for Müller]' . . . was to create a myth." He believes that in this public lecture, du Bois-Reymond "offered German physiology its first narrative . . . a history of physiology recounted in du Bois-Reymond's own image." Gabriel Ward Finkelstein, "Emil du Bois-Reymond: The Making of a Liberal German Scientist (1818–1851)" (Ph.D. diss., Princeton University, 1996), 173–74. Jardine argues that both du Bois-Reymond and Rudolf Virchow took on "the mantle and mystique of Müller by portraying him in their own image." He calls their memorial addresses for Müller "moves in a power struggle," proposing that both students were "out to advance their personal positions" and "legitimate their own programs." Nicholas Jardine, "The Mantle of Müller and the Ghost of Goethe: Interactions between the Sciences and Their Histories," in *History and the Disciplines: The Reclassification of Knowledge in Early Modern Europe*, ed. Donald R. Kelley (Rochester, NY: University of Rochester Press, 1997), 306. Nyhart observes that "du Bois-Reymond was instrumental in constructing this history [of physiology] and projecting it onto the earlier part of the century." Lynn K. Nyhart, *Biology*

Takes Form: Animal Morphology and the German Universities, 1800–1900 (Chicago: University of Chicago Press, 1995), 74, n. 24.

3. Johannes Müller, *Briefe von Johannes Müller an Anders Retzius von dem Jahre 1830 bis 1857* (Stockholm: Aftonbladets Aktiebolags Tryckeri, 1900), ii. My translation. Unless otherwise indicated, all translations in this book are my own.

4. The historian Frederic L. Holmes draws attention to this issue when he warns that Helmholtz's "retrospective account" of Müller in 1877 cannot be read as a "transparent memory of how Helmholtz had perceived the situation in the 1840s." Instead, it should be regarded as "a reconstruction that fitted an early episode in his own life into the trajectory of his prior and subsequent experiences." Frederic L. Holmes, "The Role of Johannes Müller in the Formation of Helmholtz's Physiological Career," in *Universalgenie Helmholtz: Rückblick nach 100 Jahren, ed.* Lorenz Krüger (Berlin: Akademie Verlag, 1994), 4.

5. To date, this self-serving representation of Müller has been best documented in the case of du Bois-Reymond. See Finkelstein, "Emil du Bois-Reymond," 173–174, and Jardine, "The Mantle of Müller," 306.

6. L. S. Jacyna, "Images of John Hunter in the Nineteenth Century," *History of Science* 21 (1983): 105.

7. I am grateful to the mathematicians Stanley Chang and Thomas Weissert for their advice on this mathematical metaphor.

8. On postmodern biographies, see Carole J. Lambert, "Postmodern Biography: Lively Hypotheses and Dead Certainties," *Biography* 18 (1995): 305–327. I thank Tom Couser and Craig Howes for their thoughts on the goals of postmodern biographies, on which I have relied in this paragraph.

9. Choderlos de Laclos, *Les Liaisons dangereuses* (Paris: Librairie Générale Française, 1987), 281.

10. Sigmund Freud, *Totem and Taboo: Some Points of Agreement between the Mental Lives of Savages and Neurotics*, in The Standard Edition of the Complete Psychological Works of Sigmund Freud, trans. James Strachey (1955; London: Hogarth Press, 1999), 13: 141–42.

11. Sigmund Freud, "Family Romances," in *Sigmund Freud: Collected Papers*, ed. James Strachey (New York: Basic Books, 1959), 5: 78.

12. Harold Bloom, *The Anxiety of Influence: A Theory of Poetry*, 2nd ed. (New York: Oxford University Press, 1997), 14–15.

13. Peter Loewenberg, *Decoding the Past: The Psychohistorical Approach* (New York: Alfred A. Knopf, 1983), 12.

14. Ibid.

15. Thomas A. Kohut, "Psychohistory as History," *American Historical Review* 91 (1986): 341.

16. Although psychohistorians claim that "the individual cannot be historically or psychologically 'known' without learning his or her sociocultural setting," many psychohistorical studies focus on the psychopathology of individuals, drawing attention away from the social situations they aim to elucidate. Presuming that psychological phenomena such as the Oedipus complex are universal can lead to greater distortions than ignorance of writers' emotional involvement with their material, an issue of which most historians have long been aware. Emotional bonds do drive scientists and historians, but they can be studied without reducing past relationships to twentieth-century formulations. See Peter Loewenberg, *Fantasy and Reality in History* (New York: Oxford University Press, 1995), 5; Kohut, "Psychohistory as History," 336; and Howard I. Kushner, "Taking Erikson's Identity Seriously: Psychoanalyzing the Psychohistorian," *The Psychohistory Review* 22 (1993–94): 8.

17. For a detailed study of how zoologists of the 1860s transformed the comparative anatomy forged by Müller's generation, see Nyhart, *Biology Takes Form.*

Chapter 1

1. Emil du Bois-Reymond, "Gedächtnisrede auf Johannes Müller," in his *Reden* (Leipzig: Veit, 1887), 2: 146; Rudolf Virchow, *Johannes Müller: Eine Gedächtnisrede* (Berlin: August Hirschwald, 1858), 10.
2. Max Lenz, *Geschichte der königlichen Friedrich-Wilhelms-Universität zu Berlin,* 2.1 (Halle: Buchhandlung des Waisenhauses, 1910), 456.
3. Johannes Steudel, *Le Physiologiste Johannes Müller* (Paris: Université de Paris, 1963), 9; Nelly Tsouyopoulos, "Schellings Naturphilosophie: Sünde oder Inspiration für den Reformer der Physiologie Johannes Müller?" in *Johannes Müller und die Philosophie*, ed. Michael Hagner and Bettina Wahrig-Schmidt (Berlin: Akademie Verlag, 1992), 67. In German, a *Handwerker* is a skilled craftsman such as a carpenter or plumber, not a factory worker, and *Handwerkerklasse* does not mean proletariat. I have translated Tsouyopoulos's phrase as "working class" because *Handwerker* in nineteenth-century Germany did not enjoy the same social respect as educated professionals.
4. Hajo Holborn, *A History of Modern Germany,* 2 (New York: Alfred A. Knopf, 1968), 371–372.
5. James J. Sheehan, *German History, 1770–1866* (Oxford: Clarendon Press, 1989), 292–294.
6. Holborn, *A History of Modern Germany*, 2: 382–385; Sheehan, *German History*, 296.
7. Holborn, *A History of Modern Germany*, 2: 393–395; Thomas Nipperdey, *Deutsche Geschichte 1800–1866: Bürgerwelt und starker Staat* (Munich: C. H. Beck, 1994), 34; Sheehan, *German History*, 296–301.
8. Holborn, *A History of Modern Germany*, 2: 406; Nipperdey, *Deutsche Geschichte*, 40–42; Sheehan, *German History*, 300.
9. Holborn, *A History of Modern Germany*, 2: 378; Nipperdey, *Deutsche Geschichte*, 50; Sheehan, *German History*, 305–306.
10. Holborn, *A History of Modern Germany*, 2: 416–420; Nipperdey, *Deutsche Geschichte*, 51–55; Sheehan, *German History*, 308–309.
11. Holborn, *A History of Modern Germany*, 2: 435–441, 455; Sheehan, *German History*, 393, 401–402.
12. Sheehan, *German History*, 403.
13. Holborn, *A History of Modern Germany*, 2: 469; Nipperdey, *Deutsche Geschichte*, 345–346.
14. Holborn, *A History of Modern Germany*, 2: 464–66; Sheehan, *German History*, 405–407.
15. "Burschenschaft," in *Brockhaus Enzyklopädie* (Mannheim: F. A. Brockhaus, 1987), 6: 228.
16. Sheehan, *German History*, 407.
17. Gottfried Koller, *Das Leben des Biologen Johannes Müller 1801–1858* (Stuttgart: Wissenschaftliche Verlagsgesellschaft, 1958), 26–27.
18. My discussion of German industrialization in this paragraph relies on Nipperdey, *Deutsche Geschichte*, 179–205.
19. Timothy Lenoir argues that the rapid development of German experimental physiology was closely related to industrialization. See Timothy Lenoir, *Instituting Science: The Cultural Production of Scientific Disciplines* (Stanford, CA: Stanford University Press, 1997), 79.
20. Sheehan, *German History*, 514.

21. Nipperdey, *Deutsche Geschichte*, 58.

22. Holborn, *A History of Modern Germany*, 2: 480–481.

23. Nipperdey, *Deutsche Geschichte*, 64–65; Sheehan, *German History*, 518.

24. Wilhelm Haberling, *Johannes Müller: Das Leben des rheinischen Naturforschers* (Leipzig: Akademische Verlagsgesellschaft, 1924), 14–15. See also Lenz, *Geschichte der königlichen Friedrich-Wilhelms-Universität*, 2.1: 456.

25. Gabriel Finkelstein, "Emil du Bois-Reymond: The Making of a Liberal German Scientist" (Ph.D. diss., Princeton University, 1996), 78. According to Finkelstein, Altenstein designed the Bonn University as "a showcase of Prussian national values," trying especially to make it "strong in the natural sciences."

26. Ulrich Ebbecke, *Johannes Müller: Der grosse rheinische Physiologe* (Hannover: Schmorl und von Seefeld, 1951), 29–30.

27. This question is the central issue of Michael Hagner and Bettina Wahrig-Schmidt's valuable anthology, *Johannes Müller und die Philosophie* (Berlin: Akademie Verlag, 1992). On nature philosophy at the Bonn University, see Finkelstein, "Emil du Bois-Reymond," 78–81.

28. Timothy Lenoir believes that du Bois-Reymond's and Helmholtz's popular lectures shaped our current, erroneous belief that modern science progressed when experimenters of the 1840s freed themselves from "the . . . utterly stifling influence of *Naturphilosophie*." Timothy Lenoir, *The Strategy of Life: Teleology and Mechanics in Nineteenth-Century German Biology* (Chicago: University of Chicago Press, 1989), 5.

29. Timothy Lenoir proposes that "a core of ideas" from Kant and other philosophers of the 1790s provided a "program for research" for German biologists from 1800 until at least 1850. Ibid., 2.

30. This paragraph summarizes Kant's argument in *Kritik der Urteilskraft*, ed. Heiner F. Klemme and Piero Giordanetti (Hamburg: Felix Meiner, 2001), 278–280.

31. Friedrich Wilhelm Joseph Schelling, *Einleitung zu seinem Entwurf eines Systems der Naturphilosophie*, in Schelling's *Historisch-Kritische Ausgabe, Reihe I: Werke*, ed. Manfred Durner and Wilhelm G. Jacobs, vol. 8 (Stuttgart: Frommann-Holzboog, 2004), 30.

32. Robert J. Richards, *The Romantic Conception of Life: Science and Philosophy in the Age of Goethe* (Chicago: University of Chicago Press, 2002), 128, 139.

33. Richards characterizes nature philosophy as a belief that nature contains "fundamental organic types" that can be arranged in a "progressive hierarchy." Ibid., 8.

34. Johann Wolfgang von Goethe, "Erster Entwurf einer allgemeinen Einleitung in die vergleichende Anatomie, ausgehend von der Osteologie," in *Goethe's Werke, Abteilung 2: Naturwissenschaftliche Schriften* (Weimar: Hermann Böhlau, 1893), 8: 10. On Goethe's morphology see Richards, *The Romantic Conception of Life*, 434–457; and George A. Wells, "Johann Wolfgang von Goethe," in *Dictionary of Scientific Biography*, 5: 442–446, both of whose analyses have guided my discussion in this paragraph.

35. Lorenz Oken, *Lehrbuch der Naturphilosophie*, 3rd ed. (Zurich: Friedrich Schulthess, 1843), 1–2.

36. Lorenz Oken, *Über das Universum als Fortsetzung des Sinnensystems: Ein pythagoräisches Fragment* (Jena: Friedrich Frommann, 1808).

37. Marc Klein, "Lorenz Oken," in *Dictionary of Scientific Biography*, 10: 195.

38. Readers shocked to see that Müller completed his M.D. and Ph.D. three years after starting his university studies should know that in the 1820s, this achievement was not unusual. No undergraduate degree was required, so that students could begin learning medicine as soon as they entered universities, and while a doctoral degree required independent research, a thesis consisted of ten to twenty pages, in Latin.

39. Johannes Müller, "Beobachtung über die Gesetze und Zahlenverhältnisse der Bewegung in den verschiedenen Thierclassen mit besonderer Rücksicht auf die Bewegung der Insekten und Polymerien," *Isis* 1 (1822): 61–62.

40. Edith Selow, "Friedrich Wilhelm Joseph Schelling," in *Dictionary of Scientific Biography*, 12: 155.

41. Haberling, *Johannes Müller*, 39–44; Johannes Steudel, "Johannes Peter Müller," in *Dictionary of Scientific Biography*, 9: 568.

42. Haberling, *Johannes Müller*, 54.

43. Karl E. Rothschuh, *History of Physiology*, ed. and trans. Guenter B. Risse (Huntington, NY: Robert E. Krieger, 1973), 197.

44. Johannes Müller, *Handbuch der Physiologie des Menschen* [hereafter HP], 3rd ed., 1 (Koblenz: J. Hölscher, 1838), 780, my translation and emphasis. William Baly translated Müller's *Handbook* into English, and it was published in London and Philadelphia as *Elements of Physiology* in 1842.

45. Ibid., 2 (1840): 411.

46. Rothschuh, *History of Physiology*, 199.

47. Haberling, *Johannes Müller*, 77–79.

48. Du Bois-Reymond, "Gedächtnisrede," 2: 166.

49. Walther's letter is reproduced in Haberling, *Johannes Müller*, 77–78.

50. In 1850, Helmholtz wrote to du Bois-Reymond that his colleague Kirchhoff in Königsberg was "full of indecision and hypochondria." In 1853, Carl Ludwig joked to du Bois-Reymond, "Love's sweet small talk . . . will clear up your hypochondriac horizon." Emil du Bois-Reymond and Hermann von Helmholtz, *Dokumente einer Freundschaft: Briefwechsel zwischen Hermann von Helmholtz and Emil du Bois-Reymond 1846–1894* (Berlin: Akademie Verlag, 1986), 96, letter of 22 April 1850; and Emil du Bois-Reymond and Carl Ludwig, *Two Great Scientists of the Nineteenth Century: Correspondence of Emil du Bois-Reymond and Carl Ludwig*, ed. Paul F. Cranefield, trans. Sabine Lichtner-Ayed (Baltimore: Johns Hopkins University Press, 1982), 80, letter of 30 October 1853.

51. Haberling, *Johannes Müller*, 84, 101.

52. Lenoir, *The Strategy of Life*, 103–111.

53. Koller, *Das Leben des Biologen Johannes Müller*, 70–71.

54. My discussion of Humboldt's career in this paragraph relies on Kurt R. Biermann, "Friedrich Wilhelm Heinrich Alexander von Humboldt," in *Dictionary of Scientific Biography*, 6: 549.

55. Müller called Magendie's experiments "the cruelest imaginable." Müller, HP, 1: 649–650.

56. Steudel, "Johannes Peter Müller," 570.

57. Johannes Müller, "Bestätigung des Bell'schen Lehrsatzes, dass die doppelten Wurzeln der Rückenmarksnerven verschiedene Functionen haben, durch neue und entscheidende Experimente," *Notizen aus dem Gebiete der Natur- und Heilkunde* 30 (1831): 115–116.

58. Koller, *Das Leben des Biologen Johannes Müller*, 215.

59. Gottwalt Christian Hirsch, "Die Forscherpersönlichkeit des Biologen Johannes Müller," *Sudhoffs Archiv* 26 (1933): 188.

60. Manfred Stürzbecher, "Aus dem Briefwechsel des Physiologen Johannes Müller mit dem Preussischen Kultusministerium," *Janus* 49 (1960): 280.

61. Joseph Ben-David and Awraham Zloczower have made this argument in a number of insightful essays. See Joseph Ben-David, "Scientific Productivity and Academic Organization in Nineteenth-Century Medicine," *American Sociological Review* 25 (1960): 839–40; Joseph Ben-David and Awraham Zloczower, "Universities and Academic Systems in Mod-

ern Societies," *European Journal of Sociology* 3 (1962): 51; Joseph Ben-David, *The Scientist's Role in Society* (Englewood Cliffs, NJ: Prentice-Hall, 1971), 123; and Awraham Zloczower, *Career Opportunities and the Growth of Scientific Discovery in Nineteenth-Century Germany* (New York: Arno Press, 1981), 1–2. I thank Lynn Nyhart for bringing Ben-David's and Zloczower's work to my attention.

62. Ben-David, "Scientific Productivity," 835; Ben-David and Zloczower, "Universities and Academic Systems," 51.

63. Zloczower, *Career Opportunities*, 30–33.

64. Ben-David and Zloczower, "Universities and Academic Systems," 51, 57; Zloczower, *Career Opportunities*, 1–4.

65. Zloczower, *Career Opportunities*, 36.

66. Ben-David, "Scientific Productivity," 842; Ben-David and Zloczower, "Universities and Academic Systems," 49.

67. Zloczower, *Career Opportunities*, 35.

68. The German titles of *ausserordentlicher Professor* and *ordentlicher Professor* are not equivalent to the American "associate professor" and "full professor," but I prefer these translations to the more confusing *Professor extraordinarius* and *Professor ordinarius*. Both are secure and permanent positions, distinct from untenured assistant professorships, but *ordentlicher Professor* designates a significantly higher rank than *ausserordentlicher*. The title *extraordinarius* gives the opposite impression.

69. Zloczower, *Career Opportunities*, 17–20.

70. Ibid., 57.

71. Manfred Stürzbecher, "Zur Berufung Johannes Müllers an die Berliner Universität," *Jahrbuch für die Geschichte Mittel- und Ostdeutschlands* 21 (1972): 193.

72. Ibid., 197.

73. Du Bois-Reymond, "Gedächtnisrede," 2: 185; William Clark, "On the Ministerial Archive of Academic Acts," *Science in Context* 9 (1996): 421–486.

74. Stürzbecher, "Zur Berufung Johannes Müller," 202.

75. Haberling, *Johannes Müller*, 150, letter of 7 January 1833.

76. Gabriel Finkelstein believes that hiring Müller was "probably the best decision in [Altenstein's] career as administrator." Finkelstein, "Emil du Bois-Reymond," 166.

77. Leopold Freiherr von Zedlitz, *Neustes Conversations–Handbuch für Berlin und Potsdam* (Berlin: A. W. Eisersdorff, 1834), 1. I thank Michael Hagner for bringing this source to my attention.

78. Henry Adams, *The Education of Henry Adams*, ed. Ira B. Nadel (New York: Oxford University Press, 1999), 69.

79. All information about Berlin's population in this paragraph is from Zedlitz, *Neustes Conversations-Handbuch* 164–166, 369.

80. Jean Eckenstein, *Der akademische Mentor für die Studirenden der Friedrich-Wilhelms-Universität zu Berlin* (Berlin: Wilhelm Schüppel, 1835), 396.

81. Zedlitz, *Neustes Conversations-Handbuch*, 166, 712.

82. Eckenstein, *Der akademische Mentor*, 394, 441–442.

83. Stürzbecher, "Zur Berufung Johannes Müller," 187, 213.

84. Eckenstein, *Der akademische Mentor*, 396.

85. Ibid., 397.

86. Ibid., 398, 346.

87. Ibid., 398.

88. Ibid., 398–399.

89. Ibid., 400–404.

90. Ibid., 362.
91. Ibid., 460.
92. Ibid., 267.
93. Ibid., 272–275.
94. Zedlitz, *Neustes Conversations-Handbuch*, 511, 547.
95. My discussion of the Berlin University in this paragraph draws upon Eckenstein, *Der akademische Mentor*, 86–87; Lenz, *Geschichte der königlichen Friedrich-Wilhelms-Universität*, 1: 290–291; and Zedlitz, *Neustes Conversations-Handbuch* 798–799.
96. Lenz, *Geschichte der königlichen Friedrich-Wilhelms-Universität*, 1: 300.
97. Eckenstein, *Der akademische Mentor*, 43–48, 69–70; Zedlitz, *Neustes Conversations-Handbuch* 797.
98. Zedlitz, *Neustes Conversations-Handbuch* 797–798.
99. Ben-David, *The Scientist's Role in Society*, 121–122; Zloczower, *Career Opportunities*, 20–23.
100. My outline of Müller's teaching responsibilities relies on du Bois-Reymond, "Gedächtnisrede," 2: 193–194. Since Berlin was the capital, Müller had to examine every candidate from every Prussian university. Unless the student showed real brilliance or stupidity, these exams could be deathly boring. Helmholtz wrote to his parents, "The examiners Müller and Gurlt sit there with their mouths hanging open, going gray with boredom." Hermann von Helmholtz, *Letters of Hermann von Helmholtz to His Parents: The Medical Education of a German Scientist 1837–1846*, ed. David Cahan (Stuttgart: Franz Steiner, 1993), 107, letter of 30 October 1845.
101. Lenz, *Geschichte der königlichen Friedrich-Wilhelms-Universität*, 3: 485–486.
102. Eckenstein, *Der akademische Mentor*, 120–121; Zedlitz, *Neustes Conversations-Handbuch*, 25–26. According to Eckenstein, the Anatomical Institute handled 200 cadavers a year; according to Zedlitz, 550.
103. Eckenstein, *Der akademische Mentor*, 121.
104. Friedrich Bidder, "Vor hundert Jahren im Laboratorium Johannes Müllers," *Münchener medizinische Wochenschrift* (12 January 1934): 62. This article is an extract from Friedrich Bidder, *Aus dem Leben eines Dorpater Universitätslehrers: Erinnerungen des Mediziners Prof. Dr. Friedrich v. Bidder 1810–1894* (Würzburg: Holzner-Verlag, 1959). Both Bidder's article and his published memoirs are based on a handwritten manuscript that he prepared "in retrospect [*nachträglich*]," after 1869. His daughter, Marie Lezius, prepared a printed version of this manuscript in 1896. In subsequent references, I will cite Bidder's complete memoirs rather than the article. I thank Ohad Parnes for bringing Bidder's memoirs to my attention.
105. Bidder, *Aus dem Leben eines Dorpater Universitätslehrers*, 71. The same description appears in Bidder, "Vor hundert Jahren im Laboratorium Johannes Müllers," 62.
106. Du Bois-Reymond, "Gedächtnisrede," 2: 315. Du Bois-Reymond made this comment in a note, preceding it with high praise for the new facility built along the Veterinary School garden. Added for publication and not read with his memorial address, the remark was almost certainly designed to curry favor with the Prussian Cultural Ministry.
107. Du Bois-Reymond, "Gedächtnisrede," 2: 193, quoted in Finkelstein, "Emil du Bois-Reymond," 167. Finkelstein translates the phrase differently.
108. Bidder, *Aus dem Leben eines Dorpater Universitätslehrers*, 68.
109. Stürzbecher, "Zur Berufung Johannes Müller," 195–197. See also Steudel, *Le Physiologiste Johannes Müller*, 20.
110. Hans-Jörg Rheinberger argues that "there are not any radical changes in Müller's position, but rather incessant adjustments to a model." Rheinberger, "From the 'Originary Phe-

nomenon' to the 'System of Pelagic Fishery': Johannes Müller (1801–1858) and the Relation between Physiology and Philosophy," in *From Physico-Theology to Bio-Technology: Essays in the Social and Cultural History of Biosciences. A Festschrift for Mikulás Teich*, ed. Kurt Bayertz and Roy Porter (Atlanta, GA: Rodopi, 1998), 134.

111. Rheinberger sharply criticizes Müller's biographer Gottfried Koller for claiming that Müller "had no research goal," arguing that a belief in scientific imagination unifies Müller's work. Rheinberger, "From the 'Originary Phenomenon' to the 'System of Pelagic Fishery,'" 151, n. 67.

112. Johannes Müller, "Bericht über die Fortschritte der vergleichenden Anatomie der Wirbelthiere im Jahre 1843," *Müllers Archiv* (1844): 50.

113. Johannes Müller, "Von dem Bedürfnis der Physiologie nach einer philosophischen Naturbetrachtung." Müller had this lecture published as an introduction to his first book, *Zur vergleichenden Physiologie des Gesichtssinnes des Menschen und der Thiere* (Leipzig: K. Knobloch, 1826), 34.

114. Emil du Bois-Reymond, "Gedächtnisrede," 2: 150.

115. Richards, *The Romantic Conception of Life*, 11.

116. Brigitte Lohff, "Johannes Müllers Rezeption der Zellenlehre in seinem *Handbuch der Physiologie des Menschen*," *Medizinhistorisches Journal* 13 (1978): 247.

117. Müller, HP, 2: 506.

118. Ibid., 2: 305 and 411. In German, *Werkzeug* means "tool" and is the word that carpenters use for their hammers.

119. See chapter 2 for a discussion of the microscopes used by Müller and his students.

120. Johannes Müller, *Bildungsgeschichte der Genitalien aus Untersuchungen an Embryonen des Menschen und der Thiere* (Düsseldorf: Arnz, 1830), 77.

121. Virchow, *Johannes Müller*, 33.

122. Koller, *Das Leben des Biologen Johannes Müller*, 137.

123. Claude Bernard, *An Introduction to the Study of Experimental Medicine*, trans. Henry Copely Greene (New York: Henry Schuman, 1949), 5–26.

124. Johannes Müller, *Zur vergleichenden Physiologie des Gesichtssinnes* (Leipzig: C. Knobloch, 1826), xxi.

125. Müller, *Bildungsgeschichte der Genitalien*, viii-ix.

126. Ibid., viii.

127. Johannes Müller, "Über die Erzeugung von Schnecken in Holothurien," *Bericht über die zur Bekanntmachung geeigneten Verhandlungen der Königlichen Preussischen Akademie der Wissenschaften zu Berlin* (October 1851): 645.

128. Müller, HP, 1: 19.

129. Johannes Müller, "Ueber die Metamorphose des Nervensystems in der Thierwelt," *J. F. Meckel's Archiv für Anatomie und Physiologie* (1828): 16.

130. Johannes Müller, *Briefe von Johannes Müller an Anders Retzius von dem Jahre 1830 bis 1857* (Stockholm: Aftonbladets Aktiebolags Tryckeri, 1900), 46, letter of 21 July 1843, and 52, letter of 28 March 1845.

131. Rheinberger, "From the 'Originary Phenomenon,'" 144.

132. Johannes Müller, "Ueber parasitische Bildungen," *Müllers Archiv* (1842): 203–204.

133. Thomas Hardy, *The Well-Beloved: A Sketch of a Temperament* (London: Macmillan, 1952), 7.

134. Johannes Müller, "Ueber verschiedene Formen von Seethieren," *Müllers Archiv* (1854): 88. To refer to the *Mitraria*, Müller uses the pronoun "she," since the name is a feminine word. A German reader given this sentence out of context would presume he

was talking about a woman. A German colleague who helped me translate the sentence was startled when she discovered that Müller was describing an embryonic worm.

135. Müller, "Ueber die Larven und die Metamorphose der Holothurien," *MKAWB* (1849): 301.

136. Haberling, *Johannes Müller,* 211.

137. Müller, "Über die Larven und die Metamorphose der Holothurien," 303, 328.

138. Müller, "Über den feinern Bau der krankhaften Geschwülste," *MKAWB* (1836): 110.

139. Johannes Müller, *Über die Compensation der physischen Kräfte am menschlichen Stimmorgan* (Berlin: Hirschwald, 1839), 30.

140. Bidder, *Aus dem Leben eines Dorpater Universitätslehrers,* 74–75.

141. Müller, *Über den feineren Bau und die Formen der krankhaften Geschwülste,* 1.

142. Johannes Müller and Jacob Henle, "On the Generic Characters of Cartilaginous Fishes, with Descriptions of New Genera," *The Magazine of Natural History* 2 (1838): 33. Henle, then Müller's student and closest associate, translated this article into English. It originally appeared as "Über die Gattungen der Plagiostomen," *Wiegmann's Archiv* (1837): 394–401.

143. Müller, "Bestätigung des Bell'schen Lehrsatzes," 117.

144. Müller, *Zur vergleichenden Physiologie des Gesichtsinnes,* 29.

145. Müller, HP, 1:1.

146. Müller, *Über den feineren Bau und die Formen der krankhaften Geschwülste,* 3.

147. Haberling, *Johannes Müller,* 372, letter of 4 September 1852.

148. See, for instance, Johannes Müller, "Über die Larven und die Metamorphose der Echinodermen," *MKAWB* (1848): 75; and "Ueber verschiedene Formen der Seethiere," *Müllers Archiv* (1854): 84.

149. Müller also competed for grants and specimens with the military doctors running the smaller museums of the Charité Hospital. See chapter 5 of this volume.

150. Eckenstein, *Der akademische Mentor,* 92–93; Zedlitz, *Neustes Conversations-Handbuch,* 829.

151. Eckenstein, *Der akademische Mentor,* 94.

152. Ibid.; Lenz, *Geschichte der königlichen Friedrich-Wilhelms-Universität,* 1: 40; Rolf Winau, *Medizin in Berlin* (Berlin: Walter de Gruyter, 1987), 107–108; Zedlitz, *Neustes Conversations-Handbuch,* 25. Eckenstein gives the year of purchase as 1805; Lenz, as 1802; Winau and Zedlitz, as 1803.

153. Zedlitz, *Neustes Conversations-Handbuch,* 25.

154. Lenz, *Geschichte der königlichen Friedrich-Wilhelms-Universität,* 2.1: 428–429, 438–440.

155. Müller, *Briefe an Anders Retzius,* 45, letter of 21 July 1843, and 48, letter of 3 February 1844.

156. Ibid., 35, letter of 21 October 1841; 48, letter of 3 February 1844; and 50, letter of 10 August 1844.

157. Ibid., 2, letter of 25 June 1830.

158. Johannes Müller to the Prussian Cultural Ministry, 14 January 1848, GSPK, Rep. 76Va, Sekt. 2, Tit. X, Nr. 11, Bd. VI, 142–143.

159. Stürzbecher, "Zur Berufung Johannes Müllers," 213–216. See also Finkelstein, "Emil du Bois-Reymond," 168.

160. Sven Dierig, "'Die Instrumente waren noch theuer und selten': Schiek-Mikroskope im Umfeld von Johannes Müller," *Unsichtbar—Sichtbar—Durchschaut: Das Mikroskop als Werkzeug des Lebenswissenschaftlers,* ed. Helmut Kettenmann, Jörg Zaun, and Stefanie Korthals (Berlin: Museumspädagogischer Dienst, 2001), 44.

161. Müller, *Briefe an Anders Retzius*, 37, letter of 12 January 1842.
162. Genesis 6:19–20. *The Complete Parallel Bible* (New York: Oxford University Press, 1993), 12.
163. Honoré de Balzac, "Avant-Propos," in *La Comédie humaine, Oeuvres complètes de Balzac* (Paris: Club de L'Honnête Homme, 1968), 64.
164. Emil du Bois-Reymond, *Jugendbriefe von Emil du Bois-Reymond an Eduard Hallmann*, ed. Estelle du Bois-Reymond (Berlin: Dietrich Reimer, 1918), 34, letter of 7 December 1839. The full name of Müller's museum was the Zootomisches-Anatomisches Museum.
165. Ibid., 61, letter of 2 June 1840.
166. Lenz, *Geschichte der königlichen Friedrich-Wilhelms-Universität*, 2.2: 370–371.
167. Haberling, *Johannes Müller*, 294, letter of 2 September 1846, and 333, letter of 1 September 1849.
168. Ibid., 328.
169. Müller, "Über den allgemeinen Plan in der Entwickelung der Echinodermen," *Physikalische Abhandlungen der könliglichen Akademie der Wissenschaften zu Berlin* (1852): 27.
170. Haberling, *Johannes Müller*, 391, letter of 30 August 1853, and 425, letter of 30 August 1855.
171. Stürzbecher, "Zur Berufung Johannes Müllers," 188.
172. Haberling, *Johannes Müller*, 381.
173. Ibid., 82.
174. Ibid., 307.
175. Müller's biographer Wilhelm Haberling believes that "in Müller, the feeling of responsibility was more pronounced than almost any other, making him anxious and uncertain." See ibid., 76.
176. Johannes Müller, *Ueber die phantastischen Gesichtserscheinungen: Eine physiologische Untersuchung* (Koblenz: Jacob Hölscher, 1826), reprinted in Ebbecke, *Johannes Müller: Der grosse rheinische Physiologe*, 102.
177. Haberling, *Johannes Müller*, 87.
178. Ibid., 295, letter of 2 September 1846.
179. Ibid., 340, letter of 25 September 1849.
180. Johannes Steudel writes that Müller "is reported to have wandered through back streets in Berlin, driven by inexplicable anxiety." Steudel, "Johannes Peter Müller," 569. Klaus Günther makes the same claim in "Die Gesellschaft Naturforschender Freunde zu Berlin, Johannes Müller und die Frage nach der Urzeugung," *Sitzungsberichte der Gesellschaft Naturforschender Freunde zu Berlin* 14 (1974): 32. Neither cites a primary source, but they appear to be generalizing du Bois-Reymond's claim that in the last year of Müller's life, he "wandered around in remote streets, as if driven by a deep inner fear." Du Bois-Reymond, "Gedächtnisrede," 297.
181. Roslynn D. Haynes has identified six negative stereotypes of scientists in Western literature, one of which, the Romantic scientist, "has reneged on human relationships." The popular notion of an isolated researcher has its roots in this stereotype and, as Haynes argues, derives from literary representations rather than any actual experience with scientists. See Roslynn D. Haynes, *From Faust to Strangelove: Representations of the Scientist in Western Literature* (Baltimore: Johns Hopkins University Press, 1994), 3.
182. Johannes Müller, "Zur Physiologie des Fötus," *Zeitschrift für die Anthropologie* (1824): 446.
183. Müller, *Briefe an Anders Retzius*, 39, letter of 19 May 1842.
184. Ibid., 60, letter of 2 July 1850.

185. Ibid., 67, letter of 21 May 1852.
186. Michael Hagner, "Sieben Briefe von Johannes Müller an Karl Ernst von Baer," *Medizinhistorisches Journal* 27 (1992): 149.
187. Müller, *Briefe an Anders Retzius*, 3, letter of 25 June 1830.
188. Ibid., 67, letter of 21 May 1852.
189. Johannes Steudel asserts that Müller's five-month depression in 1827 "had no connection to his way of working, to his outer or inner experiences; it was an inherited part of his nature." See Johannes Steudel, "Wissenschaftslehre und Forschungsmethodik Johannes Müllers," *Deutsche medizinsche Wochenschrift* 77 (1952): 117. He repeats this claim in *Le Physiologiste Johannes Müller*, 26, and "Johannes Peter Müller," 569. Gottwalt Christian Hirsch also attributes Müller's behavior to an inherited personality type in "Die Forscherpersönlichkeit des Biologen Johannes Müller," 168.
190. Haberling, *Johannes Müller*, 240.
191. Ibid., 236–237.
192. Du Bois-Reymond, "Gedächtnisrede," 2: 238. Steudel echoes this claim in "Johannes Peter Müller," 569.
193. Nipperdey, *Deutsche Geschichte*, 603; Jonathan Sperber, *The European Revolutions, 1848–1851* (Cambridge: Cambridge University Press, 1994), x. I thank David Cahan for bringing Sperber's book to my attention.
194. Nipperdey, *Deutsche Geschichte*, 595.
195. Ibid., 598–599, 605.
196. Ibid., 654.
197. Sperber, *The European Revolutions*, xv–xvi.
198. Nipperdey, *Deutsche Geschichte*, 659–660; Sperber, *The European Revolutions*, xvi-xvii.
199. Virchow's biographer Erwin Ackerknecht claims that Virchow "caused much trouble" for Müller "as a leading spirit in the meetings of rebellious *Privatdozenten* and associate professors." See Erwin H. Ackerknecht, *Rudolf Virchow: Doctor, Statesman, Anthropologist* (Madison: University of Wisconsin Press, 1953), 16.
200. Nipperdey, *Deutsche Geschichte*, 479.
201. Du Bois-Reymond called Müller conservative in his memorial address. See du Bois-Reymond, "Gedächtnisrede," 2: 274. Max Lenz wrote that Müller had a "conservative attitude and convictions." Lenz, *Geschichte der königlichen Friedrich-Wilhelms-Universität*, 2.2: 163.
202. Lenz, *Geschichte der königlichen Friedrich-Wilhelms-Universität*, 2.2: 198, 218. While many of the fighters were Berlin University students, the vast majority of students were not involved in the fighting.
203. Ibid., 2.2: 215–218, 231–232.
204. Ibid., 2.2: 259.
205. Ibid., 2.2: 267–69. See also Constantin Goschler, *Rudolf Virchow: Mediziner, Anthropologe, Politiker* (Cologne: Böhlau, 2002), 74.
206. Lenz, *Geschichte der königlichen Friedrich-Wilhelms- Universität*, 2.2: 250–253.
207. Haberling, *Johannes Müller*, 322–323.
208. Alexander von Humboldt and Emil du Bois-Reymond, *Briefwechsel zwischen Alexander von Humboldt und Emil du Bois-Reymond*, ed. Ingo Schwarz and Klaus Wenig (Berlin: Akademie Verlag, 1997), 83, letter of 9 April 1849.
209. Müller, *Briefe an Anders Retzius*, 68, letter of 21 May 1852.
210. Haberling, *Johannes Müller*, 368.
211. Müller, *Briefe an Anders Retzius*, 76, letter of 18 August 1854.
212. Albert E. Gunther, *A Century of Zoology at the British Museum through the Lives of Two*

Keepers 1815–1914 (London: Dawsons, 1975), 242. I thank Julia Voss for bringing this source to my attention.

213. On 14 October 1855, Helmholtz wrote to du Bois-Reymond, "Johannes Müller visited me and told me the truly hideous story of his shipwreck." *Dokumente einer Freundschaft*, 157.

214. As a young student, Müller had a special talent for classical languages, and at the time may have written better in Latin than he did in German. When Johannes Schulze reorganized the curriculum in the Rhine region, where Müller was attending high school, the Prussian reformer placed special emphasis on classical languages and rhetoric. Müller later used the argumentative styles of Roman writers in his scientific articles. See Ebbecke, *Johannes Müller: Der grosse rheinische Physiologe*, 23; Brigitte Lohff, "Johannes Müller: Von der Nervenwissenschaft zur Nervenphysiologie," in *Das Gehirn—Organ oder Seele? Zur Ideengeschichte der Neurobiologie*, ed. Ernst Florey and Olaf Breidbach (Berlin: Akademie Verlag, 1993), 40, and "Hat die Rhetorik Einfluss auf die Entstehung einer experimentellen Biologie in Deutschland gehabt?" In *Disciplinae Novae: Zur Entstehung neuer Denk- und Arbeitsrichtungen in der Naturwissenschaft* (Göttingen: Vandenhoeck and Ruprecht, 1979), 128.

215. Haberling, *Johannes Müller*, 289–290.

216. Johannes Müller, "Ueber die Erzeugung von Schnecken in Holothurien," 2.

217. Müller, HP, 2: 526.

218. Haberling, *Johannes Müller*, 97–98.

219. Müller, *Ueber die phantastischen Gesichtserscheinungen*, 104, 124.

220. Du Bois-Reymond found Müller's book on fantasy images highly amusing. When he was first getting to know Müller, he wrote to his friend Eduard Hallmann, "I'm reading a funny thing of his now: *On Fantasy Images*. To compare our stiff, rigid, imperious, get-on-with-the-job (standard line in the dissecting room) [scientist] to that enthusiastic, lovable, devoted, Goethe-crazed, practical doctor from Bonn decked out in Hegelian expressions is in fact quite entertaining." Du Bois-Reymond, *Jugendbriefe*, 41, letter of 13 January 1840.

221. Müller, "Zur Physiologie des Fötus," 461–462.

222. Müller, "Über den feinern Bau der krankhaften Geschwülste," 111, and "Ueber parasitische Bildungen," 206.

223. Müller, "Ueber parasitische Bildungen," 200.

224. Müller, "Über die Larven und die Metamorphose der Holothurien," 302.

225. Du Bois-Reymond looked to the classical period for inspiration and built his own "aesthetics of experimentation" around the ideals of Greek art and athleticism. His reference to "horrible memories of the impure forms of rococo architecture" may express his frustration with Müller's ongoing love of echinoderm embryos instead of experimental physiology. See du Bois-Reymond, "Naturwissenschaft und bildende Kunst," in his *Reden*, 2nd ed. (Leipzig: Veit, 1912), 2: 397; and Sven Dierig, "'Die Kunst des Versuchens': Emil du Bois-Reymonds *Untersuchungen über thierische Elektricität*," in *Kultur im Experiment*, ed. Henning Schmidgen, Peter Geimer, and Sven Dierig (Berlin: Kadmos, 2004), 123–146. See also Norton Wise, "What's in a Line? The 3 D's: Dürer, Dirichlet, Du Bois," presented as the Rothschild Lecture at Harvard University, April 2001.

226. Müller, "Über die Larven und die Metamorphose der Echinodermen," 85.

227. Johannes Müller and Theodor Schwann, "Versuche über die künstliche Verdauung des geronnenen Eiweisses," *Müllers Archiv* (1836): 73–74. *Eiweiss* means both egg white and protein in general. In the article's title, *Eiweiss* denotes protein, but in the description quoted, it probably refers to egg white, since Müller compares it to cooked meat.

228. Müller, "Ueber parasitische Bildungen," 194.

229. Müller, HP 1: 255.
230. Müller, "Über die Erzeugung von Schnecken in Holothurien," 7.
231. Two excellent studies of the close relationship between nineteenth-century realism and scientific writing are Richard Menke, "Fiction as Vivisection: G. H. Lewes and George Eliot," *English Literary History* 67 (2000): 617–653; and Lawrence Rothfield, *Vital Signs: Medical Realism in Nineteenth-Century Fiction* (Princeton, NJ: Princeton University Press, 1992).
232. Müller, "Über die Erzeugung von Schnecken in Holothurien," 628.
233. Haberling, *Johannes Müller*, 427.

Chapter 2

1. In this letter, Emil du Bois-Reymond was complaining to his friend Carl Ludwig about his longtime assistant Isidor Rosenthal (1836–1915). Du Bois-Reymond continued: "These are things that make me wish with all my heart that we part company, but not at all things that speak against him in the least. They are grievances grown out of the inconvenience that he has for so long remained stuck in a position suitable only for a younger, still flexible man who is subservient as far as necessary." Emil du Bois-Reymond and Carl Ludwig, *Two Great Scientists of the Nineteenth Century: Correspondence of Emil du Bois-Reymond and Carl Ludwig*, ed. Paul F. Cranefield, trans. Sabine Lichtner-Ayed (Baltimore: Johns Hopkins University Press, 1982), 104–105, letter of 17 July 1868.
2. Mannfred Stürzbecher, "Zur Berufung Johannes Müllers an die Berliner Universität," *Jahrbuch für die Geschichte Mittel- und Ostdeutschlands* 21 (1972): 188.
3. Johannes Müller, *Briefe von Johannes Müller an Anders Retzius* (Stockholm: Aftonbladets Aktiebolags Tryckeri, 1900), 23–24, letter of 24 November 1834; 32, letter of 12 September 1839; and 74, letter of 1 July 1854.
4. Historians disagree about the extent to which Müller's thinking merged with that of his students. Frederic L. Holmes believes that Müller, Henle, and Schwann "worked together so closely that it is difficult entirely to separate their respective contributions." See Frederic L. Holmes, "The Role of Johannes Müller in the Formation of Helmholtz's Physiological Career," in *Universalgenie Helmholtz: Rückblick nach 100 Jahren*, ed. Lorenz Krüger (Berlin: Akademie Verlag, 1994), 9. Timothy Lenoir argues most strongly for dissent between Müller and the students, calling Helmholtz and du Bois-Reymond "the rebellious students of Johannes Müller" who "set themselves up in conscious opposition to their great master." Timothy Lenoir, *The Strategy of Life: Teleology and Mechanics in Nineteenth-Century German Biology* (Chicago: University of Chicago Press, 1989), 195. L. J. Rather proposes that the relationship between Müller, Henle, and Schwann was only "loosely collaborative." See L. J. Rather, Patricia Rather, and John B. Frerichs, *Johannes Müller and the Nineteenth-Century Origins of Tumor Cell Theory* (Canton, MA: Science History Publications, 1986), 2. Hans-Jörg Rheinberger writes that Müller's "lab" depended on "the creativity of many minds united loosely by a shared heuristic maxim." See Hans-Jörg Rheinberger, "From the 'Originary Phenomenon' to the 'System of Pelagic Fishery': Johannes Müller (1801–1858) and the Relation between Physiology and Philosophy." In *From Physico-Theology to Bio-Technology: Essays in the Social and Cultural History of Biosciences. A Festschrift for Mikulás Teich*, ed. Kurt Bayertz and Roy Porter (Atlanta, GA: Rodopi, 1998), 144.
5. Friedrich Merkel, *Jacob Henle: Ein deutsches Gelehrtenleben* (Brunswick: Friedrich Vieweg, 1891), 1–4. Since Merkel was Henle's son-in-law and dedicated the biography to his wife, one must presume that he depicted Henle in as favorable a light as possible, and

therefore should read his account with caution. Merkel's is the only existing biography of Henle.

6. Ibid., 21.

7. Ibid., 173. See also Erich Hintzsche, "Friedrich Gustav Jacob Henle," in *Dictionary of Scientific Biography,* 6: 268.

8. Sander Gilman, *Jewish Self-Hatred: Anti-Semitism and the Hidden Language of the Jews* (Baltimore: Johns Hopkins University Press, 1986), 189.

9. Hermann Hoepke, "Jakob Henle's Briefe aus seiner Heidelberger Studentenzeit (26. April 1830–Januar 1831)," *Heidelberger Jahrbücher* 11 (1967): 54. In 1961, Hoepke obtained Henle's letters to his family (1830–49) from Henle's grandson, Werner Henle, of Philadelphia, which he published in this article and two others cited below.

10. Friedrich Merkel, *Jacob Henle: Gedächtnisrede* (Brunswick: Friedrich Vieweg, 1909), 1.

11. Hermann Hoepke, "Jakob Henles Briefe aus Berlin 1834–40," *Heidelberger Jahrbücher* 8 (1964): 57, quoted with kind permission of Springer Science and Business Media. See also Merkel, *Jacob Henle,* 101. Henle's religious identity is difficult to define. In his letters, he writes about Christmas and seems to regard himself as a Christian. Müller, Schwann, and Bidder never mention Henle's original faith, yet their letters cannot fully reveal the way that Berlin society regarded Henle. His friendship with Felix Mendelssohn-Bartholdy suggests that at least to some degree, he sought companions who shared his early background. At the time, Judaism was understood as a cultural, not a racial category, so that a convert was not seen as bearing un-German blood. There is no evidence that anti-Semitism ever motivated anyone's treatment of Henle.

12. Wilhelm Haberling, *Johannes Müller: Das Leben des rheinischen Naturforschers* (Leipzig: Akademische Verlagsgesellschaft, 1924), 86.

13. Hintzsche, "Friedrich Gustav Jakob Henle," 268.

14. Merkel, *Jacob Henle,* 39, letter of 11 November 1827.

15. Ibid., 41, letter of 26 June 1829.

16. Hoepke, "Jakob Henles Briefe aus seiner Heidelberger Studentenzeit," 46–47.

17. Henle's frequent use of this expression suggests the different outlooks of Müller and his wealthier students. Emil du Bois-Reymond, who like Henle came from a well-to-do family, complained to his friend Eduard Hallmann, "I'm working like an ox (*Ich ochse wie ein Vieh*)." The expression did not deprecate physical labor, since du Bois-Reymond specified, "To me, working like an ox (*ochsen*) means any work that one can't see, hear, feel, and do with one's own hands." See Emil du Bois-Reymond, *Jugendbriefe von Emil du Bois-Reymond an Eduard Hallmann,* ed. Estelle du Bois-Reymond (Berlin: Dietrich Reimer, 1918), 116, letter of 3 September 1844, and 29, letter of 15 October 1839. Instead, it seems to have meant brute memorization, work that demeaned a person because it demanded no active, creative thought. Since du Bois-Reymond used the expression a decade or more later, it cannot have been short-lived slang. Müller never used this expression, nor did Rudolf Virchow or Hermann von Helmholtz, who attended medical school on military scholarships.

18. Merkel, *Jacob Henle,* 45–46, undated letter.

19. Hoepke,"Jakob Henles Briefe aus seiner Heidelberger Studentenzeit," 52.

20. The historian Max Lenz states that in the 1820s, *Burschenschaften* attracted "the best of German youth," the sons of civil servants, pastors, theologians, and aristocrats. See Max Lenz, *Geschichte der königlichen Friedrich-Wilhelms-Universität zu Berlin* (Halle: Buchhandlung des Waisenhauses, 1910), 2.1: 53.

21. Merkel, *Jacob Henle,* 53–55.

22. I am grateful to Sander Gilman for informing me about *Burschenschaft* practices.

23. In his 1891 biography of Henle, Merkel states that Henle was cut on his right cheek.

In his 1909 address in honor of Henle's hundredth birthday, he claims it was his left. In the portrait included in the biography, Henle's face is turned so that only the left (unscarred) cheek can be seen. The scar is hidden in shadows on his right cheek. See Merkel, *Jacob Henle*, 55, and *Jacob Henle: Gedächtnisrede*, 1.

24. Hoepke, "Jakob Henles Briefe aus seiner Heidelberger Studentenzeit," 51, letter of 21 July 1830.

25. Ibid., 53, letter of 21 August 1830 and an undated letter, probably of September 1830.

26. Ibid., 52, letter of 20 May 1830.

27. Hermann Hoepke, "Der Bonner Student Jakob Henle in seinem Verhältnis zu Johannes Müller," *Sudhoffs Archiv* 53 (1969): 198. According to Friedrich Merkel, Müller, not Henle, discovered the pupil membrane in June 1831: "Müller first noticed the pupil membrane when Henle was injecting fetuses for other reasons; he [then] induced him to pursue the matter." Merkel, *Jacob Henle*, 92.

28. Hoepke, "Der Bonner Student Jakob Henle," 197, letter of 10 June 1831.

29. Merkel, *Jacob Henle*, 78–81; Hoepke, "Der Bonner Student Jakob Henle," 198–199. Merkel gives the date of the letter as 20 June 1831; Hoepke, as 30 June. In their search for worthy mates, du Bois-Reymond and Henle again expressed themselves in similar terms. In a letter to Helmholtz, du Bois-Reymond bragged, "Riding for my wife is second nature." Emil du Bois-Reymond and Hermann von Helmholtz, *Dokumente einer Freundschaft: Briefwechsel zwischen Hermann von Helmholtz und Emil du Bois-Reymond 1846–1894* (Berlin: Akademie Verlag, 1986), 226, letter of 1 June 1867.

30. Merkel, *Jacob Henle*, 81.

31. Haberling, *Johannes Müller*, 144.

32. In his published letters to Retzius, Müller called the Swedish scientist *Sie* until 30 November 1832. He first called him *Du* on 24 August 1834, so that the switch occurred sometime between these dates. Müller had known Retzius since September 1828. Müller, *Briefe an Anders Retzius.*

33. In his memorial address for Müller, du Bois-Reymond recalls a conversation about the second part of his *Animal Electricity* in which he told Müller how much he owed him scientifically. Müller remarked, "Oh, come on [*gehen Sie doch*], you just see things from a completely different point of view!" Du Bois-Reymond began working for Müller in early 1840; the second part of *Animal Electricity* appeared in 1849. Emil du Bois-Reymond, "Gedächtnisrede auf Johannes Müller," in his *Reden* (Leipzig: Veit, 1887), 2: 222.

34. Hoepke, "Der Bonner Student Jakob Henle," 200, letter of 20 August 1831.

35. Müller, *Briefe an Anders Retzius*, 5, letter of 14 November 1831.

36. Merkel, *Jacob Henle*, 84–85. In his biography of Müller, Haberling quotes Henle's letter and states that the visit occurred on 13 September 1831. Haberling, *Johannes Müller*, 128–29.

37. Hermann Hoepke, "Der Bonner Student Jakob Henle," 209, letter of 1 October 1831.

38. Ibid., 211, letter of 19 November 1831.

39. Merkel, *Jacob Henle*, 89–91.

40. Ibid., 98.

41. Jakob Henle, *Theodor Schwann: Nachruf* (Bonn: Max Cohen, 1882), 3.

42. Du Bois-Reymond, "Gedächtnisrede," 2: 303. Du Bois-Reymond quotes a letter from Schwann of 22 December 1858, describing his relationship with Müller. See also Marcel Florkin, *Naissance et déviation de la théorie cellulaire dans l'oeuvre de Théodore Schwann* (Paris: Hermann, 1960), 29.

43. Arleen Marcia Tuchman, *Science, Medicine, and the State in Germany: The Case of Baden 1815–1871* (New York: Oxford University Press, 1993), 64–65. I thank Volker Hess for introducing me to Tuchman's work.

44. Hoepke, "Jakob Henles Briefe aus seiner Heidelberger Studentenzeit," 55, letter of 7 February 1831.

45. Rembert Watermann, *Theodor Schwann: Leben und Werk* (Düsseldorf: L. Schwann, 1960), 17–18.

46. Merkel, *Jacob Henle*, 98.

47. Ibid., 103–109.

48. Tuchman reports that "Henle had written home to his parents more than once complaining about the competition among the *Privatdozenten* [and] his frustration with the university's policy of dividing up newly vacated positions among many individuals." Tuchman, *Science, Medicine, and the State in Germany*, 58.

49. Hoepke, "Jakob Henles Briefe aus Berlin," 66, letter of 25 March 1835.

50. Secondary sources disagree about the time of Schwann's appointment as museum helper. Schwann's biographers Marcel Florkin and Rembert Watermann give it as October 1834; Henle's biographer Merkel and Müller's biographer Haberling, as October 1835. The evidence supports Florkin and Watermann. Friedrich Bidder, who worked with Müller in the fall of 1834, described Schwann in his memoirs. See Florkin, *Naissance et déviation de la théorie cellulaire*, 35; Watermann, *Theodor Schwann*, 19; Merkel, *Jacob Henle*, 120; and Haberling, *Johannes Müller*, 176.

51. Hoepke, "Jakob Henles Briefe aus Berlin," 61, letter of October 1835.

52. Ibid., 68.

53. Friedrich Bidder, *Aus dem Leben eines Dorpater Universitätslehrers: Erinnerungen des Mediziners Prof. Dr. Friedrich v. Bidder 1810–1894* (Würzburg: Holzner Verlag, 1959), 70–71.

54. Hoepke, "Jakob Henles Briefe aus Berlin," 61, letter of 25 June 1840.

55. Ibid., 61.

56. Merkel, *Jacob Henle*, 137–139.

57. Henle, *Theodor Schwann: Nachruf*, 1–2.

58. Merkel, *Jacob Henle*, 146.

59. Hoepke, "Jakob Henles Briefe aus Berlin," 65, undated letter, and 67, letter of November 1835.

60. Henle, *Theodor Schwann: Nachruf*, 2.

61. Merkel, *Jacob Henle*, 144–145; Henle, *Theodor Schwann: Nachruf*, 2.

62. Bidder, *Aus dem Leben eines Dorpater Universitätslehrers*, 74.

63. Hoepke, "Jakob Henles Briefe aus Berlin," 61–62, letter of March 1834.

64. Bidder, *Aus dem Leben eines Dorpater Universitätslehrers*, 90, 81.

65. Merkel, *Jacob Henle*, 116.

66. Hoepke, "Jakob Henles Briefe aus Berlin," 63–64, letter of 19 July 1835.

67. "Burschenschaft," in *Brockhaus Enzyklopädie* (Mannheim: F. A. Brockhaus, 1987), 6: 228.

68. Merkel, *Jacob Henle*, 119.

69. Hoepke, "Jakob Henles Briefe aus Berlin," 65, letter of 28 September 1835.

70. Ibid., 71, letter of 19 July 1835. Hoepke uses an ellipsis to show that he has omitted a phrase from Henle's letter, possibly to hide incriminating words: "I had been under investigation for my participation in . . . ."

71. Ibid., 70, letter of 9 July 1835, and 72, letter of 2 August 1835.

72. Merkel, *Jacob Henle*, 129.

73. Ibid., 129–130.

74. Ibid., 133.

75. Hoepke, "Jakob Henles Briefe aus Berlin," 72, letter of 2 August 1835.

76. Haberling, *Johannes Müller*, 187, letter of 29 August 1837.

77. Ibid., 196, letter of 25 September 1837.

78. Jakob Henle, *Vergleichend-anatomische Beschreibung des Kehlkopfs mit besonderer Berücksichtigung des Kehlkopfs der Reptilien* (Leipzig: Leopold Voss, 1839), 1.

79. P. Harting, *Theorie und allgemeine Beschreibung des Mikroskopes* (Brunswick: Friedrich Vieweg, 1866), 3: 132–142.

80. Tuchman, *Science, Medicine, and the State in Germany*, 57.

81. Haberling, *Johannes Müller*, 54, 81.

82. Bidder, *Aus dem Leben eines Dorpater Universitätslehrers*, 70.

83. Johannes Müller, *Über den feineren Bau und die Formen der krankhaften Geschwülste* (Berlin: G. Reimer, 1838), 5.

84. Henle, *Theodor Schwann: Nachruf*, 2.

85. Albert Kölliker, *Erinnerungen aus meinem Leben* (Leipzig: Wilhelm Engelmann, 1899), 9. I thank Ohad Parnes for bringing Kölliker's memoirs to my attention.

86. Sven Dierig, "'Die Instrumente waren noch theuer und selten': Schiek-Mikroskope im Umfeld von Johannes Müller," in *Unsichtbar—Sichtbar—Durchschaut: Das Mikroskop als Werkzeug des Lebenswissenschaftlers*, ed. Helmut Kettenmann, Jörg Zaun, and Stefanie Korthals (Berlin: Museumspädagogischer Dienst, 2001), 44.

87. The correspondence between Müller's assistants and foreign scientists who worked with them suggests that the Berlin physiologists became a clearinghouse, of sorts, for supplying isolated microscopists with Schiek instruments. In 1837, the Swiss histologist Friedrich Miescher-His (1811–1887) sent Müller a shopping list, explaining that the magnifying power of his colleague's Fraunhofer microscope was insufficient and that at high magnifications it produced blurry images. After using a Schiek instrument, he wrote, "One cannot be even remotely satisfied with that one." From Müller he asked for a Schiek microscope "of the biggest and best kind," with a screw micrometer (a device for measuring the size of specimens while observing them), a compressor and ocular system like the one that the anatomist Jan Purkinje used, a steel needle for mercury injections, and two pairs of forceps for opening frog spines. Since he had heard that Schiek never sent anything without receiving a deposit, he included 100 crowns. In a letter to Miescher-His the next year, Henle wrote that he would soon be sending a Hirschmann microscope to the Swiss scientist's colleague Dr. Streckeisen, at a cost of 18 taler. Possibly Henle was deducting a down payment from the price of the Hirschmann instrument, since the cost seems very low. He also promised Miescher-His a new Schiek microscope "as the most pleasant Christmas gift," since Schiek had repeatedly promised to deliver it before 25 December. Manfred Frey, *Friedrich Miescher-His (1811–1887) und sein Beitrag zur Histopathologie des Knochens* (Basel: Benno Schwabe, 1962), 60–61, 65–67.

88. Ibid., 10.

89. Jakob Henle, *Allgemeine Anatomie: Lehre von den Mischungs- und Formbestandtheilen des menschlichen Körpers* (Leipzig: Leopold Voss, 1841), 134.

90. Ibid., 136–140.

91. Jakob Henle, "Ueber die Ausbreitung des Epithelium im menschlichen Körper," *Müllers Archiv* (1838): 115.

92. Jakob Henle, "Ueber Schleim- und Eiterbildung und ihr Verhältnis zur Oberhaut," *Hufelands Journal der praktischen Heilkunde* 86 (1838): 52.

93. Thomas D. Brock, *Robert Koch: A Life in Medicine and Bacteriology* (Madison, WI: Science Tech Publishers, 1988), 11, 28–29. Henle taught Koch anatomy at the University of Göttingen. Koch claimed that Henle's lectures did not inspire him to investigate microorganisms, since they were about general human anatomy, but they did awaken his interest in

scientific research. See Christoph Gradmann, *Krankheit im Labor: Robert Koch und die medizinische Bakteriologie* (Göttingen: Wallstein, 2005), 39.

94. Jakob Henle, *Von den Miasmen und Kontagien und von den miasmatisch-kontagiösen Krankheiten* (1840; reprint, Leipzig: Barth, 1910), 26. I have relied on Thomas Brock's translation of Henle's argument, *Robert Koch*, 28.

95. Henle, *Theodor Schwann: Nachruf*, 2, 19.

96. Florkin, *Naissance et déviation de la théorie cellulaire*, 31.

97. Theodor Schwann, "Ueber die Nothwendigkeit der atmosphärischen Luft zur Entwickelung des Hühnchens in dem bebrüteten Ei," *Notizen aus dem Gebiete der Natur- und Heilkunde* 41 (1834): 241–245.

98. Marcel Florkin, "Theodor Ambrose Hubert Schwann," in *Dictionary of Scientific Biography*, 12: 240.

99. Florkin, *Naissance et déviation de la théorie cellulaire*, 38–40. See also Florkin, "Theodor Ambrose Hubert Schwann," 240–241.

100. Florkin, *Naissance et déviation de la théorie cellulaire*, 41–46.

101. The original titles are "Versuche über die künstliche Verdauung des geronnenen Eiweisses" and "Ueber das Wesen des Verdauungsprocesses," *Müllers Archiv* (1836): 66–89 and 90–139, respectively. I am grateful to Frederic L. Holmes for directing my attention to Schwann's work on protein digestion.

102. Müller and Schwann, "Versuche über die künstliche Verdauung," 66.

103. Schwann, "Ueber das Wesen des Verdauungsprocesses," 90.

104. Müller and Schwann, "Versuche über die künstliche Verdauung," 81.

105. Theodor Schwann, "Vorläufige Mittheilung, betreffend Versuche über die Weingährung und Fäulniss," *Annalen der Physik und Chemie* 41 (1837): 184–193.

106. Florkin, *Naissance et déviation de la théorie cellulaire*, 53.

107. Brian J. Ford argues that new technology was not the decisive factor leading to the idea that all living things are made of cells. See Ford's *Single Lens: The Story of the Simple Microscope* (New York: Harper & Row, 1985).

108. The historian of science Ohad Parnes writes that "the envisioning of cells . . . was . . . not a result of a simple gaze upon the microscopical. Cells were not the objective, but the means for Schwann's investigation. His aim was to deliver an account of the life of animal tissue—a lawful account that did not take recourse in vital force." See Ohad Parnes, "The Envisioning of Cells," *Science in Context* 13 (2000): 86.

109. Florkin, *Naissance et déviation de la théorie cellulaire*, 62, speech of 23 June 1878.

110. Marc Klein, "Jakob Mathias Schleiden," in *Dictionary of Scientific Biography*, 12: 173. See also Rather et al., *Johannes Müller and the Nineteenth-Century Origins of Tumor Cell Theory*, 2, 33.

111. Theodor Schwann, *Microscopical Researches into the Accordance in the Structure and Growth of Animals and Plants*, trans. Henry Smith (London: Sydenham Society, 1847), 7, xi.

112. Ibid., ix.

113. Ibid., 167.

114. Ibid., 190.

115. Theodor Schwann to Cultural Minister von Altenstein, letter of 6 December 1838, GSPK, Rep. 76Va, Sekt. 2, Tit. X, Nr. 11, Bd. VI, "Das anatomische Museum der Universität zu Berlin 1836–1858," 47, reproduced in Theodor Schwann, *Lettres de Théodore Schwann*, ed. Marcel Florkin (Liège: Société Royale des Sciences de Liège, 1961), 36.

116. Ibid., 48.

117. Watermann, *Theodor Schwann*, 24.

118. Hajo Holborn, *A History of Modern Germany* (New York: Alfred A. Knopf, 1968), 2: 506.
119. Henle, *Theodor Schwann: Nachruf,* 41.
120. Florkin, *Naissance et déviation de la théorie cellulaire,* 88–89; Rather et al., *Johannes Müller and the Nineteenth-Century Origins of Tumor Cell Theory,* 35.
121. Florkin, *Naissance et déviation de la théorie cellulaire,* 87; Holborn, *A History of Modern Germany,* 2: 505.
122. Schwann, *Lettres de Théodore Schwann,* 32, letter of 7 March 1835.
123. Schwann, *Microscopical Researches,* 189.
124. Florkin, "Theodor Ambrose Hubert Schwann," 243.
125. Theodor Schwann, "Versuche, um auszumitteln, ob die Galle im Organismus eine für das Leben wesentliche Rolle spielt," *Müllers Archiv* (1844): 127–159.
126. Rather et al., *Johannes Müller and the Nineteenth-Century Origins of Tumor Cell Theory,* 35, 33.
127. Henle, *Theodor Schwann: Nachruf,* 24.
128. Du Bois-Reymond, "Gedächtnisrede," 2: 304.
129. Ibid.
130. Henle, *Theodor Schwann: Nachruf,* 44–45.
131. Merkel, *Jacob Henle,* 95.
132. Hoepke, "Jakob Henles Briefe aus Berlin," 82–83, letter of August 1834. See also Bidder, *Aus dem Leben eines Dorpater Universitätslehrers,* 72.
133. Manfred Stürzbecher, "Aus dem Briefwechsel des Physiologen Johannes Müller mit dem Preussischen Kultusministerium," *Janus* 49 (1960): 277–278.
134. Du Bois-Reymond, "Gedächtnisrede," 2: 304.
135. Hoepke, "Jakob Henles Briefe aus Berlin," 82, letter of 16 February 1834. According to Friedrich Bidder, Henle himself had no qualms about using assistants. At first, seeing Bidder's skill at dissection, Henle began asking him for help with the preparations for Müller's classes. When Henle saw that Bidder could do the work on his own, he simply let him do it and concentrated on his own studies, so that Müller began bringing his requests directly to Bidder. "In fact, I was functioning fully as [Müller's] prosector," Bidder remembers with pride. There is no indication that he received the prosector's salary. Bidder, *Aus dem Leben eines Dorpater Universitätslehrers,* 69.
136. Müller, *Über den feinern Bau und die Formen der krankhaften Geschwülste,* 3.
137. Johannes Müller, *Handbuch der Physiologie* [hereafter HP], 3rd ed., 2 (Koblenz: J. Hölscher, 1840), 758.
138. Du Bois-Reymond, *Jugendbriefe,* 51, letter of 24 April 1840.
139. Ibid., 36, letter of 7 December 1839.
140. Watermann, *Theodor Schwann,* 176. See also Brigitte Lohff, "Johannes Müllers Rezeption der Zellenlehre in seinem *Handbuch der Physiologie des Menschen,*" *Medizinhistorisches Journal* 13 (1978): 247–258.
141. Arleen Tuchman argues that Henle "broke with tradition and brought the microscope into the classroom." Before 1838, some German medical courses had offered microscopical demonstrations, but this does not necessarily mean that the students had much opportunity to look through the microscopes themselves. Tuchman, *Science, Medicine, and the State in Germany,* 56.
142. Hermann Hoepke, "Jakob Henle's Gutachten zur Besetzung des Lehrstuhls für Anatomie an der Universität Berlin 1883," *Anatomischer Anzeiger* 120 (1967): 228.
143. Tuchman emphasizes the difference between Henle and Müller as teachers, credit-

ing Henle with introducing the microscope as a teaching tool as opposed to Müller, who reserved it for gifted researchers. See Tuchman, *Science, Medicine, and the State in Germany*, 58.

144. Du Bois-Reymond, *Jugendbriefe*, 6.

145. Kölliker, *Erinnerungen aus meinem Leben*, 8. See also François Duchesneau, "Kölliker and Schwann's Cell Theory," in *La storia della medicina e della scienza tra archivio e laboratorio*, ed. Guido Cimino, Carlo Maccagni, and Luigi Belloni (Florence: L. S. Olschki, 1994), 104.

146. Kölliker, *Erinnerungen aus meinem Leben*, 8, quoted in Tuchman, *Science, Medicine, and the State in Germany*, 58. Tuchman's translation differs slightly from mine.

147. Wilhelm Waldeyer, "J. Henle: Nachruf," *Archiv für mikroskopische Anatomie* 26 (1886): 3.

148. Hoepke, "Jakob Henles Briefe aus Berlin," 83, letter of 1 February 1839.

149. Ibid., 62, letter of 12 June 1835.

150. Ibid., 73, letter of 1 March 1836.

151. Ibid., 83–84, letter of 30 April 1840. Arleen Tuchman argues that Müller took away Henle's prosector and editor jobs, and consequently his salaries, because the success of Henle's microscopic anatomy course "aroused his suspicions." Having studied Henle's letters in the Göttingen University Archive, Tuchman includes a sentence of his 30 April 1840 letter to Schöll that Hoepke does not: "Müller had long ago divided this money [for the prosector- and editorships] among individuals whom he hoped to mother and from whom he hoped he would have less to fear." Tuchman, *Science, Medicine, and the State in Germany*, 58–59.

152. Du Bois-Reymond, *Jugendbriefe*, 79, letter of 26 December 1840. Du Bois-Reymond's daughter, Estelle, who edited his letters, omitted passages she considered damaging to her father's reputation. An ellipsis follows this sentence.

153. Du Bois-Reymond, *Jugendbriefe*, 83, letter of 29 March 1841. An ellipsis precedes this passage.

154. Müller, *Briefe an Anders Retzius*, 33, letter of 21 October 1841.

155. Letter of Gustav Magnus to Jakob Henle, 14 May 1858, NL Hermann Hoepke, KE 74, UH. I thank Christoph Gradmann for bringing Magnus's letters to Henle to my attention.

156. Du Bois-Reymond and Helmholtz, *Dokumente einer Freundschaft*, 144, letter of 30 May 1853.

157. Jakob Henle, *Handbuch der rationellen Pathologie* (Brunswick: Friedrich Vieweg, 1846), 1: 97.

158. Hintzsche, "Friedrich Gustav Jakob Henle," 268.

159. One can infer Henle's statements about the Berlin job from Gustav Magnus's letter to him of 9 June 1858. Magnus expresses great disappointment that Henle is not coming to Berlin but writes, "Even so, I can understand that when someone feels well and satisfied in a position, he will not gladly exchange it for another which . . . may be still more pleasant but in no way nourishes the inner peace and contented satisfaction that I imagine to be one of the loveliest aspects of a Göttingen professorship." NL Hermann Hoepke, KE 74, UH.

160. In their biographies of Henle, Merkel and Hintzsche both state that after the *Burschenschaft* incident, Henle preferred to stay out of Prussia. Hinzsche writes that Henle "was unable to become reconciled to Prussia's domestic politics." See Merkel, *Jacob Henle*, 129; and Hintzsche, "Friedrich Gustav Jakob Henle," 269.

161. Du Bois-Reymond, *Jugendbriefe*, 88, letter of 2 April 1841.

162. Du Bois-Reymond and Ludwig, *Two Great Scientists of the Nineteenth Century*, 52, letter of 29 December 1849.

163. Du Bois-Reymond, *Jugendbriefe*, 114, letter of March 1843.

164. Du Bois-Reymond, "Gedächtnisrede," 2: 302.
165. Ibid., 304.
166. Ibid.
167. Ibid., 305.
168. Ibid., 306.
169. Florkin, *Naissance et déviation de la théorie cellulaire*, 74.
170. Haberling, *Johannes Müller*, 324, letter of 6 November 1848.

Chapter 3

1. The historians Gabriel Finkelstein and Nicholas Jardine have both argued that du Bois-Reymond used the eulogy for Müller as an opportunity to express his own scientific views. See Gabriel Ward Finkelstein, "Emil du Bois-Reymond: The Making of a Liberal German Scientist (1818–1851)" (Ph.D. diss., Princeton University, 1996), 173–174; and Nicholas Jardine, "The Mantle of Müller and the Ghost of Goethe: Interactions between the Sciences and Their Histories," in *History and the Disciplines: The Reclassification of Knowledge in Early Modern Europe*, ed. Donald R. Kelley (Rochester, NY: University of Rochester Press, 1997), 306.

2. Emil du Bois-Reymond, *Jugendbriefe von Emil du Bois-Reymond an Eduard Hallmann*, ed. Estelle du Bois-Reymond (Berlin: Dietrich Reimer, 1918), 86, letter of 29 March 1841, quoted in Finkelstein, "Emil du Bois-Reymond," 203. Finkelstein translates the passage slightly differently. He has also pointed out du Bois-Reymond's focus on action during his early years as a scientist. Finkelstein, "Emil du Bois-Reymond," 28, 96.

3. Hermann von Helmholtz, *Letters of Hermann von Helmholtz to His Parents: The Medical Education of a German Scientist 1837–1846*, ed. David Cahan (Stuttgart: Franz Steiner, 1993), 42, n. 5, and 60, letter of 15 May 1839.

4. William Shakespeare, *Hamlet* (Harmondsworth, UK: Penguin Popular Classics, 1994), 82 (III, i).

5. I agree with Timothy Lenoir that du Bois-Reymond was one of those students who "set themselves up in conscious opposition" to Müller. Timothy Lenoir, *The Strategy of Life: Teleology and Mechanics in Nineteenth-Century German Biology* (Chicago: University of Chicago Press, 1989), 195.

6. Du Bois-Reymond, *Jugendbriefe*, 93, letter of 25 May 1841, quoted in Finkelstein, "Emil du Bois-Reymond," 203–4. My translation differs slightly from Finkelstein's.

7. Gabriel Finkelstein has suggested another pair of literary works that helped du Bois-Reymond to develop his identity as a scientist: Goethe's *Wilhelm Meister's Apprenticeship* and *Wilhelm Meister's Travels*, to which Finkelstein refers frequently in his analysis of du Bois-Reymond's education and early life. Goethe's writing would have been better known to a young German scholar than Shakespeare's, but I believe that in the case of du Bois-Reymond, both *Wilhelm Meister* and *Hamlet* served as inspirational narratives. See Finkelstein, "Emil du Bois-Reymond," 20–24.

8. Shakespeare, *Hamlet*, 105 (III, iv).

9. Ibid., 154 (V, ii).

10. See David Cahan, "Introduction," in Hermann von Helmholtz, *Science and Culture: Popular and Philosophical Essays*, ed. David Cahan (Chicago: University of Chicago Press, 1995), ix-xi; Finkelstein, "Emil du Bois-Reymond," 174; Peter W. Ruff, *Emil du Bois-Reymond* (Leipzig: B. G. Teubner, 1981), 79; and Jochen Zwick, "Akademische Erinnerungskultur, Wissenschaftsgeschichte und Rhetorik im 19. Jahrhundert: Über Emil du Bois-Reymond

als Festredner," *Scientia Poetica: Jahrbuch für Geschichte der Literatur und der Wissenschaften* 1 (1997): 120–139.

11. Emil du Bois-Reymond, *Untersuchungen über thierische Elektricität* (Berlin: Reimer, 1848), xix.

12. Emil du Bois-Reymond, "Über Geschichte der Wissenschaft," in his *Reden* (Leipzig: Veit, 1887), 2: 350, 354.

13. Emil du Bois-Reymond and Hermann von Helmholtz, *Dokumente einer Freundschaft: Briefwechsel zwischen Hermann von Helmholtz und Emil du Bois-Reymond 1846–1894* (Berlin: Akademie Verlag, 1986), 113, letter of 16 May 1851.

14. Sven Dierig has pointed out du Bois-Reymond's use of *zweckmässig* as an aesthetic term to describe the beauty of functional, streamlined instruments. See Sven Dierig, "'Die Kunst des Versuchens': Emil du Bois-Reymonds *Untersuchungen über thierische Elektricität*," in *Kultur im Experiment*, ed. Henning Schmidgen, Peter Geimer, and Sven Dierig (Berlin: Kulturverlag Kadmos, 2004), 136.

15. Emil du Bois-Reymond, Laboratory Diary VIII, Experiments 1 January 1850–March 1856, posted for public use in *The Virtual Laboratory: Essays and Resources on the Experimentalization of Life*, Max Planck Institute for the History of Science, Berlin, http://vlp.mpiwg-berlin.mpg.de. In English prose, the lines read: "May the daily work of my hands bring great happiness, so that I may complete it! Oh, don't let me get discouraged! No, these are not empty dreams: now mere saplings, these trees will someday give fruit and shade." I am grateful to Sven Dierig for bringing this poem to my attention and to Christoph Hoffmann and Killian Nauhaus for identifying it. In "What's in a Line?" Norton Wise shows the connection between the root and branch imagery in Goethe's poem and du Bois-Reymond's drawing of a tree on the membership certificate of the Berlin Physical Society, arguing that in both cases, du Bois-Reymond has in mind a tree of knowledge. Norton Wise, "What's in a Line? The 3 D's: Dürer, Dirichlet, Du Bois," Rothschild Lecture, Harvard University, April 2001, 31.

16. Gabriel Finkelstein calls hope a "secular religion," tying it to du Bois-Reymond's cultured Huguenot background and liberal politics as a motivating force in his career. See Finkelstein, "Emil du Bois-Reymond," 202.

17. My discussion of du Bois-Reymond's early life in this paragraph draws on Paul F. Cranefield, "Carl Ludwig and Emil du Bois-Reymond: A Study in Contrasts," *Gesnerus* 45 (1988): 271–272; Paul Diepgen, "Foreword," in Emil du Bois-Reymond and Carl Ludwig, *Two Great Scientists of the Nineteenth Century: Correspondence of Emil du Bois-Reymond and Carl Ludwig*, ed. Paul F. Cranefield, trans. Sabine Lichtner-Ayed (Baltimore: Johns Hopkins University Press, 1982), xiv; Finkelstein, "Emil du Bois-Reymond," 12–24; K. E. Rothschuh, "Emil Heinrich du Bois-Reymond," in *Dictionary of Scientific Biography*, 4: 200; and Ruff, *Emil du Bois-Reymond*, 8–9. Finkelstein offers the most detailed information.

18. Ruff, *Emil du Bois-Reymond*, 84. Ruff does not cite any primary source to support this claim.

19. Du Bois-Reymond and Ludwig, *Two Great Scientists*, 58, letter of 9 April 1850.

20. Alfred E. Hoche, *Jahresringe: Innenansicht eines Menschenlebens* (Munich: J. F. Lehmans, 1940), 84–85, quoted in Finkelstein, "Emil du Bois-Reymond," 19, n. 39.

21. Estelle du Bois-Reymond, "Einleitung," in Emil du Bois-Reymond, *Jugendbriefe*, 9; Finkelstein, "Emil du Bois-Reymond," 24–25; Rothschuh, "Emil Heinrich du Bois-Reymond," 200; Ruff, *Emil du Bois-Reymond*, 9.

22. My discussion of Félix du Bois-Reymond's life relies on Diepgen, "Foreword," xiv; Estelle du Bois-Reymond, "Einleitung," 8; Finkelstein, "Emil du Bois-Reymond," 13–17;

Rothschuh, "Emil Heinrich du Bois-Reymond," 200; and Ruff, *Emil du Bois-Reymond,* 8. Finkelstein offers the most informative account.

23. Finkelstein has pointed out how Félix du Bois-Reymond's ability to "cultivate benefactors" helped him to survive. See Finkelstein, "Emil du Bois-Reymond," 106.

24. Eugénie Rosenberger, *Félix du Bois-Reymond, 1782–1865* (Berlin: Meyer and Jessen, 1912), quoted in Finkelstein, "Emil du Bois-Reymond," 46.

25. Du Bois-Reymond, *Jugendbriefe,* 52, letter of 24 April 1840, quoted in Finkelstein, "Emil du Bois-Reymond," 147. Finkelstein translates the passage slightly differently.

26. Du Bois-Reymond and Helmholtz, *Dokumente einer Freundschaft,* 88, letter of 14 October 1849.

27. Finkelstein, "Emil du Bois-Reymond," 126–130.

28. Helmholtz, *Letters of Helmholtz to His Parents,* 110, n. 6.

29. Gabriel Finkelstein, "The Ascent of Man? Emil du Bois-Reymond's Reflections on Scientific Progress," *Endeavour* 24 (2000): 129.

30. On du Bois-Reymond's studies, see Finkelstein, "Emil du Bois-Reymond," 24–195; Rothschuh, "Emil Heinrich du Bois-Reymond," 200, and "Emil du Bois-Reymond (1818–1896) und die Elektrophysiologie der Nerven," in *Von Boerhaave bis Berger: Die Entwicklung der kontinentalen Physiologie im 18. und 19. Jahrhundert mit besonderer Berücksichtigung der Neurophysiologie, ed. Karl E. Rothschuh* (Stuttgart: Gustav Fischer, 1964), 86; and Ruff, *Emil du Bois-Reymond,* 10.

31. Finkelstein, "Emil du Bois-Reymond," 54, 81. Albert E. Gunther, who studied with Müller in the early 1850s, recalled that "the custom of the 'Wanderjahre' (during which a student migrated to another university to continue his studies) was general in German student life." Albert E. Gunther, *A Century of Zoology at the British Museum through the Lives of Two Keepers, 1815–1914* (London: Dawsons of Pall Mall, 1975), 234.

32. Finkelstein, "Emil du Bois-Reymond," 103, 112.

33. Du Bois-Reymond, *Jugendbriefe,* 66, letter of 19 August 1840, and 79, letter of 26 December 1840. See also Finkelstein, "Emil du Bois-Reymond," 189.

34. Du Bois-Reymond and Ludwig, *Two Great Scientists,* 21, letter of 9 February 1849. See also Finkelstein, "Emil du Bois-Reymond," 260.

35. Alexander von Humboldt and Emil du Bois-Reymond, *Briefwechsel zwischen Alexander von Humboldt und Emil du Bois-Reymond,* ed. Ingo Schwarz and Klaus Wenig (Berlin: Akademie Verlag, 1997), 83, letter of 15 April 1849. See also Finkelstein, "Emil du Bois-Reymond," 278.

36. Humboldt and du Bois-Reymond, *Briefwechsel,* 85, n. 1, letter of 17 April 1849.

37. Rosenberg, *Félix du Bois-Reymond,* 273, quoted in Finkelstein, "Emil du Bois-Reymond," 24.

38. Du Bois-Reymond and Helmholtz, *Dokumente einer Freundschaft,* 199–200, letters of 15 and 25 March 1862.

39. Du Bois-Reymond, "Vorrede," in his *Thierische Elektricität,* lv-lvi.

40. In *The Anxiety of Influence,* Harold Bloom quotes Kierkegaard and Nietzsche to explain how poets re-create their precursors. According to Kierkegaard, "He who is willing to work gives birth to his own father." According to Nietzsche, "When one hasn't had a good father, it is necessary to invent one." Since Bloom gives no sources, these statements should be taken as paraphrases rather than direct quotations. Harold Bloom, *The Anxiety of Influence: A Theory of Poetry* (New York: Oxford University Press, 1997), 56. Both epigrams shed light on du Bois-Reymond's representations of his father and Müller.

41. Du Bois-Reymond, *Jugendbriefe,* 94, letter of 9 August 1841.

42. For a description of the *Bildungsbürgertum* in the early nineteenth century, see Finkelstein, "Emil du Bois-Reymond," 3–5, 42–43.

43. Du Bois-Reymond, *Jugendbriefe*, 80, letter of 26 December 1840.

44. Ibid., 35, letter of 7 December 1839.

45. Du Bois-Reymond and Ludwig, *Two Great Scientists*, 20, letter of 23 October 1848.

46. Ibid., 78, letter of 9 January 1853.

47. Du Bois-Reymond, *Jugendbriefe*, 42, letter of 3 February 1840.

48. Ibid., 47, letter of 19 February 1840. In German, Thiele is that "undurchdringliches kieselartiges und rochenstacheliges Hautskelett unseres Museums."

49. Ibid., 91, 25 May 1841; du Bois-Reymond and Helmholtz, *Dokumente einer Freundschaft*, 100, letter of 25 August 1850.

50. Emil du Bois-Reymond to Henry Bence Jones, 31 October-7 November 1859, no. 37, S. 63, SBPK, HA, SD, NL Emil du Bois-Reymond, 3k 1852 (4). I am grateful to Gabriel Finkelstein for bringing the du Bois-Reymond-Bence Jones correspondence to my attention.

51. Ruff, *Emil du Bois-Reymond*, 82.

52. Gabriel Finkelstein proposes that du Bois-Reymond's commitment to liberal views shaped his scientific career. See Finkelstein, "Emil du Bois-Reymond." I cite no page since this is a central argument of Finkelstein's dissertation.

53. Du Bois-Reymond and Ludwig, *Two Great Scientists*, 9, 13, letter of 22 April 1848.

54. Du Bois-Reymond, *Jugendbriefe*, 128, letter of 6 January 1849, quoted in Finkelstein, "Emil du Bois-Reymond," 259. My translation varies slightly from Finkelstein's.

55. Du Bois-Reymond and Ludwig, *Two Great Scientists*, 11, letter of 22 April 1848.

56. Du Bois-Reymond, *Jugendbriefe*, 86, letter of 29 March 1841.

57. Ibid., 69, letter of 19 August 1840.

58. Ibid., 109, letter of May 1842.

59. Edwin Clarke and L. S. Jacyna, *Nineteenth-Century Origins of Neuroscientific Concepts* (Berkeley: University of California Press, 1987), 163–180.

60. Du Bois-Reymond, "Vorrede," in his *Untersuchungen über thierische Elektricität*, v.

61. Finkelstein, "The Ascent of Man," 129.

62. Emil du Bois-Reymond to Henry Bence Jones, 31 January 1858, no. 30, S. 52, SBPK, HA, SD, NL Emil du Bois-Reymond, 3k 1852 (4).

63. Gabriel Finkelstein emphasizes that du Bois-Reymond's "equipment virtually established electrophysiology as a modern discipline." Finkelstein, "Emil du Bois-Reymond," 308. See also Rothschuh, "Emil du Bois-Reymond und die Elektrophysiologie der Nerven," 99–103.

64. Finkelstein, "Emil du Bois-Reymond," 6.

65. Ibid., 264–265.

66. Ibid., 75–77, 113. Finkelstein offers extensive evidence that Steffens shaped du Bois-Reymond's early scientific thinking.

67. Du Bois-Reymond, "Der physiologische Unterricht sonst und jetzt," in his *Reden* (Leipzig: Veit, 1887), 2: 364, quoted in Finkelstein, "Emil du Bois-Reymond," 72, n. 57. My italics.

68. Du Bois-Reymond, "Vorrede," in his *Thierische Elektricität*, xl.

69. Ibid., xxvii, xxxix.

70. Ibid., 1.

71. Many historians have described du Bois-Reymond's introduction to *Animal Electricity* as a manifesto. Sven Dierig calls it an "anti-vitalistic manifesto." See Dierig, "'Die Kunst des Versuchens,'" 141. Finkelstein characterizes it as "a manifesto of mechanist physiology,"

"Emil du Bois-Reymond," 264, and "a kind of scientific manifesto," "The Ascent of Man," 130.

72. Ruff, *Emil du Bois-Reymond*, 87.

73. Karl Marx and Friedrich Engels, "Manifest der Kommunistischen Partei," in their *Ausgewählte Schriften in zwei Bänden* (Berlin: Dietz, 1970), 1: 25.

74. Du Bois-Reymond, "Vorrede," in his *Thierische Elektricität*, xxxviii.

75. Ibid., xxxvii.

76. Ibid., xlix.

77. Bloom, *The Anxiety of Influence*, 14.

78. Dierig, "'Die Kunst des Versuchens,'" 133–140.

79. Du Bois-Reymond, "Vorrede," in his *Thierische Elektricität*, xlviii.

80. Emil du Bois-Reymond, "Ueber die Lebenskraft," in his *Reden* (Leipzig: Veit, 1887), 2: 26, n. 1.

81. Timothy Lenoir has noted that "the term 'zweckmässig' was carefully expunged from [Helmholtz's] scientific lexicon." See Lenoir, *The Strategy of Life*, 196.

82. Johannes Müller, HP 1: 67, quoted in Ruff, *Emil du Bois-Reymond*, 36.

83. Du Bois-Reymond, "Vorrede," in his *Thierische Elektricität*, xxiv.

84. Norton Wise argues that du Bois-Reymond's imagery represents life force as a "dangerously feminine sea of romanticism" associated with the spring that reduced Hermaphroditus to half a man. Such threats were to be fought with spearlike laboratory instruments. See Wise, "What's in a Line?" 30.

85. On 18 December 1859, du Bois-Reymond wrote to Ludwig, "If it were ten years ago and I were sitting as a simple doctor over the stable in Carlstrasse . . . I would be clear about these things in a couple of weeks' time." Du Bois-Reymond and Ludwig, *Two Great Scientists*, 101.

86. Du Bois-Reymond, "Vorrede," in his *Thierische Elektricität*, xliii.

87. Du Bois-Reymond, "Der physiologische Unterricht sonst und jetzt," 363. Rothschuh quotes this passage in *History of Physiology*, 223, but translates it quite differently.

88. Finkelstein, "Emil Du Bois-Reymond," 253; Ruff, *Emil du Bois-Reymond*, 91.

89. Du Bois-Reymond, *Jugendbriefe*, 48, letter of 31 March 1840.

90. Hermann von Helmholtz, Speech honoring the fiftieth anniversary of Emil du Bois-Reymond's doctoral degree, *Neue Freie Presse* (Vienna), 23 September 1893, quoted in Dierig, "'Die Kunst des Versuchens,'" 123.

91. Du Bois-Reymond, *Jugendbriefe*, 95, letter of 9 August 1841, quoted in Finkelstein, "Emil du Bois-Reymond," 206. Finkelstein translates the passage slightly differently. I have relied in part on Finkelstein's translation.

92. Du Bois-Reymond and Ludwig, *Two Great Scientists*, 51, letter of 29 December 1849.

93. Du Bois-Reymond, *Jugendbriefe*, 103, letter of 25 October 1841.

94. Du Bois-Reymond and Ludwig, *Two Great Scientists*, 20, letter of 23 October 1848.

95. Ibid., 13, letter of 22 April 1848.

96. Du Bois-Reymond, *Thierische Elektricität*, 1: 459.

97. Du Bois-Reymond and Ludwig, *Two Great Scientists*, 107, letter of 18 January 1871.

98. Ibid., 114–115, letters of 27, 28, and 30 November 1875.

99. Du Bois-Reymond, *Thierische Elektricität*, 2:26, quoted in Finkelstein, "Emil du Bois-Reymond," 236. My translation varies slightly from Finkelstein's.

100. Johannes Müller, "Bestätigung des Bell'schen Lehrsatzes," *Notizen aus dem Gebiete der Natur- und Heilkunde* 30 (1831): 115.

101. Humboldt and du Bois-Reymond, *Briefwechsel*, 90, n. 9, letter of 12 May 1849. See also Finkelstein, "Emil du Bois-Reymond," 282–283.

102. Humboldt and du Bois-Reymond, *Briefwechsel*, 74, letter of 20 May 1845, unsent.
103. Ibid., 85, letter of 17 April 1849.
104. Du Bois-Reymond and Ludwig, *Two Great Scientists*, 44–45, letter of 7 August 1849.
105. Humboldt and du Bois-Reymond, *Briefwechsel*, 101, letter of 18 January 1850.
106. Ibid., 103, letter of 7 February 1850.
107. Du Bois-Reymond and Helmholtz, *Dokumente einer Freundschaft*, 107, letter of 18 March 1851.
108. Du Bois-Reymond, "Der physiologische Unterricht sonst und jetzt," 378.
109. I agree with Norton Wise that the motivations of Berlin's physiologists "had deep roots in the local culture, particularly in the attempts of [the] educated elite (*Bildungsbürgertum*) to interrelate neo-humanist ideals with new industrial realities." See Wise, "What's in a Line?" 5.
110. Sven Dierig argues that du Bois-Reymond viewed the experimental scientist as a classical athlete, so that manual labor done for science's sake became part of his personal cultivation, or *Bildung*. See Dierig, "'Die Kunst des Versuchens,'" 124–132.
111. Du Bois-Reymond to Bence Jones, 17 February 1851, no. 1, s. 1, SBPK, HA, SD, NL Emil du Bois-Reymond, 3k 1852 (4).
112. Du Bois-Reymond, "Der physiologische Unterricht sonst und jetzt," 364, quoted in Finkelstein, "Emil du Bois-Reymond," 207; and Rothschuh, *History of Physiology*, 223. Both translate it differently.
113. Dierig, "'Die Kunst des Versuchens,'" 124–132.
114. Ibid., 138–140. Norton Wise argues that to artists, artisans, and scientists of the 1840s, "sharp outlines and smooth surfaces symbolized definiteness, unity, and above all, rationality, seen in sharp contrast to an over-embellished, sentimental Rococo style." Wise, "What's in a Line?" 9.
115. Du Bois-Reymond and Ludwig, *Two Great Scientists*, 4, letter of 4 January 1848.
116. Finkelstein, "Emil du Bois-Reymond," 128.
117. Emil du Bois-Reymond, "Naturwissenschaft und bildende Kunst," in his *Reden*, 2nd ed. (Leipzig: Veit, 1912), 2: 397.
118. Finkelstein offers good explanations of how du Bois-Reymond's galvanometers worked in "Emil du Bois-Reymond," 222–223, and "M. du Bois-Reymond Goes to Paris," *British Journal for the History of Science* 36 (2003): 264–266.
119. Rothschuh, "Emil du Bois-Reymond und die Elektrophysiologie der Nerven," 88.
120. Ruff, *Emil du Bois-Reymond*, 30; Rothschuh, "Emil du Bois-Reymond und die Elektrophysiologie der Nerven," 95.
121. Müller, *Briefe an Anders Retzius*, 56, letter of 24 March 1847.
122. Emil du Bois-Reymond and Karl Ludwig, *Zwei grosse Naturforscher des 19. Jahrhunderts: Ein Briefwechsel zwischen Emil du Bois-Reymond und Karl Ludwig*, ed. Estelle du Bois-Reymond (Leipzig: Johann Ambrosius Barth, 1927), 185, n. 10. I am grateful to Sven Dierig for locating this source.
123. Du Bois-Reymond, *Jugendbriefe*, 93, letter of 25 May 1841. In German, du Bois-Reymond calls the person or people rolling the galvanometer coil "die G.," indicating either a woman or a feminine noun like "die [physikalische] Gesellschaft."
124. Emil du Bois-Reymond, "Über die Lebenskraft," 1.
125. In this quotation from *King Henry IV, Part I* (I, iii), du Bois-Reymond has omitted a line. Shakespeare's text reads, "And now I will unclasp a secret book,/And to your quick-conceiving discontents/I'll read you matter deep and dangerous,/As full of peril, and advent'rous spirit,/As to o'erwalk a current, roaring loud,/On the unsteadfast footing of a spear." http://www.onlineliterature.com/shakespeare/henryIV1.

126. The image of two electrodes poised over a river again suggests the association of life force with "a dangerously feminine sea of romanticism," a tendency toward emasculating, uncontrolled thought that must be defined and disciplined with laboratory instruments. See Wise, "What's in a Line?" 30.

127. Finkelstein argues that du Bois-Reymond's friendships followed a pattern of "enthusiasm, annoyance, and indifference." See Finkelstein, "Emil du Bois-Reymond," 180–182.

128. Du Bois-Reymond, *Jugendbriefe*, 60, letter of 27 May 1840. Timothy Lenoir believes that Reichert rejected du Bois-Reymond, telling him that he did not "penetrate and think in the spirit of nature." Timothy Lenoir, *Instituting Science: The Cultural Production of Scientific Disciplines* (Stanford, CA: Stanford University Press, 1997), 139. Lenoir quotes a letter from du Bois-Reymond to Hallmann of 25 May 1841, *Jugendbriefe*, 89.

129. On du Bois-Reymond's interest in gymnastics and the social role of gyms in the 1830s, see Dierig, "'Die Kunst des Versuchens,'" 126–128; and Finkelstein, "Emil du Bois-Reymond," 48–49.

130. Estelle du Bois-Reymond, "Einleitung," in du Bois-Reymond's *Jugendbriefe*, 2–3; Finkelstein, "Emil du Bois-Reymond," 135–139.

131. Estelle du Bois-Reymond, "Einleitung," in *Jugendbriefe*, 3–6.

132. Ibid., 7.

133. Humboldt and du Bois-Reymond, *Briefwechsel*, 65, n. 1, letter of 13 February 1840.

134. Estelle du Bois-Reymond, "Einleitung," in *Jugendbriefe*, 2; Finkelstein, "Emil du Bois-Reymond," 140.

135. Du Bois-Reymond, *Jugendbriefe*, 18, letter of 26 September 1839, quoted in Finkelstein, "Emil du Bois-Reymond," 144. My translation differs from Finkelstein's, but his rendering of "Der Unbefangene" as "the ingénue" is particularly apt because of du Bois-Reymond's French-language background.

136. Finkelstein proposes that du Bois-Reymond projected himself onto Müller, writing that "he characterized Müller in terms that he wished to apply to himself, as a man whose decision to investigate the material world followed from his strength of will." Finkelstein, "Emil du Bois-Reymond," 163. I agree with this view only in part. While du Bois-Reymond depicted the experimental physiologist Müller of the 1820s as the scientist he wanted to be, he defined himself in opposition to the comparative anatomist Müller of the 1840s and 1850s. If the idealized physiologist was a projection, the self-crippling anatomist was a straw man who showed the superiority of du Bois-Reymond's own scientific assumptions.

137. Du Bois-Reymond, *Jugendbriefe*, 13, letter of 26 September 1839.

138. Finkelstein believes that du Bois-Reymond had to "court" Müller, and did so by "concealing his intimidation through bluster." See Finkelstein, "Emil du Bois-Reymond," 176.

139. Du Bois-Reymond, *Jugendbriefe*, 27–28, letter of 15 October 1839.

140. Ibid., 35, letter of 7 December 1839.

141. Ibid., 41, letter of 13 January 1840.

142. Ibid., 42, letter of 3 February 1840.

143. Ibid., 44, letter of 19 February 1840.

144. Finkelstein notes that when du Bois-Reymond describes his standing with Müller during this period, "His letters to Hallmann read like military dispatches." Finkelstein, "Emil du Bois-Reymond," 178.

145. Du Bois-Reymond, *Jugendbriefe*, 56–57, letter of 27 May 1840, and 63, letter of 27 July 1840, quoted in Finkelstein, "Emil du Bois-Reymond," 179. My translation differs from Finkelstein's.

146. Du Bois-Reymond, *Jugendbriefe*, 68, letter of 19 August 1840.

147. Ibid., 70, letter of 19 August 1840.
148. Ibid., 77, letter of 17 October 1840.
149. Ibid., 102, letter of 25 October 1841.
150. Ibid., 48, letter of 31 March 1840.
151. Du Bois-Reymond, "Der physiologische Unterricht sonst und jetzt," 361.
152. Du Bois-Reymond, "Gedächtnisrede," 289.
153. Finkelstein believes that the "consistency of [Müller's] success in finding happy matches [of students and projects] suggests more than coincidence." See Finkelstein, "Emil du Bois-Reymond," 195.
154. Ibid., 227.
155. Du Bois-Reymond, *Jugendbriefe*, 111, letter of March 1843.
156. Du Bois-Reymond and Ludwig, *Two Great Scientists*, 41, letter of 26 June 1849.
157. Ibid., 45, letter of 7 August 1849.
158. Ibid., 90–91, letter of 7 October 1855.
159. Ibid., 87, letter of 27 December 1854.
160. Du Bois-Reymond and Helmholtz, *Dokumente einer Freundschaft*, 158, letter of 27 April 1856.
161. Emil du Bois-Reymond to Henry Bence Jones, 11 February-19 March 1857, no. 24, s. 42, SBPK, HA, SD, NL Emil du Bois-Reymond, 3k 1852 (4).
162. Du Bois-Reymond and Helmholtz, *Dokumente einer Freundschaft*, 175, letter of 26 July 1857.
163. Ibid., 146, letter of 21 June 1854; 147, letter of 7 July 1854; and 149, letter of 16 August 1854.
164. Ibid., 186, letters of 6 and 29 May 1858.
165. Du Bois-Reymond and Ludwig, *Two Great Scientists*, 97–98, letter of 7 November 1858.
166. Du Bois-Reymond to Bence Jones, 19 May 1858, no. 31, s. 54, SBPK, HA, SD, NL Emil du Bois-Reymond, 3k 1852 (4).
167. Du Bois-Reymond to Bence Jones, 2 May 1854, no. 14, s. 27, SBPK, HA, SD, NL Emil du Bois-Reymond, 3k 1852 (4).
168. Humboldt and du Bois-Reymond, *Briefwechsel*, 85, n. 1, letter of 17 April 1849.
169. Cahan, "Introduction," in Helmholtz, *Science and Culture*, xi.
170. Keith M. Anderton believes that du Bois-Reymond became "a paramount, highly visible symbol of the German scientific enterprise." Keith M. Anderton, "The Limits of Science: A Social, Political, and Moral Agenda for Epistemology in Nineteenth-Century Germany," Ph.D. diss., Harvard University, 1993, 4.
171. Finkelstein, "Emil du Bois-Reymond," 107.
172. Ruff, *Emil du Bois-Reymond*, 75. On du Bois-Reymond as orator, see W. Kloppe, "Du Bois-Reymonds Rhetorik im Urteil einiger seiner Zeitgenossen," *Deutsches medizinisches Journal* 27 (1958): 80–82.
173. Du Bois-Reymond and Helmholtz, *Dokumente einer Freundschaft*, 206, letter of 9 December 1863.
174. Ibid., 238, letter of 15 May 1870.
175. I am grateful to Kristin Asdal for urging me to think more about the social and political dimensions of du Bois-Reymond's memorial address for Müller in her commentary on my talk "The Lab as a Literary Creation," in Oslo, 6 September 2005.
176. Du Bois-Reymond, "Gedächtnisrede," 144. In German, the word *Geschichte* can mean either story or history, depending on context. In this sentence, du Bois-Reymond

is paraphrasing a line from the French physiologist Marie Jean Pierre Flourens's (1794–1867) memorial address for Georges Cuvier. Dorinda Outram has written that the memorial addresses read before the French Academy of Sciences often had political agenda, justifying the activities of living scientists by selectively praising those of dead ones. See Dorinda Outram, "The Language of Natural Power: The 'Éloges' of Georges Cuvier and the Public Language of Nineteenth-Century Science," *History of Science* 16 (1978): 153–178.

177. Finkelstein writes that "Müller's obituary offered German physiology its first narrative, . . . a history of physiology recounted in du Bois-Reymond's own image." See Finkelstein, "Emil du Bois-Reymond," 174. Nicholas Jardine argues that in du Bois-Reymond's memorial address for Müller, the physiologist is "out to advance [his] personal position, taking on the mantle and mystique of Müller by portraying him in [his] own image." Nicholas Jardine, "The Mantle of Müller and the Ghost of Goethe," 306.

178. In this description I rely on Frederic L. Holmes's assessment of Helmholtz's recollection of Müller, "a reconstruction that fitted an early episode in his own life into the trajectory of his prior and subsequent experiences." See Frederic L. Holmes, "The Role of Johannes Müller in the Formation of Helmholtz's Physiological Career," in *Universalgenie Helmholtz: Rückblick nach 100 Jahren,* ed. Lorenz Krüger (Berlin: Akademie Verlag, 1994), 4.

179. Emil du Bois-Reymond to Henry Bence Jones, undated letter, no. 32, s. 54, SBPK, HA, SD, NL Emil du Bois-Reymond, 3k 1852 (4). I thank Gabriel Finkelstein for alerting me to these comments on the "Gedächtnisrede" in du Bois-Reymond's letters to Bence Jones.

180. Du Bois-Reymond and Helmholtz, *Dokumente einer Freundschaft,* 187, letter of 14 July 1858.

181. Du Bois-Reymond to Bence Jones, undated letter, no. 32, s. 55, SBPK, HA, SD, NL Emil du Bois-Reymond, 3k 1852 (4).

182. Du Bois-Reymond to Bence Jones, 4 March 1860, no. 38, s. 64–65, SBPK, HA, SD, NL Emil du Bois-Reymond, 3k 1852 (4).

183. Du Bois-Reymond's notes and correspondence for the memorial address for Müller have been gathered by the SBPK and can be found in HA, SD, NL Emil du Bois-Reymond, Kasten 2, Mappe 3. His list of German, Austrian, French, English, and Italian obituaries to consult is on s. 1.

184. Du Bois-Reymond's notes on Müller's Cultural Ministry file constitute Bl. 1–15 of Kasten 2, Mappe 3, SBPK, HA, SD, NL du Bois-Reymond.

185. Max Müller to Emil du Bois-Reymond, 2 December 1858, SBPK, HA, SD, NL du Bois-Reymond, Kasten 2, Mappe 3, Bl. 28–30.

186. Du Bois-Reymond, "Gedächtnisrede," 299, n. 1.

187. Ibid., 146, 149.

188. Ibid., 159.

189. Ibid., 159–160.

190. Ibid., 196.

191. Ibid., 167.

192. Ibid., 203.

193. Ibid., 185.

194. Ibid., 215–216.

195. Rheinberger believes that in the memorial address, du Bois-Reymond tried to depict Müller's move from confused vitalism to mature empiricism as a sort of religious conversion, but ran into trouble because he was also arguing that Müller remained a vitalist. See Hans-Jörg Rheinberger, "From the 'Originary Phenomenon' to the 'System of Pelagic Fish-

ery': Johannes Müller (1801–1858) and the Relation between Physiology and Philosophy," in *From Physico-Theology to Bio-Technology: Essays in the Social and Cultural History of Biosciences. A Festschrift for Mikulás Teich*, ed. Kurt Bayertz and Roy Porter (Atlanta, GA: Rodopi, 1998), 151–52, n. 71.

196. The first half of the address supports Finkelstein's claim that "du Bois-Reymond cast Müller as an emblematic physiologist, but making a common Romantic error, he sympathized with the figure too much." See Finkelstein, "Emil du Bois-Reymond," 174. The second half points to a different reading, however. Once du Bois-Reymond turns critical, his chief narrative purpose is to show how Müller's philosophical beliefs prevented him from achieving his scientific potential. In the first half, Müller comes across as an idealized physiologist; in the second, as an example of how a scientist can go wrong.

197. Du Bois-Reymond, "Gedächtnisrede," 217.

198. Ibid., 218. Here again, du Bois-Reymond practices what Harold Bloom has called "tessera," appropriating the *Zweckmässigkeit* of Kant and nature philosophers and giving it a new meaning.

199. Ibid., 219.

200. Ibid., 222.

201. Ibid., 282.

202. Ibid., 238.

203. "By then," writes du Bois-Reymond, "having to compete with so many others, it must have seemed impossible to him always to be number one." Ibid., 238.

204. Ibid., 244.

205. Ibid., 245.

206. Rheinberger believes that Müller fished rather than raising embryos in tanks because without modern heating and aeration systems, it was too hard to keep them alive. Rheinberger, personal communication.

207. Du Bois-Reymond, "Gedächtnisrede," 290–291.

208. Ibid., 293.

209. Ibid., 265.

210. Ibid., 267.

211. Ibid., 271.

212. Ibid., 274–75.

213. Ibid., 275.

214. Ibid., 278–81.

215. Du Bois-Reymond to Bence Jones, 23 July 1860, no. 39, s. 69, SBPK, HA, SD, NL Emil du Bois-Reymond, 3k 1852 (4).

216. Gordon S. Haight, *George Eliot: A Biography* (London: Penguin, 1992), 171.

217. Bence Jones to du Bois-Reymond, 23 November 1860, s. 310, SBPK, HA, SD, NL Emil du Bois-Reymond, 3k 1852 (4).

218. Keith M. Anderton has proposed that du Bois-Reymond, Helmholtz, and Virchow "encountered in Müller . . . a hindering epistemological limit to investigation" because Müller would not "accept the complete explicability of the human organism in physical terms." Anderton, "The Limits of Science," 17.

219. Bloom, *Anxiety of Influence*, 14. Bloom has called this maneuver "Clinamen." I have replaced the word "poem" with the word "science."

220. Ibid., 15. I have again replaced the word "poem" with the word "science."

221. Ibid., 16. I have replaced the words "poem" and "poet" with "science" and "scientist." Here Bloom's insight coincides with Finkelstein's idea that in writing the memorial

address for Müller, du Bois-Reymond projected himself onto his mentor. Finkelstein, "Emil du Bois-Reymond," 174.

222. Shakespeare, *Hamlet,* 86 (III, ii).

Chapter 4

1. Gabriel Ward Finkelstein, "Emil du Bois-Reymond: The Making of a Liberal German Scientist (1818–1851)" (Ph.D. diss., Princeton University, 1996), 208; Erna Lesky, "Ernst Wilhelm von Brücke," in *Dictionary of Scientific Biography,* 2: 530.

2. David Cahan, "Introduction," in *Letters of Hermann von Helmholtz to His Parents: The Medical Education of a German Scientist 1837–1846,* ed. David Cahan (Stuttgart: Franz Steiner, 1993), 26–27; Finkelstein, "Emil du Bois-Reymond," 213; Timothy Lenoir, *Instituting Science: The Cultural Production of Scientific Disciplines* (Stanford, CA: Stanford University Press, 1997), 139; Peter W. Ruff, *Emil du Bois-Reymond* (Leipzig: B. G. Teubner, 1981), 49–50.

3. Lenoir, *Instituting Science,* 140.

4. My discussion of Helmholtz's early life draws upon Cahan "Introduction," 4–8; and R. Steven Turner, "Hermann von Helmholtz," in *Dictionary of Scientific Biography,* 6: 241.

5. According to Helmholtz's biographer Leo Koenigsberger, "The more the young man's thoughts, the direction of his labors, and his whole scientific attitude . . . took him away from metaphysical speculation, the stronger and for some time the more irreconcilable became the contrast with the wholly speculative philosophy of his father." See Leo Koenigsberger, *Hermann von Helmholtz,* trans. Frances A. Welby (1906; New York: Dover, 1965), 30. See also Lenoir, *Instituting Science,* 144.

6. Cahan, "Introduction," 7.

7. Ibid., 8–9; Turner, "Hermann von Helmholtz," 241.

8. Cahan, "Introduction," 14.

9. Jean Eckenstein, *Der akademische Mentor für die Studirenden der Friedrich-Wilhelms-Universität zu Berlin* (Berlin: Wilhelm Schüppel, 1835), 119; Leopold Freiherr von Zedlitz, *Neustes Conversations—Handbuch für Berlin und Potsdam* (Berlin: A. W. Eisersdorff, 1834), 576–577.

10. Cahan, "Introduction," 19–20.

11. Ibid., 20.

12. Hermann von Helmholtz, *Letters of Hermann von Helmholtz to His Parents,* ed. David Cahan (Stuttgart: Franz Steiner, 1993), 44–45, letter of 31 October 1838 and 51, letter of 5 November 1838.

13. Cahan, "Introduction," 10, and 36, n. 7.

14. Helmholtz, *Letters of Hermann von Helmholtz to His Parents,* 45–46, n. 8, letter of 31 October 1838.

15. Cahan, "Introduction," 17.

16. Helmholtz, *Letters of Hermann von Helmholtz to His Parents,* 63, letter of 15 May 1839.

17. Helmholtz's notes on Müller's lectures can be found at the ABBAW, NL Hermann von Helmholtz, Nr. 538.

18. Cahan, "Introduction," 20.

19. Hermann von Helmholtz, "Das Denken in der Medicin," in his *Vorträge und Reden,* 4th ed. (Brunswick: Friedrich Vieweg, 1896), 2: 180, quoted in Cahan, "Introduction," 20. See also Hermann von Helmholtz, "Thought in Medicine," in *Science and Culture: Popular*

and Philosophical Essays, ed. David Cahan (Chicago: University of Chicago Press, 1995), 319. I have relied in part on Cahan's translation in his introduction to Helmholtz's letters, which is better than the nineteenth-century translation in *Science and Culture.*

20. Koenigsberger, *Hermann von Helmholtz*, 24.

21. Helmholtz, *Letters of Hermann von Helmholtz to His Parents*, 91, letter of 1 August 1842, quoted in Frederic L. Holmes, "The Role of Johannes Müller in the Formation of Helmholtz's Physiological Career," in *Universalgenie Helmholtz: Rückblick nach 100 Jahren*, ed. Lorenz Krüger (Berlin: Akademie Verlag, 1994), 14. I have relied in part on Holmes's translation.

22. Historians disagree about how closely Helmholtz worked with Müller on his thesis project. Leo Koenigsberger depicts a close collaboration, claiming—with no apparent evidence—that from early 1841 onward, Helmholtz "lived entirely in the circle of Müller's pupils." Koenigsberger, *Hermann von Helmholtz*, 21. Bruno Kisch, determined to uphold Remak's priority for the discovery that nerve fibers emerged from ganglion cells, wrote that Müller "persuaded one of his most brilliant younger students, Hermann Helmholtz, to investigate the problem in invertebrates. Helmholtz . . . confirmed Remak's findings convincingly." Bruno Kisch, "Robert Remak, 1815–1865," *Transactions of the American Philosophical Society* 44 (1954): 249. Frederic L. Holmes's assessment of Müller and Helmholtz's relationship seems the most reasonable: it is likely that Müller helped Helmholtz with his research from the start, but no one can exclude the possibility that Helmholtz designed it entirely on his own. See Holmes, "The Role of Johannes Müller," 17.

23. Johannes Müller, *Handbuch der Physiologie des Menschen* [hereafter HP], 3rd ed. (Koblenz: J. Hölscher, 1838), 1: 612, quoted in Holmes, "The Role of Johannes Müller," 16. My translation differs slightly from Holmes's.

24. On 25–26 December 1845, du Bois-Reymond wrote to Hallmann, "In the meantime, I've made Helmholtz's acquaintance, which has in fact brought me great pleasure." Emil du Bois-Reymond, *Jugendbriefe von Emil du Bois-Reymond an Eduard Hallmann*, ed. Estelle du Bois-Reymond (Berlin: Dietrich Reimer, 1918), 122, quoted in Cahan, "Introduction," 27. Cahan's translation differs slightly from mine.

25. Gabriel Finkelstein has pointed out the differences between du Bois-Reymond's and Helmholtz's scientific styles. See Finkelstein, "Emil du Bois-Reymond," 262.

26. Du Bois-Reymond's Berlin University transcript, reproduced in Gabriel Finkelstein's dissertation, indicates that he took Müller's theoretical and practical anatomy in the winter of 1839–1840; physiology and comparative anatomy in the summer of 1840; and pathological anatomy in the summer of 1842. Finkelstein, "Emil du Bois-Reymond," 335–336.

27. Cahan, "Introduction," 15.

28. Rolf Winau, *Medizin in Berlin* (Berlin: Walter de Gruyter, 1987), 76–81.

29. Information about the Charité in the remainder of this paragraph is from Eckenstein, *Der akademische Mentor*, 108–110; and Zedlitz, *Neustes Conversations—Handbuch*, 125–129.

30. Koenigsberger, *Hermann von Helmholtz*, 23.

31. Cahan, "Introduction," 25; Koenigsberger, *Hermann von Helmholtz*, 28–29.

32. Koenigsberger, *Hermann von Helmholtz*, 17.

33. Ibid., 52. I have taken this comment out of context. In this lecture, Helmholtz was "auditioning" for a job teaching anatomy at the Art Academy and was arguing that knowledge of anatomy is essential for producing realistic art.

34. Helmholtz's study appeared in Müller's *Archive* in 1843 as "Über das Wesen der Fäulnis und Gährung." Helmholtz's methods and findings are described in Lenoir, *The Strategy of Life*, 197–99; and Kathryn M. Olesko and Frederic L. Holmes, "Experiment, Quantification, and Discovery: Helmholtz's Early Physiological Researches, 1843–1850," in

Hermann von Helmholtz and the Foundations of Nineteenth-Century Science, ed. David Cahan (Berkeley: University of California Press, 1993), 53–54.

35. For fuller descriptions of Helmholtz's muscle chemistry experiments, see Lenoir, *The Strategy of Life*, 200–202; and Olesko and Holmes, "Experiment, Quantification and Discovery," 54–59.

36. Helmholtz's equipment and strategies in his physiological heat experiments are described in Lenoir, *The Strategy of Life*, 202–209; and Olesko and Holmes, "Experiment, Quantification and Discovery," 59–74.

37. Lenoir (*The Strategy of Life*, 211) and Olesko and Holmes ("Experiment, Quantification and Discovery," 66–67) suggest that Helmholtz's thought-intensive design of his setups for his animal heat studies led him to the Law of Conservation of Force.

38. The historians Timothy Lenoir and Frederic L. Holmes differ considerably in their representations of Helmholtz's relationship with Müller. Lenoir depicts Helmholtz's work of the 1840s as an aggressive attempt to disprove Müller's notion of life force, whereas Holmes points to Müller's inspiration of these early studies and his enthusiastic promotion of Helmholtz's work. Müller's proactive attempt to free Helmholtz so that he could perform more research, and his powerful recommendations of the young experimenter, support Holmes's view of Müller as an admiring mentor. See Holmes, "The Role of Johannes Müller," 20–21; Lenoir, *The Strategy of Life*, 195; and Olesko and Holmes, "Experiment, Quantification and Discovery," 82.

39. Quoted in Holmes, "The Role of Johannes Müller," 19. I have used Holmes's translation.

40. Cahan, "Introduction," 29.

41. The nineteenth-century translator of Helmholtz's popular lectures rendered *Studiengenossen* as "fellow student," falsely implying that Ludwig studied with Müller in Berlin. See Helmholtz, "Thought in Medicine," 320, and "Das Denken in der Medicin," 182. Holmes believes that in light of Helmholtz's "glaring error" of identifying Ludwig as Müller's student, we should be cautious in reading Helmholtz's "distant recollections" of Müller in this essay. See Holmes, "The Role of Johannes Müller," 18. *Studiengenossen* does not necessarily mean "fellow student," however.

42. Karl E. Rothschuh, *History of Physiology*, ed. and trans. Guenter B. Risse (Huntington, NY: Robert E. Krieger, 1973), 206.

43. George Rosen, "Carl Friedrich Wilhelm Ludwig," in *Dictionary of Scientific Biography*, 8: 540.

44. Paul F. Cranefield, "Carl Ludwig and Emil du Bois-Reymond: A Study in Contrasts," *Gesnerus* 45 (1988): 275; Rosen, "Carl Friedrich Wilhelm Ludwig," 540.

45. Rothschuh, *History of Physiology*, 206.

46. Emil du Bois-Reymond and Carl Ludwig, *Two Great Scientists of the Nineteenth Century: Correspondence of Emil du Bois-Reymond and Carl Ludwig*, ed. Paul F. Cranefield, trans. Sabine Lichtner-Ayed (Baltimore: Johns Hopkins University Press, 1982), 24, letter of 12 February 1849.

47. Ibid., 30, letter of 30 March 1849. Ludwig was staying at his friend Volkmann's house, and Volkmann had offered to put du Bois-Reymond up as well.

48. Emil du Bois-Reymond and Hermann von Helmholtz, *Dokumente einer Freundschaft: Briefwechsel zwischen Hermann von Helmholtz und Emil du Bois-Reymond 1846–1894* (Berlin: Akademie Verlag, 1986), 89, letter of 30 December 1849.

49. Du Bois-Reymond and Ludwig, *Two Great Scientists*, 23, letter of 12 February 1849.

50. Ibid., 29, letter of 25 Febrary 1849, and 34, letter of 17 May 1849.

51. Ibid., 41, letter of 26 June 1849.

52. Ibid., 30, letter of 7 April 1849.

53. Ibid., 32, letter of 17 May 1849.

54. Ibid., 32–34, letter of 17 May 1849.

55. Wilhelm Haberling, *Johannes Müller: Das Leben des rheinischen Naturforschers* (Leipzig: Akademische Verlagsgesellschaft, 1924), 329–330, letter of summer 1849.

56. Finkelstein, "Emil du Bois-Reymond," 260.

57. Du Bois-Reymond and Helmholtz, *Dokumente einer Freundschaft*, 158, letter of 27 April 1856.

58. Gabriel Finkelstein believes that du Bois-Reymond "did not feel perfectly at ease in [Helmholtz's] company," probably because he and Helmholtz had such different personalities and scientific styles. Whereas du Bois-Reymond stuck closely to electrophysiology, Helmholtz kept up with many fields at once and used discoveries from one to advance the experimental technology of another. Finkelstein concludes that "a good deal of the distance du Bois-Reymond felt toward Helmholtz probably derived from jealousy." Finkelstein, "Emil du Bois-Reymond," 261–262.

59. Du Bois-Reymond and Ludwig, *Two Great Scientists*, 42, letter of 26 June 1849.

60. Olesko and Holmes, "Experiment, Quantification and Discovery," 87–88.

61. Ibid., 89, n. 86.

62. Du Bois-Reymond and Helmholtz, *Dokumente einer Freundschaft*, 92, letter of 19 March 1850.

63. Olesko and Holmes, "Experiment, Quantification and Discovery," 90. Olesko and Holmes quote an unpublished letter from Müller to Helmholtz of 7 February 1850, ABBAW, NL Helmholtz 318.

64. Du Bois-Reymond and Helmholz, *Dokumente einer Freundschaft*, 86, letter of 14 October 1849.

65. Ibid., 128, letter of 24 March 1852.

66. Ibid., 116, letter of 12 June 1851.

67. Turner, "Hermann von Helmholtz," 246.

68. Ibid., 242.

69. Du Bois-Reymond and Helmholtz, *Dokumente einer Freundschaft*, 155, letter of 16 March 1855.

70. Ibid., 157, letter of 14 October 1855.

71. Ibid., 168, letter of 18 May 1857.

72. Ibid., 177, letter of 5 March 1858.

73. Turner, "Hermann von Helmholtz," 242.

74. On 13 June 1857, du Bois-Reymond wrote to Henry Bence Jones, "There is much talk now about getting up a new professorship for experimental physiology in Heidelberg. There are four candidates in view, Ludwig, Brücke, Helmholtz, and myself. Brücke and Helmholtz seem little disposed to leave their present situation, and as for Ludwig it is almost certain that the Heidelberg people cannot give him as much as he will ask and as he wants. It thus becomes very likely that I shall remain the only candidate, and I am resolved to go and live at Heidelberg with Bunsen and Kirchhoff unless the Prussian government grant me several demands which it is exceedingly improbable they will do." SBPK, HA, SD, NL du Bois-Reymond, 3k 1852 (4), no. 25, s. 44.

75. Du Bois-Reymond and Ludwig, *Two Great Scientists*, 70, letter of 6 February 1852.

76. Du Bois-Reymond and Helmholtz, *Dokumente einer Freundschaft*, 178, letter of 15 March 1858.

77. Ibid., 209, letter of 15 May 1864.

78. Turner, "Hermann von Helmholtz," 243.

79. Ruff, *Emil du Bois-Reymond*, 63–64.
80. Rosen, "Carl Friedrich Wilhelm Ludwig," 540.
81. Du Bois-Reymond and Ludwig, *Two Great Scientists*, 66, letter of 5 August 1851.
82. Ibid., 95, letter of 5 January 1857.
83. Finkelstein, "Emil du Bois-Reymond," 302.
84. Karl E. Rothschuh, "Emil Heinrich du Bois-Reymond,"in *Dictionary of Scientific Biography*, 4: 201.
85. Ruff, *Emil du Bois-Reymond*, 71.
86. Rosen, "Karl Friedrich Wilhelm Ludwig," 540–541.
87. Timothy Lenoir has argued that Helmholtz accepted the Law of Specific Sense Energies as a "fundamental law of sensory physiology," but he saw it as the starting point for investigations, not the final word on how much people could know. Timothy Lenoir, "Helmholtz, Müller, und die Erziehung der Sinne," in *Johannes Müller und die Philosophie*, ed. Michael Hagner and Bettina Wahrig-Schmidt (Berlin: Akademie Verlag, 1992), 215.
88. Hermann von Helmholtz, "Ueber die Natur der menschlichen Sinnesempfindungen," in his *Gesammelte Schriften*, ed. Jochen Brüning (Hildesheim: Olms-Weidmann, 2003), 1.2.2: 593.
89. Ibid., 605.
90. Hermann von Helmholtz, *Helmholtz's Treatise on Physiological Optics*, ed. James P. C. Southall (Bristol, UK: Thoemmes Press, 2000), 2: 309.
91. Ibid., 12.
92. Ibid., 19–20.
93. Ibid., 3: 228.
94. Lenoir argues that "vision for Helmholtz is a purely psychological phenomenon in which the brain uses the eye as a measuring device for the purpose of constructing a practically efficient map of the external world." Timothy Lenoir, "The Eye as Mathematician: Clinical Practice, Instrumentation, and Helmholtz's Construction of an Empiricist Theory of Vision," in *Hermann Helmholtz and the Foundations of Nineteenth-Century Science*, ed. David Cahan (Berkeley: University of California Press, 1993), 123. See also Lenoir, "Helmholtz, Müller, und die Erziehung der Sinne," 216, and *Instituting Science*, 144.
95. Timothy Lenoir contends that in Helmholtz's discussions of spatial perception and arguments against nativism, he raises philosophical issues that go far beyond binocular vision. See Lenoir, "Helmholtz, Müller, und die Erziehung der Sinne," 208. The driving force of Helmholtz's visual studies, Lenoir proposes, is the desire to learn "how we get outside the world of our retinas and into the real world of things." Lenoir, "The Eye as Mathematician," 119.
96. Helmholtz, *Optics*, 3: 190.
97. Ibid., 543.
98. Helmholtz, "Das Denken in der Medicin," 2: 181.
99. Ibid., 181–82.
100. Ibid., 182.
101. Helmholtz, "Erinnerungen," 1: 9. Compare Helmholtz's "An Autobiographical Sketch," in *Science and Culture: Popular and Philosophical Essays*, ed. David Cahan (Chicago: University of Chicago Press, 1995), 385. Frederic L. Holmes has noticed that here, as in "Thought in Medicine," Helmholtz incorrectly depicts Müller as Ludwig's teacher, but he concludes that "distant memories can often be incorrect in details, yet accurate in their central meaning." When one considers that Müller inspired Ludwig through du Bois-Reymond and Helmholtz, it may be true that Müller "introduced" Ludwig to physiology. Holmes, "The Role of Johannes Müller," 18.

Chapter 5

1. My discussion of Virchow's early life in this paragraph relies on Erwin H. Ackerknecht, *Rudolf Virchow: Doctor, Statesman, Anthropologist* (Madison: University of Wisconsin Press, 1953), 3–5; Christian Andree, *Rudolf Virchow: Leben und Ethos eines grossen Arztes* (Munich: Langen Müller, 2002), 30–31; Byron Albert Boyd, "Rudolf Virchow: The Scientist as Citizen" (Ph.D. diss., University of North Carolina at Chapel Hill, 1981), 2; and Constantin Goschler, *Rudolf Virchow: Mediziner-Anthropologe-Politiker* (Cologne: Böhlau, 2002), 26–28.

2. Virchow's daughter, Marie Rabl, edited and published some of his letters to his parents: *Rudolf Virchow: Briefe an seine Eltern 1839 bis 1864* (Leipzig: Wilhelm Engelmann, 1906). Erwin Ackerknecht believes that these letters were "badly mutilated by the conservative father and later by the editing daughter" (Ackerknecht, *Rudolf Virchow*, 16). Recently, Christian Andree has prepared a complete volume of Virchow's letters to his parents, restoring the passages that Rabl excluded: Christian Andree, *Rudolf Virchow: Sämtliche Werke*, 59.4, *Briefe: Der Briefwechsel mit den Eltern 1839–1864* (Berlin: Blackwell Wissenschafts-Verlag, 2001). All subsequent references to Virchow's letters home are to Andree's edition. In 1990, L. J. Rather translated the letters in Rabl's edition: *Rudolf Virchow: Letters to His Parents 1839 to 1864* (Canton, MA: Science History Publications, 1990). Where indicated, I rely on Rather's translation; otherwise, all translations of Virchow's letters are my own. The letter mentioning the crops is in Andree, *Der Briefwechsel mit den Eltern*, 289, letter of 13 August 1846. The Virchow farm consisted of fifty *Morgen* (a *Morgen* is .6–.9 acre).

3. Thomas Nipperdey, *Deutsche Geschichte 1800–1866: Bürgerwelt und starker Staat* (Munich: C. H. Beck, 1994), 476.

4. Boyd, "Rudolf Virchow," 6.

5. On Virchow's uncles and his acceptance to the Pépinière, see Andree, *Rudolf Virchow*, 35; and Goschler, *Rudolf Virchow*, 37.

6. Andree, *Der Briefwechsel mit den Eltern*, 221, letter of 14 May 1843.

7. Ibid., 273, letter of 15 October 1845.

8. Ibid., 158, letter of 22 February 1842.

9. Ibid., 208, letter of 26 January 1843. The original meaning of *Knecht*, the word Virchow uses for "slave," is "farmhand."

10. Ibid., 258, letter of 15 December 1844.

11. Ibid., 225, letter of 30 July 1843.

12. Ibid., 264, letter of 9 May 1845.

13. Ibid., 265, letter of 24 July 1845. Marie Rabl expurgated Virchow's passage attacking the Prussian monarchy and police state; it has been restored in Andree's edition.

14. Virchow lists the times and locations of his first semester's classes in ibid., 33–34, letter of 18 November 1839.

15. Ibid., 37, letter of 5 December 1839.

16. Although he had not received the prestigious Berlin anatomy and physiology chair, Schlemm had stayed on as Müller's colleague, teaching the clinically oriented anatomy courses that Müller chose not to supervise.

17. Andree, *Der Briefwechsel mit den Eltern*, 33–34, letter of 18 November 1839.

18. Rudolf Virchow, "Erinnerungsblätter," *Virchows Archiv* 4 (1852): 543.

19. Virchow's notes on Müller's physiology course can be found at the ABBAW, NL R. Virchow, nr. 2803.

20. Ibid., nr. 2804.

21. Ibid., nr. 2805.

22. Andree, *Der Briefwechsel mit den Eltern*, 33, letter of 18 November 1839.
23. Ibid., 38, letter of 5 December 1839.
24. Ibid., 246, letter of 4 July 1844.
25. Rudolf Virchow, "Über die Reform der pathologischen und therapeutischen Anschauungen durch die mikroskopischen Untersuchungen," *Virchows Archiv* 1 (1847): 208.
26. Rudolf Virchow, "Prospectus," *Virchows Archiv* 1 (1847).
27. Andree, *Rudolf Virchow*, 39.
28. Andree, *Der Briefwechsel mit den Eltern*, 222, letter of 3 June 1843.
29. Ibid., 267, letter of 24 July 1845.
30. Ibid., 214, letter of 7 April 1843; 221, letter of 3 June 1843; and 297, letter of 1 May 1847.
31. Virchow describes these negotiations in a letter to his father on 17 March 1843. Ibid., 211.
32. Ibid., 217–218, letter of 14 May 1843.
33. Ibid., 218, letter of 14 May 1843.
34. Ibid., 229, letter of 30 August 1843.
35. Ibid., 242, letter of 8 March 1844.
36. Goschler, *Rudolf Virchow*, 41.
37. Rolf Winau, "Rudolf Virchow und die Berliner Medizin," *Notabene medici* 12 (1994): 433.
38. Wilhelm Waldeyer, "Der Unterricht in den anatomischen Wissenschaften an der Universität Berlin im ersten Jahrhundert ihres Bestehens," *Berliner klinische Wochenschrift* (10 October 1910): 1863. I thank Volker Hess for bringing this source to my attention.
39. For a history of the power struggles between Charité and university instructors, see Manfred Stürzbecher, "Zur Geschichte der Instruktionen des Prosektors der Charité 1831–1833," *Janus* 52 (1965): 40–75, and "Die Prosektur der Berliner Charité im Briefwechsel zwischen Robert Froriep und Rudolf Virchow," in *Beiträge zur Berliner Medizingeschichte: Quellen und Studien zur Geschichte des Gesundheitswesens vom 17. bis zum 19. Jahrhundert* (Berlin: Walter de Gruyter, 1966), 156–221.
40. Cay-Rüdiger Prüll, "Zwischen Krankenversorgung und Forschungsprimat: Die Pathologie an der Berliner Charité im 19. Jahrhundert," *Jahrbuch für Universitätsgeschichte* 3 (2000): 91–92.
41. Stürzbecher, "Zur Geschichte der Instruktionen des Prosektors," 43.
42. Stürzbecher, "Die Prosektur der Berliner Charité," 164.
43. Prüll, "Zwischen Krankenversorgung und Forschungsprimat," 88.
44. Stürzbecher, "Zur Geschichte der Instruktionen des Prosektors," 51.
45. Prüll, "Zwischen Krankenversorgung und Forschungsprimat," 89; Stürzbecher, "Die Prosektur der Berliner Charité," 173.
46. On Müller's support of Froriep's appointment, see Prüll, "Zwischen Krankenversorgung und Forschungsprimat," 92; and Stürzbecher, "Die Prosektur der Berliner Charité," 167.
47. Stürzbecher, "Zur Geschichte der Instruktionen des Prosektors," 70.
48. Prüll, "Zwischen Krankenversorgung und Forschungsprimat," 90.
49. Stürzbecher, "Die Prosektur der Berliner Charité," 169.
50. Prüll, "Zwischen Krankenversorgung und Forschungsprimat," 93.
51. Andree, *Der Briefwechsel mit den Eltern*, 275, letter of 14 December 1845.
52. Stürzbecher, "Die Prosektur der Berliner Charité," 160–161.
53. Stürzbecher, "Zur Geschichte der Instruktionen des Prosektors," 43.
54. Ibid., 48, 59.
55. Stürzbecher, "Die Prosektur der Berliner Charité," 171–172.

56. Ibid., 170, and 190, letter of Froriep to Virchow, 29 March 1847.

57. Ibid., 172.

58. Prüll, "Zwischen Krankenversorgung und Forschungsprimat," 92–93.

59. Andree, *Der Briefwechsel mit den Eltern*, 285–286, letter of 25 May 1846. I have relied partly on L. J. Rather's translation, *Virchow: Letters to His Parents*, 67.

60. Stürzbecher, "Die Prosektur der Berliner Charité," 179–80, letter of Virchow to Froriep, 2 August 1846, and 187, letter of Virchow to Froriep, 2 March 1847.

61. On Froriep's role in introducing Virchow to contemporary research, see Ackerknecht, *Rudolf Virchow*, 8; Goschler, *Rudolf Virchow*, 48; and P. Krietsch, "Zur Geschichte der Prosektur der Charité Berlin," *Zentralblatt für Pathologie* 137 (1991): 534.

62. Stürzbecher, "Die Prosektur der Berliner Charité," 207, letter of Froriep to Virchow, 21 April 1852.

63. Andree, *Der Briefwechsel mit den Eltern*, 254, letter of 29 November 1844.

64. For a fuller discussion of Virchow's studies of pus and blood, see Ackerknecht, *Rudolf Virchow*, 59–70, from which I have drawn this brief summary.

65. In the same letter in which Virchow described his new research theme, he wrote that Froriep had said it was "abolutely necessary" to learn English, and asked his father for money to buy French and English journals. Andree, *Der Briefwechsel mit den Eltern*, 255, letter of 29 November 1844.

66. Ibid., 270, letter of 27 August 1845. In German, Virchow calls Froriep his *liebenwürdigen Vorgesetzten*.

67. Stürzbecher, "Die Prosektur der Berliner Charité," 186, letter of Virchow to Froriep, 2 March 1847.

68. Virchow, "Erinnerungsblätter," 543–544.

69. Andree, *Der Briefwechsel mit den Eltern*, 246, letter of 4 July 1844.

70. Ibid., 254, letter of 29 November 1844.

71. Ibid., 268, letter of 24 July 1845.

72. Stürzbecher, "Die Prosektur der Berliner Charité," 175.

73. Arleen Marcia Tuchman, *Science, Medicine, and the State in Germany: The Case of Baden 1815–1871* (New York: Oxford University Press, 1993), 64–65; Rolf Winau, *Medizin in Berlin* (Berlin: Walter de Gruyter, 1987), 139–140.

74. Krietsch, "Zur Geschichte der Prosektur der Charité Berlin," 535, 532.

75. Stürzbecher, "Die Prosektur der Berliner Charité," 178–179, letter of Virchow to Froriep, 2 August 1846.

76. Boyd, "Rudolf Virchow," 18–19.

77. Stürzbecher, "Die Prosektur der Berliner Charité," 179, letter of Virchow to Froriep, 2 August 1846.

78. Andree, *Der Briefwechsel mit den Eltern*, 288, letter of 13 August 1846.

79. On the relation between Virchow's science and politics, see Ackerknecht, *Rudolf Virchow*, 43; Winau, "Rudolf Virchow und die Berliner Medizin," 435; and Woodruff D. Smith, *Politics and the Sciences of Culture in Germany 1840–1920* (Oxford: Oxford University Press, 1991), 53.

80. Andree, *Der Briefwechsel mit den Eltern*, 347, letter of 1 May 1848. I have relied partly on L. J. Rather's translation, *Rudolf Virchow: Letters to His Parents*, 88.

81. Andree, *Der Briefwechsel mit den Eltern*, 370, letter of 19 December 1848.

82. Ibid., 263, letter of 9 May 1845.

83. Boyd, "Rudolf Virchow," 14–15.

84. Andree, *Der Briefwechsel mit den Eltern*, 266–67, letter of 24 July 1845.

85. Kurd Schulz, "Rudolf Virchow und die Oberschlesische Typhusepidemie von 1848," *Jahrbuch des Schlesischen Friedrich-Wilhelms-Universität zu Breslau* 19 (1978): 116.
86. Boyd, "Rudolf Virchow," 21–22.
87. Andree, *Der Briefwechsel mit den Eltern,* 316, letter of 20 February 1848.
88. Ibid., 312, letter of 13 February 1848.
89. Ibid., 317–18, letter of 24 February 1848. I have relied partly on Rather's translation, *Letters to His Parents,* 77.
90. Schulz, "Rudolf Virchow und die Oberschlesische Typhusepidemie," 111.
91. Boyd, "Rudolf Virchow," 10–11.
92. Andree, *Der Briefwechsel mit den Eltern,* 226, letter of 30 July 1843.
93. Schulz, "Rudolf Virchow und die Oberschlesische Typhusepidemie," 111.
94. Andree, *Der Briefwechsel mit den Eltern,* 317, letter of 24 February 1848.
95. Virchow, "Erinnerungsblätter," 541.
96. Ackerknecht, *Rudolf Virchow,* 149–150.
97. Carl Ludwig Schleich, *Besonnte Vergangenheit: Lebenserinnerungen 1859–1919* (Berlin: Ernst Rowohlt, 1921), 62, quoted in Ackerknecht, *Rudolf Virchow,* 15–16. My translation differs from Ackerknecht's.
98. Andree, *Der Briefwechsel mit den Eltern,* 330, 334. I have relied in part on Rather's translation, *Letters to His Parents,* 82, 84.
99. Ibid., 345, letter of 1 May 1848.
100. On Virchow's political activity in 1848–1849, see Boyd, "Rudolf Virchow," 28; Goschler, *Rudolf Virchow,* 64–72; and Dieter Schott, "'Die Medizin ist eine sociale Wissenschaft'—Rudolf Virchow und die 'Medicinische Reform' in der Revolution 1848/49," in *Die Revolution hat Konjunktur: Soziale Bewegung, Alltag und Politik in der Revolution 1848/49* (Münster: Westfälisches Dampfboot, 1999), 95.
101. Andree, *Der Briefwechsel mit den Eltern,* 346, letter of 1 May 1848. I have relied partly on Rather's translation, *Letters to His Parents,* 87–88.
102. Ibid., 335–36, letter of 24 March 1848.
103. Ibid., 365, letter of 29 September 1848.
104. Boyd, "Rudolf Virchow," 40–41.
105. Goschler, *Rudolf Virchow,* 74.
106. Boyd, "Rudolf Virchow," 42.
107. Andree, *Der Briefwechsel mit den Eltern,* 381, letter of 8 March 1849.
108. Ibid., 393, letter of 6 April 1849.
109. Stürzbecher, "Die Prosektur der Berliner Charité," 197, letter of Virchow to Froriep, 18 December 1849.
110. Andree, *Der Briefwechsel mit den Eltern,* 416, letter of 6 August 1849.
111. Andree, *Rudolf Virchow,* 78.
112. Ackerknecht, *Rudolf Virchow,* 11; Boyd, "Rudolf Virchow," 19–20; Goschler, *Rudolf Virchow,* 52–53.
113. Boyd, "Rudolf Virchow," 43–44.
114. Andree, *Rudolf Virchow,* 78–79; Goschler, *Rudolf Virchow,* 90, 115.
115. Schleich, *Besonnte Vergangenheit,* 181.
116. Goschler, *Rudolf Virchow,* 157.
117. On Müller's involvement in Virchow's call to Berlin, see Ackerknecht, *Rudolf Virchow,* 22–23; Andree, *Rudolf Virchow,* 66; and Goschler, *Rudolf Virchow,* 157–158.
118. Prüll, "Zwischen Krankenversorgung und Forschungsprimat," 95–97.
119. Andree, *Rudolf Virchow,* 66.

120. Goschler, *Rudolf Virchow*, 158.

121. For a summary of Virchow's political career, see Ackerknecht, *Rudolf Virchow*, 25–27, on which I rely in this paragraph.

122. For a detailed account of the duel challenge and responses throughout the German territories, see Hella Machetanz, "Trichinen und die Duell-Forderung Bismarcks an Virchow im Jahre 1865," *Medizinhistorisches Journal* 13 (1978): 297–306.

123. Virchow campaigned actively against anti-Semitism throughout the 1870s and 1880s. See Roger Thomas, "Rudolf Virchow: Pathologist, Public Health Physician, Liberal Politician, Anthropologist, and Opponent of Anti-Semitism," *Journal of Medical Biography* 7 (1999): 204; and Andrew Zimmerman, "Anti-Semitism as Skill: Rudolf Virchow's *Schulstatistik* and the Racial Composition of Germany," *Central European History* 32 (1999): 409–429. I thank Sander Gilman for informing me about Adolf Stoecker.

124. *Schlesische Zeitung* nr. 274 (15 June 1865), quoted in Machetanz, "Trichinen und die Duell-Forderung Bismarcks," 303.

125. Andree, *Der Briefwechsel mit den Eltern*, 271, letter of 15 October 1845. I have relied on Rather's translation, *Letters to His Parents*, 63.

126. Schott, "Die Medizin ist eine sociale Wissenschaft," 103.

127. Ackerknecht, *Rudolf Virchow*, 27.

128. Christian Andree, "Welches Verhältnis hatte Rudolf Virchow zu zeitgenössischen Dichtern, Künstlern, Verlegern und Editoren? Versuch einer Annäherung über die Korrespondenzpartner," *Würzburger medizinhistorische Mitteilungen* 12 (1994): 259–286.

129. Rudolf Virchow, *Cellular Pathology*, trans. Frank Chance (1858; reprint Ann Arbor, MI: Edwards Brothers, 1940), viii.

130. Letter of Virchow to Alexander Frantzius, 19 February 1850, quoted in Goschler, *Rudolf Virchow*, 187.

131. Goschler, *Rudolf Virchow*, 187.

132. Rudolf Virchow, "Zur pathologischen Physiologie des Bluts. Die Bedeutung der Milz- und Lymphdrüsen-Krankheiten für die Blutmischung (Leukaemie)," *Virchows Archiv* 5 (1853): 60–62.

133. My summary of Virchow's writing achievements in this paragraph relies on Ackerknecht, *Rudolf Virchow*, 17, 21, 25, 30, and 97–98.

134. Ibid., 35.

135. Schleich, *Besonnte Vergangenheit*, 183–184.

136. Andree, *Der Briefwechsel mit den Eltern*, 254–255, letter of 29 November 1844.

137. Ibid., 353–354, letter of 18 May 1848.

138. Keith Anderton has pointed out the parallels among du Bois-Reymond, Helmholtz, and Virchow as three students of Müller who worked actively to represent science to the public in particular ways. See Keith M. Anderton, "The Limits of Science: A Social, Political, and Moral Agenda for Epistemology in Nineteenth-Century Germany" (Ph.D. diss., Harvard University, 1993).

139. Andree, *Rudolf Virchow*, 130.

140. Ackerknecht, *Rudolf Virchow*, 146–47.

141. Nicholas Jardine, "The Mantle of Müller and the Ghost of Goethe: Interactions between the Sciences and Their Histories," in *History and the Disciplines: The Reclassification of Knowledge in Early Modern Europe*, ed. Donald R. Kelley (Rochester, NY: University of Rochester Press, 1997), 301–302.

142. Andree, *Der Briefwechsel mit den Eltern*, 225, letter of 30 July 1843.

143. Ibid., 232, letter of 25 October 1843.

144. Rudolf Virchow, *Disease, Life and Man: Selected Essays by Rudolf Virchow*, trans. L. J. Rather (Stanford, CA: Stanford University Press, 1958), 43.
145. Rudolf Virchow, "Ueber die Reform der pathologischen und therapeutischen Anschauungen," 213, 224.
146. Rudolf Virchow, "Die endogene Zellenbildung beim Krebs," *Virchows Archiv* 3 (1851): 219.
147. Rudolf Virchow, "Ueber Perlgeschwülste (Cholesteatoma Joh. Müllers)," *Virchows Archiv* 8 (1855): 373, 414–415.
148. Virchow, *Cellular Pathology*, 101. The German term is *Bindegewebe*.
149. Ibid., 399, 480–481.
150. Andree, *Der Briefwechsel mit den Eltern*, 811–816. Almost all of this letter was omitted from Marie Rabl's edition, but Andree has restored it in full. At the top of the first page Rose Virchow wrote "to be burned after reading."
151. Jardine, "The Mantle of Müller and the Ghost of Goethe," 297–298.
152. Rudolf Virchow, *Johannes Müller: Eine Gedächtnisrede* (Berlin: August Hirschwald, 1858), 14.
153. Ibid., 8, 10; Jardine, "The Mantle of Müller and the Ghost of Goethe," 303.
154. Virchow, *Johannes Müller: Eine Gedächtnisrede*, 3–4.
155. Ibid., 5.
156. Ibid., 44, n. 44.
157. Ibid., 34.
158. Ibid., 8–9.
159. Ibid., 20.
160. Ibid., 44, n. 44.
161. Ibid., 30.
162. Ibid., 8, 21.
163. Ibid., 43, n. 42.
164. Ibid., 27.
165. Ibid., 28.
166. Ibid., 28–29.
167. Ibid., 37.
168. Ibid., 46–47, n. 66.
169. Ibid., 47–48, n. 66.
170. Virchow, *Disease, Life and Man*, 184.
171. Ibid., 197.
172. Ibid., 204–205.
173. My summary of Virchow's work on blood in this paragraph draws upon Erwin Ackerknecht's in *Rudolf Virchow*, 60–66.
174. Ibid., 82–83.
175. Ibid., 109–110.
176. Ibid., 132–135.
177. For an evaluation of Virchow's opposition to the teaching of evolutionary theory in Prussian schools, see Jutta Kolkenbrock-Netz, "Wissenschaft als nationaler Mythos: Anmerkungen zur Haeckel-Virchow-Kontroverse auf der 50. Jahresversammlung deutscher Naturforscher und Ärzte in München (1877)," in *Nationale Mythen und Symbole in der zweiten Hälfte des 19. Jahrhunderts: Strukturen und Funktionen von Konzepten nationaler Identität*, ed. Jürgen Link and Wulf Wülfing (Stuttgart: Klett-Kotta, 1991), 212–236; and Peter Zigman, "Ernst Haeckel und Rudolf Virchow: Der Streit um den Charakter der Wissenschaft in der Auseinandersetzung um den Darwinismus," *Medizinhistorisches Journal* 35 (2000): 263–302.

178. Stürzbecher, "Die Prosektur der Berliner Charité," 181, letter of Virchow to Froriep, 2 August 1846.
179. Ibid., 199–200, letter of Virchow to Froriep, 18 December 1849.
180. Ackerknecht, *Rudolf Virchow*, 92.
181. Virchow, *Cellular Pathology*, 17.
182. Sven Dierig, "Rudolf Virchow und das Nervensystem: Zur Begründung der zellulären Neurobiologie," in *Das Gehirn—Organ oder Seele?*, ed. Ernst Florey and Olaf Breidbach (Berlin: Akademie Verlag, 1993), 58.
183. Virchow, *Disease, Life and Man*, 204.
184. Lazare Benaroyo, "Rudolf Virchow and the Scientific Approach to Medicine," *Endeavour* 22 (1998): 115. Benaroyo argues that Virchow's vision of the cell as the basic living unit helped him to escape materialism, with which he had never been comfortable. See also Boyd, "Rudolf Virchow," 58–60.
185. Ackerknecht, *Rudolf Virchow*, 56.
186. Thomas, "Rudolf Virchow," 206.
187. Ackerknecht, *Rudolf Virchow*, 147.

Chapter 6

1. Friedrich Merkel, *Jacob Henle: Ein Deutsches Gelehrtenleben* (Brunswick: Friedrich Vieweg, 1891), 173, quoted in Bruno Kisch, "Robert Remak, 1815–1865," *Transactions of the American Philosophical Society* 44 (1954): 232. In German, *filzig* means fuzzy like an old, balled-up sweater, implying filth only indirectly. Kisch reads Henle's remark as rampant hypocrisy and does not consider Henle's irony. Probably the anatomist was laughing at himself, as he so often did.
2. Emil du Bois-Reymond and Carl Ludwig, *Two Great Scientists of the Nineteenth Century: Correspondence of Emil du Bois-Reymond and Carl Ludwig*, ed. Paul F. Cranefield, trans. Sabine Lichtner-Ayed (Baltimore: Johns Hopkins University Press, 1982), 9, letter of 22 April 1848. In German, du Bois-Reymond's exact words are "eine fatale Clique jüdischer Literaten." Emil du Bois-Reymond and Carl Ludwig, *Zwei grosse Naturforscher des 19. Jahrhunderts: Ein Briefwechsel zwischen Emil du Bois-Reymond und Karl Ludwig*, ed. Estelle du Bois-Reymond (Leipzig: Johann Ambrosius Barth, 1927), 12.
3. Hermann von Helmholtz and Emil du Bois-Reymond, *Dokumente einer Freundschaft: Briefwechsel zwischen Hermann von Helmholtz und Emil du Bois-Reymond 1846–1894* (Berlin: Akademie Verlag, 1986), 216, letter of 18 February 1865.
4. Hermann von Helmholtz, *Letters of Hermann von Helmholtz to His Parents: The Medical Education of a German Scientist 1837–1846*, ed. David Cahan (Stuttgart: Franz Steiner Verlag, 1993), 79, letter of 1 September 1840.
5. Christian Andree, ed., *Rudolf Virchow: Sämtliche Werke* (Berlin: Blackwell, 2001), 59: 733, letter of 15 February 1856, quoted in Kisch, "Robert Remak," 280. Kisch translates the passage slightly differently.
6. Ernst Haeckel, *Entwicklungsgeschichte einer Jugend: Briefe an die Eltern 1852–1856* (Leipzig: K. F. Koehler, 1921), 40, letter of 17 February 1853.
7. Peter W. Ruff, *Emil du Bois-Reymond* (Leipzig: B. G. Teubner, 1981), 85.
8. All information in this paragraph derives from Kisch, "Robert Remak," 229–230; and Heinz-Peter Schmiedebach, *Robert Remak (1815–1865): Ein jüdischer Arzt im Spannungsfeld von Wissenschaft und Politik* (Stuttgart: Gustav Fischer, 1995), 13–15, 24.

9. Kisch, "Robert Remak," 230; Schmiedebach, *Robert Remak*, 150–154.
10. Kisch, "Robert Remak," 262–63.
11. Schmiedebach, *Robert Remak*, 157–158.
12. Thomas Nipperdey, *Deutsche Geschichte 1800–1866: Bürgerwelt und starker Staat* (Munich: C. H. Beck, 1994), 479.
13. Kisch, "Robert Remak," 231.
14. Anonymous, "Nachruf," *Deutsche Klinik* no. 44 (4 November 1865): 413, quoted in Kisch, "Robert Remak," 232. Kisch translates the passage differently and omits the words *hervorstechenden Gesichtszugen* (prominent features), which may be a coded anti-Semitic reference.
15. Schmiedebach, *Robert Remak*, 28.
16. Erich Hintzsche, "Robert Remak," in *Dictionary of Scientific Biography*, 11: 368; Kisch, "Robert Remak," 237.
17. Kisch, "Robert Remak," 277.
18. Schmiedebach, *Robert Remak*, 42–45.
19. Kisch, "Robert Remak," 277; Schmiedebach, *Robert Remak*, 45–50.
20. Kisch, "Robert Remak," 239–240; Schmidebach, *Robert Remak*, 28.
21. Kisch, "Robert Remak," 241; Schmiedebach, *Robert Remak*, 252–254.
22. Schmiedebach, *Robert Remak*, 41.
23. Kisch, "Robert Remak," 276–277; Schmiedebach, *Robert Remak*, 249–250.
24. J. M. S. Pearce, "Remak, Father and Son," *The Lancet* 347 (1996): 1669.
25. Schmiedebach, *Robert Remak*, 71.
26. Robert Remak, "Ueber die physiologische Bedeutung des organischen Nervensystems, besonders nach anatomischen Thatsachen," *Monatschrift für Medizin, Augenheilkunde und Chirurgie* 3 (1840): 228–229.
27. Edwin Clarke and L. S. Jacyna, *Nineteenth-Century Origins of Neuroscientific Concepts* (Berkeley: University of California Press, 1987), 84–85; Kisch, "Robert Remak," 234.
28. Robert Remak, "Ueber die Verrichtungen des organischen Nervensystems," *Neue Notizen aus dem Gebiete der Natur- und Heilkunde* 7 (1838): 69.
29. Robert Remak, "Neue Beiträge zur Kenntnis vom organischen Nervensystem," *Medicinische Zeitung* 9 (1840): 7.
30. Robert Remak, "Ueber die Ganglien der Herznerven des Menschen und deren physiologsische Bedeutung," *Wochenschrift für die gesammte Heilkunde* no. 10 (9 March 1839): 152.
31. Robert Remak, "Ueber extracellulare Entstehung thierischer Zellen und über Vermehrung derselben durch Theilung," *Müllers Archiv* (1852): 49, quoted in Kisch, "Robert Remak," 258, but translated differently.
32. Bruno Kisch insists that Remak should have priority for this discovery, arguing that Virchow unjustly took credit for Remak's idea. Kisch, "Robert Remak," 258–260.
33. Robert Remak, "Ein Beitrag zur Entwickelungsgeschichte der krebshaften Geschwülste," *Deutsche Klinik* 6 (1854): 173.
34. Robert Remak, "Ueber die Anzeigen zur Ausrottung krankhafter Geschwülste," *Medicinische Zeitung* 10 (1841): 132.
35. Remak, "Ueber extracellulare Entstehung thierischer Zellen," 57.
36. These investigations were published together as *Untersuchungen über die Entwickelung der Wirbelthiere* (Berlin: G. Reimer, 1855).
37. I rely here on Schmiedebach's summary, *Robert Remak*, 177–178.
38. Kisch, "Robert Remak," 284; Schmiedebach, *Robert Remak*, 213.
39. Kisch, "Robert Remak," 233.

40. Ibid., 237.
41. Ibid., 233; Schmiedebach, *Robert Remak*, 84.
42. Schmiedebach, *Robert Remak*, 91.
43. Robert Remak, "Vorläufige Mittheilung microscopischer Beobachtungen über den innern Bau der Cerebrospinalnerven und über die Entwickelung ihrer Formelemente," *Müllers Archiv* (1836): 159.
44. Robert Remak, "Über gangliöse Nervenfasern beim Menschen und bei den Wirbelthieren," MKAWB (1853): 294.
45. Robert Remak, "Einleitung," in *Ueber ein selbständiges Darmnervensystem* (Berlin: G. Reimer, 1847). Remak's introduction to this work has no page numbers.
46. Robert Remak, "Ueber die Entwickelung des Hühnchens im Ei," *Müllers Archiv* (1843): 484.
47. Remak, "Einleitung."
48. Robert Remak, "Weitere mikroscopische Beobachtungen über die Primitivfasern des Nervensystems der Wirbelthiere," *Neue Notizen aus dem Gebiete der Natur- und Heilkunde* 3 (1837): 36, quoted in Kisch, "Robert Remak," 233, but translated differently.
49. Robert Remak, *Galvanotherapie der Nerven- und Muskelkrankheiten* (Berlin: August Hirschwald, 1858), xi.
50. Remak, "Einleitung."
51. Kisch, "Robert Remak," 255, 234, 242.
52. Ibid., 240, 261; Schmiedebach, *Robert Remak*, 29.
53. Kisch, *Robert Remak*, 257; Schmiedebach, *Robert Remak*, 115.
54. Kisch, "Robert Remak," 265; Schmiedebach, *Robert Remak*, 117–118.
55. Hintzsche, "Robert Remak," 369.
56. Wilhelm Haberling, *Johannes Müller: Das Leben des rheinischen Naturforschers* (Leipzig: Akademische Verlagsgesellschaft, 1924), 330, letter of 7 May 1849. See also Kisch, "Robert Remak," 275–276.
57. Haberling, *Johannes Müller*, 330, letter of 7 May 1849.
58. Quoted in Schmiedebach, *Robert Remak*, 257.
59. Remak, Dedication, in his *Untersuchungen über die Entwickelung der Wirbelthiere.* Remak's dedication has no page numbers.
60. Ludwig Geiger, "Briefe Alexander von Humboldts an Dr. Robert Remak 1839 bis 1855," *Jahrbuch für jüdische Geschichte und Literatur* (1916): 116–117, letter of 8 September 1839. I have relied in part on Kisch's translation of this letter, "Robert Remak," 240.
61. Geiger, "Briefe Alexander von Humboldts an Dr. Robert Remak," 124, letter of 23 January 1847, quoted in Kisch, "Robert Remak," 269. Kisch's translation differs from mine.
62. Remak, Dedication.
63. Geiger, "Briefe Alexander von Humboldts an Dr. Robert Remak," 130, letter of 21 June 1847.
64. Max Lenz, *Geschichte der königlichen Friedrich-Wilhelms-Universtät zu Berlin* (Halle: Verlag der Buchhandlung des Waisenhauses, 1910), 3: 494–495, 490–491; Schmiedebach, *Robert Remak*, 160–161.
65. M. Kalisch, *Die Judenfrage in ihrer wahren Bedeutung für Preussen* (Leipzig: Veit, 1860), 21–22. I have relied in part on Kisch's translation of this letter of 14 March 1843, Kisch, "Robert Remak," 264.
66. Schmiedebach, *Robert Remak*, 162.
67. Kisch, "Robert Remak," 270.
68. Kalisch, *Die Judenfrage*, 23–24, letter of 27 February 1847. I have relied in part on Kisch's translation, "Robert Remak," 271.

69. Kalisch, *Die Judenfrage*, 23, letter of 24 February 1847. I have relied in part on Kisch's translation, "Robert Remak," 270.

70. Kisch, "Robert Remak," 272.

71. Geiger, *Briefe Alexander von Humboldts an Dr. Robert Remak*, 129, letter of 10 March 1847. I have relied in part on Kisch's translation, "Robert Remak," 272.

72. Schmiedebach, *Robert Remak*, 163–166.

73. Kisch, "Robert Remak," 273–274. Manfred Stürzbecher claims that in 1810, before the Edict of Emancipation, the obstetrician's assistant (*Geburtshelfer*) Nathan Friedländer, an unbaptized Jew, was appointed lecturer at the medical faculty of the Berlin University. If this is true, then Remak was not the first unbaptized Jew to teach on the Berlin medical faculty. Manfred Stürzbecher, "Die Prosektur der Berliner Charité im Briefwechsel zwischen Robert Froriep und Rudolf Virchow," in *Beiträge zur Berliner Medizingeschichte: Quellen und Studien zur Geschichte des Gesundheitswesens vom 17. bis zum 19. Jahrhundert* (Berlin: Walter de Gruyter, 1966), 174, n. 52.

74. Kisch, "Robert Remak," 265; Schmiedebach, *Robert Remak*, 247.

75. Wilhelm Waldeyer, "Der Unterricht in den anatomischen Wissenschaften an der Universität Berlin im ersten Jahrhundert ihres Bestehens," *Berliner klinische Wochenschrift* (10 October 1910): 1864.

76. Schmiedebach, *Robert Remak*, 249–250.

77. Kisch, "Robert Remak," 256.

78. Anonymous, Announcement, *Allgemeine medizinische Zentral-Zeitung* 20 (1851): 46, quoted in Kisch, "Robert Remak," 256. My translation differs from Kisch's.

79. Du Bois-Reymond and Helmholtz, *Dokumente einer Freundschaft*, 107, letter of 18 March 1851.

80. Schmiedebach, *Robert Remak*, 258–259.

81. A. Kölliker, *Erinnerungen aus meinem Leben* (Leipzig: Wilhelm Engelmann, 1899), 9–10, quoted in Kisch, "Robert Remak," 256, but translated slightly differently.

82. Wilhelm His der Ältere, *Lebenserinnerungen und ausgewählte Schriften* (Bern: H. Huber, 1965), 32, quoted in Kisch, "Robert Remak," 276, but translated differently. I thank Ohad Parnes for bringing His's memoirs to my attention.

83. Schmiedebach, *Robert Remak*, 293.

84. Kisch, "Robert Remak," 278.

85. Schmiedebach, *Robert Remak*, 170–172.

86. Ibid., 255–257.

87. Lenz, *Geschichte der königlichen Friedrich-Wilhelms-Universtät zu Berlin*, 3: 491; Schmiedebach, *Robert Remak*, 260.

88. All information in this paragraph derives from Schmiedebach, *Robert Remak*, 261–267.

89. Kisch, "Robert Remak," 282.

90. Schmiedebach, *Robert Remak*, 278–281.

91. Kisch, "Robert Remak," 243, 237; Schmiedebach, *Robert Remak*, 28.

92. I am grateful to Michael Berkowitz of the Institute of Jewish Studies, University College, London, for explaining how Remak's need to keep kosher would have created distance between him and Müller's other assistants.

93. Du Bois-Reymond and Ludwig, *Two Great Scientists*, 34, letter of 17 May 1849.

94. Erwin Ackerknecht describes Remak as a "perennial competitor" of Virchow. Erwin Ackerknecht, *Rudolf Virchow: Doctor, Statesman, Anthropologist* (Madison: University of Wisconsin Press, 1953), 108.

95. Letter of Remak to Virchow, 3 October 1850, quoted in Schmiedebach, *Robert Remak*, 251.

96. Kisch, "Robert Remak," 260.

97. Letter of Remak to Virchow, 9 July 1855, quoted in Schmiedebach, *Robert Remak*, 190.

98. Kisch, "Robert Remak," 280–281; Schmiedebach, *Robert Remak*, 189. Kisch represents Virchow as an anti-Semite who actively tried to hurt Remak's chances of getting the job. I agree with Schmiedebach that this is a "grotesque distortion" of the facts.

99. Quoted in Schmiedebach, *Robert Remak*, 238.

100. Remak, "Ueber extracellulare Entstehung thierischer Zellen," 49.

101. Remak, "Ueber die Anzeigen zur Ausrottung krankhafter Geschwülste," 126, 131.

102. Remak, "Ueber die Entwickelung des Hühnchens im Ei," 482.

103. Remak, *Untersuchungen über die Entwickelung der Wirbelthiere*, 178.

104. Remak, "Ueber extracellulare Entstehung thierischer Zellen," 52–53.

105. Remak, *Galvanotherapie der Nerven- und Muskelkrankheiten*, viii.

106. Schmiedebach, *Robert Remak*, 79–80, n. 27.

107. Remak, "Ueber die physiologische Bedeutung des organischen Nervensystems," 228.

108. Remak, *Untersuchungen über die Entwickelung der Wirbelthiere*, 164.

109. Robert Remak, "Ueber die sogenannten Blutkörperchen haltenden Zellen," *Müllers Archiv* (1851): 483.

110. Robert Remak, "Ueber Theilung thierischer Zellen," *Müllers Archiv* (1854): 376.

111. Robert Remak, "Ueber die embryologische Grundlage der Zellenlehre," *Müllers Archiv* (1862): 237.

112. Remak, "Ueber die Entwickelung des Hühnchens im Ei," 484.

113. Remak, "Vorläufige Mittheilung microscopischer Beobachtungen," 152.

114. Robert Remak, "Neurologische Erläuterungen," *Müllers Archiv* (1844): 468.

115. Robert Remak, "Ueber peripherische Ganglien in den Nerven des Nahrungsrohrs," *Müllers Archiv* (1858): 192.

116. Robert Remak, "Anatomische Beobachtungen über das Gehirn, das Rückenmark und die Nervenwürzeln," *Müllers Archiv* (1841): 506, 518.

117. Remak, *Galvanotherapie der Nerven- und Muskelkrankheiten*, vii.

118. Remak, "Vorläufige Mittheilung microscopischer Beobachtungen," 145–146.

119. Remak, *Ueber ein selbständiges Darmnervensystem*, 33.

120. Remak, "Ueber gangliöse Nervenfasern," 298.

121. Remak, "Neurologische Erläuterungen," 468.

122. Remak, *Galvanotherapie der Nerven- und Muskelkrankheiten*, ix.

123. Remak, Introduction.

124. Robert Remak, "Ueber die genetische Bedeutung des oberen Keimblattes im Eie der Wirbelthiere," *Müllers Archiv* (1849): 210, and "Ueber die Entstehung des Bindegewebes und des Knorpels," *Müllers Archiv* (1852): 67.

125. Remak, "Anatomische Beobachtungen über das Gehirn," 512–513.

126. Remak, "Ueber extracellulare Entstehung thierischer Zellen," 50.

127. Robert Remak, "Ueber die Theilung der Blutzellen beim Embryo," *Müllers Archiv* (1858): 178.

128. Robert Remak, "Ueber vielkernige Zellen der Leber," *Müllers Archiv* (1854): 101.

129. Remak, *Galvanotherapie der Nerven- und Muskelkrankheiten*, xi.

130. Remak, "Ueber die Anzeigen zur Ausrottung krankhafter Geschwülste," 125.

131. Remak, "Neue Beiträge zur Kenntnis vom organischen Nervensystem," 8.

132. Remak, "Ueber die Anzeigen zur Ausrottung krankhafter Geschwulste," 136.

133. Remak, "Weitere microscopische Beobachtungen," 37. One line (*Linie*) in mid-nineteenth century Prussia was equivalent to .0858 English inches, in today's measurement, 2.18

millimeters. See Savile Bradbury, *Basic Measurement Techniques for Light Microscopy* (Oxford:Oxford University Press, 1991), 7.

134. Robert Remak, "Ueber den Inhalt der Nervenprimitivröhren," *Müllers Archiv* (1843): 198. Wilhelm His recalls that "innumerable crawfish had also to be sacrificed, since Remak cleansed the embryos with the fresh blood of crawfish." His, *Lebenserinnerungen*, 32, quoted in Kisch, "Robert Remak," 276, Kisch's translation.

135. Remak, "Ueber die Theilung der Blutzellen beim Embryo," 179.

136. Remak, "Neurologische Erläuterungen," 464.

137. Robert Remak, "Ueber runde Blutgerinnsel und über pigmentkugelhaltige Zellen," *Müller's Archiv* (1852): 119.

138. C. Thomas Anderson has pointed out that "Remak's scientific style was not flamboyant enough to attract notice." C. Thomas Anderson, "Robert Remak and the Multinucleated Cell: Eliminating a Barrier to the Acceptance of Cell Division," *Bulletin of the History of Medicine* 60 (1986): 525.

139. Remak, "Weitere microscopische Beobachtungen," 36.

140. Remak, "Ueber die Verrichtungen des organischen Nervensystems," 66.

141. Remak, "Ueber den Inhalt der Nervenprimitivröhren," 199, n. 1.

142. Ibid., 199; Remak, "Ueber die Theilung der Blutzellen beim Embryo," 181, 185.

143. Robert Remak, "Ueber die Enden der Nerven im elektrischen Organ der Zitterrochen," *Müllers Archiv* (1856): 472.

144. Remak, "Ueber vielkernige Zellen der Leber," 99.

145. Remak, *Galvanotherapie der Nerven- und Muskelkrankheiten*, 186–188.

146. Remak, "Neurologische Erläuterungen," 472.

147. Ibid., 472, n. 1.

Chapter 7

1. Lynn Nyhart argues persuasively that historians should never presume "that generation gaps inevitably cause conflict." Nyhart analyzes six generations of German morphologists, only some of whom rebelled against their predecessors. See Lynn K. Nyhart, *Biology Takes Form: Animal Morphology and the German Universities 1800–1900* (Chicago: University of Chicago Press, 1995), 31.

2. Ernst Haeckel, *Entwicklungsgeschichte einer Jugend: Briefe an die Eltern 1852–1856* (Leipzig: K. F. Koehler, 1921), 183, letter of 23 April 1856.

3. Ernst Haeckel, "Ernst Haeckel: Eine autobiographische Skizze," in *Ernst Haeckel: Gemeinverständliche Werke* (Leipzig: Alfred Kröner, 1924), 1: xx.

4. Erika Krausse has noted that Haeckel uses Müller as the "principal witness [*Kronzeuge*]" for his theories. Erika Krausse, "Johannes Müller und Ernst Haeckel: Erfahrung und Erkenntnis," in *Johannes Müller und die Philosophie*, ed. Michael Hagner and Bettina Wahrig-Schmidt (Berlin: Akademie Verlag, 1992), 237.

5. Erika Krausse, *Ernst Haeckel* (Leipzig: B. G. Teubner, 1984), 11, 18.

6. Haeckel, *Entwicklungsgeschichte einer Jugend*, 127, letter of 25 April 1855, and 175, letter of 13 January 1856. In both instances Haeckel writes that he finds the book inspiring. For a discussion of the way Goethe's works shaped the attitudes of young men from the *Bildungsbürgertum*, see Gabriel Ward Finkelstein, "Emil du Bois-Reymond: The Making of a Liberal German Scientist (1818–1851)," (Ph.D. diss., Princeton University, 1996), 20–22.

7. Krausse, *Ernst Haeckel*, 39–40.

8. In September 1851, Haeckel's father retired and moved to Berlin. Haeckel had planned to study botany at Jena with Matthias Schleiden but had to begin learning medicine in Berlin because a rheumatic knee forced him to return home. See Krausse, *Ernst Haeckel*, 18.
9. Ibid., 18–31.
10. Haeckel, *Entwicklungsgeschichte einer Jugend*, 56, letter of 1 June 1853.
11. Ibid., 103, letter of 9 March 1854.
12. Ernst Haeckel, *The History of Creation* (New York: Appleton, 1880), 1: 4.
13. Haeckel, *Entwicklungsgeschichte einer Jugend*, 4–5, letter of 31 October 1852, and 73, letter of 1 November 1853.
14. Ibid., 78, letter of 16 November 1853. The numbers in parentheses indicate the total number of hours per week that Haeckel spent in each class.
15. Ibid., 178, letter of 2 February 1856.
16. Ibid., 57, letter of 1 June 1853.
17. Ibid., 186, letter of 8 May 1856.
18. Ibid., 6, letter of 31 October 1852.
19. Ibid., 10, letter of 6 November 1852.
20. Ibid., 193, letter of 11 June 1856.
21. Ibid., 186, letter of 8 May 1856.
22. Haeckel calls a course on surgical instruments "the most horrible thing [*das Greulichste*] imaginable." Ibid., 53, letter of 14 May 1853.
23. Ibid., 76, 79, letter of 16 November 1853.
24. Ibid., 85–86, letter of 4 December 1853.
25. Ibid., 101, letter of 17 February 1854.
26. Ibid., 169, letter of 3 December 1855.
27. Ibid., 179–80, letter of 17 February 1856.
28. Haeckel, "Autobiographische Skizze," xv.
29. Haeckel, *Entwicklungsgeschichte einer Jugend*, 14, letter of 19 November 1852.
30. Ibid., 27, letter of 25 December 1852.
31. Ibid., 44, letter of 27 February 1853.
32. Ibid., 46, letter of 10 March 1853.
33. Krausse, *Ernst Haeckel*, 49.
34. Haeckel, *Entwicklungsgeschichte einer Jugend*, 26, letter of 21 December 1852.
35. Ibid., 60, letter of 18 June 1853.
36. Ibid., 68, letter of 30 August 1853.
37. Ibid., 74, letter of 1 November 1853.
38. Ibid., 68, letter of 18 August 1853.
39. Ibid., 84, letter of 4 December 1853, and 88, letter of 21 December 1853.
40. Ibid., 71, letter of 13 October 1853.
41. Ibid., 101, letter of 17 February 1854. Haeckel expresses the same wish in his letter of 21 December 1853 but does not call his microscope his wife.
42. Ibid., 85, letter of 4 December 1853.
43. Ibid., 50, letter of 4 May 1853.
44. Ibid., 117–118, letter of 22 August 1854.
45. Ernst Haeckel, *Die Radiolarien* (Berlin: Georg Reimer, 1862), vi.
46. Ernst Haeckel, *Generelle Morphologie der Organismen* (Berlin: Georg Reimer, 1866), 3.
47. Haeckel, *Entwicklungsgeschichte einer Jugend*, 187, letter of 8 May 1856.
48. Ibid., 109, letter of 25 March 1854.
49. Ibid., 142–43, letter of 17 June 1855. Haeckel uses a popular expression difficult to translate: "mit dem man 'Herz in Herz und Seel' in Seele drängen kann."

50. For a brief discussion of Goethe's comparative anatomy, see chapter 1.

51. Robert J. Richards, *The Romantic Conception of Life: Science and Philosophy in the Age of Goethe* (Chicago: University of Chicago Press, 2002), 435.

52. Gerd Rehkämpfer has written that Haeckel's juxtaposition of Goethe and Darwin suggests "how little Haeckel recognized the really revolutionary character of Darwinism," proposing that in biology, one can hardly imagine a pair of thinkers more opposed than Goethe and Darwin. See "Zur frühen Rezeption von Darwins Selektionstheorie und deren Folgen für die vergleichende Morphologie heute," *Sudhoffs Archiv* 81 (1997): 178–179, quoted in Peter Zigman, "Ernst Haeckel und Rudolf Virchow: Der Streit um den Charakter der Wissenschaft in der Auseinandersetzung um den Darwinismus," *Medizinhistorisches Journal* 35 (2000): 281, n. 69.

53. Krausse, *Ernst Haeckel*, 17–18.

54. Haeckel, *Entwicklungsgeschichte einer Jugend*, 39, letter of 17 February 1853.

55. Ibid., 103–4, letter of 9 March 1854.

56. Erika Krausse, "Zum Verhältnis von Carl Gegenbauer (1826–1903) und Ernst Haeckel (1834–1919): Generelle und spezielle Morphologie," in *Miscellen zur Geschichte der Biologie*, ed. Armin Geus, Wolfgang Friedrich Gutmann, and Michael Weingarten (Frankfurt am Main: Waldemar Kramer, 1994), 84.

57. Nyhart, *Biology Takes Form*, 146–147.

58. William Coleman, "Carl Gegenbaur," in *Dictionary of Scientific Biography*, 5: 165–66.

59. Haeckel, *Entwicklungsgeschichte einer Jugend*, 22–23, letter of 12 December 1852.

60. Nyhart, *Biology Takes Form*, 148.

61. Ernst Haeckel, *Über die Biologie in Jena während des 19. Jahrhunderts* (Jena: Gustav Fischer, 1905), 12–13.

62. Krausse, "Zum Verhältnis von Carl Gegenbauer und Ernst Haeckel," 94. Krausse specifies that Haeckel learned from Gegenbaur how to use morphological comparisons as a research method as opposed to a philosophical or theoretical tool.

63. Haeckel, *Entwicklungsgeschichte einer Jugend*, 94, letter of 7 February 1854.

64. Krausse, *Ernst Haeckel*, 41, and "Zum Verhältnis von Carl Gegenbauer und Ernst Haeckel," 85.

65. Nyhart, *Biology Takes Form*, 152. Nyhart does not speculate about Gegenbaur's or Haeckel's choices of research topics but points out that such divisions of turf were "customary for anatomists and zoologists in the 1860s."

66. Krausse, *Ernst Haeckel*, 43.

67. Krausse, "Zum Verhältnis von Carl Gegenbauer und Ernst Haeckel," 85; Nyhart, *Biology Takes Form*, 146.

68. Haeckel, "An Carl Gegenbaur," in his *Generelle Morphologie*. The dedication has no page numbers.

69. All information on Kölliker's career in this paragraph derives from Erich Hintzsche, "Rudolf Albert von Koelliker," in *Dictionary of Scientific Biography*, 7: 437–438.

70. Krausse, *Ernst Haeckel*, 20. The enrollment did not increase steadily, since Haeckel wrote to his parents that in the fall of 1853, it had dropped below 300. Although Virchow and Kölliker offered cutting-edge science, the professors of therapy and surgery taught poorly. See Haeckel, *Entwicklungsgeschichte einer Jugend*, 74, letter of 5 November 1853.

71. Haeckel, *Entwicklungsgeschichte einer Jugend*, 9, letter of 6 November 1852.

72. Ibid., 14, letter of 14 November 1852.

73. Ibid., 65, letter of 18 July 1853.

74. Ibid., 67, letter of 18 July 1853.

75. Ibid., 87–88, letter of 21 December 1853.

76. Ibid., 103, letter of 9 March 1854.
77. Ibid., 141–142, letter of 17 June 1855.
78. Ibid., 165, letter of 2 November 1855.
79. Luigi Belloni, "Haeckel als Schüler und Assistent von Virchow und sein Atlas der pathologischen Histologie bei Prof. Rudolf Virchow, Würzburg, Winter 1855–1856," *Physis* 15 (1973): 6.
80. Haeckel, *Entwicklungsgeschichte einer Jugend*, 32, letter of 20 January 1853.
81. Ibid., 80, letter of 16 November 1853.
82. Ibid., 89, letter of 21 December 1853.
83. Ibid., 136–137, letter of 17 May 1855.
84. Ibid., 137, letter of 17 May 1855.
85. Ibid., 180, letter of 17 February 1856.
86. Belloni, "Haeckel als Schüler und Assistent von Virchow," 12.
87. Haeckel, *Entwicklungsgeschichte einer Jugend*, 182, letter of 20 February 1856.
88. Ibid., 187, letter of 8 May 1856.
89. Ibid., 192–193, letter of 21 May 1856.
90. Ibid., 194, letter of 11 June 1856.
91. Ibid., 205, letter of 4 August 1856.
92. Zigman, "Ernst Haeckel und Rudolf Virchow," 278. See also Krausse, *Ernst Haeckel*, 93–95.
93. D. V. Engelhardt, "Virchow et Haeckel et la liberté de la science," in *Médecine et philosophie à la fin du XIXème siècle* (Paris: Université Paris—Val de Marne, 1981), 119.
94. Johannes Hemleben, *Ernst Haeckel in Selbstzeugnissen und Bilddokumenten* (Reinbek bei Hamburg: Rowohlt, 1964), 30, quoted in Zigman, "Ernst Haeckel und Rudolf Virchow," 272.
95. Ernst Haeckel, *Der Kampf um den Entwickelungsgedanken* (Berlin: Georg Reimer, 1905), 29.
96. Peter Zigman has written that although they were contemporaries, Virchow and Haeckel "seemed to belong to different worlds. They spoke to each other as though they didn't listen to or understand each other." Zigman, "Ernst Haeckel und Rudolf Virchow," 293.
97. My summary of Haeckel's contact with Müller relies partly on Krausse's in "Johannes Müller und Ernst Haeckel," 223.
98. Haeckel, *Entwicklungsgeschichte einer Jugend*, 49–50, letter of 4 May 1853.
99. Ibid., 52, letter of 14 May 1853.
100. Ibid., 59, letter of 1 June 1853, and 109, letter of 25 March 1854.
101. This diary (*Tagebuch*) entry is quoted in Krausse, *Ernst Haeckel*, 25; and Georg Uschmann, *Ernst Haeckel: Biographie in Briefen* (Leipzig: Urania, 1983), 32–33. The Haeckel archive at the Haeckel-Haus in Jena has no such diary.
102. Haeckel, *Entwicklungsgeschichte einer Jugend*, 123, letter of 30 August 1854.
103. Ibid., 124, letter of 10 September 1854.
104. Haeckel writes in his letter of 30 August (a Wednesday) that Müller had arrived "Tuesday afternoon." Since his last letter, written on Tuesday, 22 August, makes no mention of Müller, Müller probably arrived on 29 August. In his letter of 10 September, Haeckel writes that Müller is due to depart the following Thursday, 14 September. This would mean that Müller and Haeckel had a maximum of fifteen days to fish.
105. Haeckel, *Entwicklungsgeschichte einer Jugend*, 185, letter of 8 May 1856.
106. Ibid., 209, letter of 22 August 1856.
107. Krausse, *Ernst Haeckel*, 33.
108. Ibid., 35.

109. Haeckel, *Die Radiolarien*, vi. Müller's article was titled "Über die Thalassicollen, Polycystinen und Acanthometren des Mittelmeeres."

110. Krausse, "Johannes Müller und Ernst Haeckel," 226–227.

111. Haeckel, *Die Radiolarien*, xii.

112. Lynn Nyhart has shown how Bronn's translation, which promoted his own view of development as an "ascent to perfection," influenced German readers' understanding of evolution. Nyhart, *Biology Takes Form*, 110–121. On Haeckel and Darwin, see Nyhart, *Biology Takes Form*, 129–138; and Krausse, "Johannes Müller und Ernst Haeckel," 226–229. Krausse points out that although Haeckel wanted to continue using Müller's methods, he had to rethink the relationships he was seeing among radiolaria in terms of evolutionary theory, since he read *The Origin of Species* after conducting the research but before completing the book.

113. Haeckel, *Generelle Morphologie*, xvii.

114. Ibid., 6.

115. Haeckel does not refer directly here to du Bois-Reymond's comparison of Müller to Alexander the Great ("Gedächtnisrede auf Johannes Müller," in du Bois-Reymond's *Reden* [Leipzig: Veit, 1887], 2: 145), but he does so frequently in his later works. See, for example, *The Riddle of the Universe*, 47; *Über die Biologie in Jena während des 19. Jahrhunderts*, 11; and *Der Kampf um den Entwickelungsgedanken*, 24.

116. Haeckel, *The History of Creation*, 2: 203, and *Anthropogenie* (Leipzig: Wilhelm Engelmann, 1874), 298.

117. Haeckel, *The History of Creation*, 1: 51, 1: 312.

118. Krausse, *Ernst Haeckel*, 107–108.

119. Haeckel, *The Riddle of the Universe*, 45.

120. Ibid., 45–46.

121. Ibid., 99–100.

122. Haeckel's "Autobiographical Sketch" was published only in 1924, when his former student Heinrich Schmidt completed it and included it as a preface to Haeckel's *Gemeinverständliche Werke*. Schmidt notes that "[the sketch] extends up to the beginning of the seventies," suggesting that it was written at that time. Haeckel, "Autobiographische Skizze," xxvii. However, the undated manuscript 311 in the Haeckel archive in Jena, on which Schmidt's version appears to be based, is in the handwriting of Haeckel's later years, probably 1900 or later. I am extremely grateful to Dr. Thomas Bach for transcribing for me the passages in ms. 311 relating to Müller.

123. Haeckel, "Autobiographische Skizze," xii. The corresponding passage in ms. 311 says that they fished together for "fünf Wochen" (five weeks).

124. Ibid., xv. Here the wording of Schmidt's published sketch follows that of ms. 311 almost exactly. See the Afterword for a discussion of Haeckel's repeated reference to himself as one of Müller's "closest students."

125. Haeckel, *Der Kampf um den Entwickelungsgedanken*, 23–24.

126. Ibid., 25.

127. After carefully studying many generations of German morphologists, Lynn Nyhart has concluded that for "scholars born around 1800, . . . Darwin's theory represented only the most recent in a long series of dramatic proposals to redraw the shape of life science." Nyhart, *Biology Takes Form*, 107.

128. Haeckel, *The Riddle of the Universe*, 49.

129. Krausse, *Ernst Haeckel*, 104–105.

130. In 1874 Haeckel wrote, "Against this '*Ignoramibus*,' which has brought the worthy investigator of nerve and muscle electricity the unanimous thanks of the *Ecclesia militans*, we

must decidedly protest in the name of progressive knowledge of nature and science capable of development!" Haeckel, *Anthropogenie*, xii.

131. Haeckel, *The Riddle of the Universe*, 180, 94. Haeckel continued this pattern in his later public lectures, calling du Bois-Reymond "the brilliant orator of the Berlin Academy" in *Der Kampf um den Entwickelungsgedanken*, 28.

132. Emil du Bois-Reymond and Hermann von Helmholtz, *Dokumente einer Freundschaft: Briefwechsel zwischen Hermann von Helmholtz und Emil du Bois-Reymond 1846–1894* (Berlin: Akademie Verlag, 1986), 312, n. 1, annotation to letter of 7 May 1881.

133. Ibid., 264, letter of 7 May 1881. In *The Perigenesis of the Plastudules* (1876), Haeckel had proposed that individual cells could store hereditary information as patterns of molecular vibration, arguing for the unity of matter and mind.

134. Haeckel, *The Riddle of the Universe*, 102–103.

135. In German, *etwas Ordentliches* does not often literally mean "something orderly." Usually, it conveys something substantial, respectable, or real. In Haeckel's writing, however, I am convinced that it forms part of a trend upholding orderly narratives as the highest form of thought and expression.

136. Haeckel, *Entwicklungsgeschichte einer Jugend*, 16, letter of 19 November 1852. The lines quoted were written by Haeckel's neighbor, Dr. Gsell-Fels, but Haeckel wrote that his experience with Schenk had fully confirmed them.

137. Ibid., 101, letter of 17 February 1854.

138. Ibid., 125, letter of 10 September 1854. Haeckel explains that he wants to organize his specimens as fast as he can because once he puts them in alcohol, their appearance is greatly altered.

139. Ibid., 79–80, letter of 16 November 1853.

140. Ibid., 80, letter of 16 November 1853.

141. Haeckel, *Anthropogenie*, 9.

142. Krausse, "Johannes Müller und Ernst Haeckel," 230.

143. Haeckel, *Entwicklungsgeschichte einer Jugend*, 50, letter of 4 May 1853.

144. Ibid., 81, letter of 16 November 1853.

145. Ernst Haeckel, "Freie Wissenschaft und freie Lehre. Eine Entgegnung auf Rudolf Virchow's Münchener Rede über 'Die Freiheit der Wissenschaft in der modernen Staat'" (1878; reprint Leipzig: A. Kröner, 1908), 52, quoted in Zigman, "Ernst Haeckel und Rudolf Virchow," 293.

146. Haeckel, *Entwicklungsgeschichte einer Jugend*, 200, letter of 21 July 1856.

147. Haeckel, *Generelle Morphologie*, xix.

148. Haeckel, "Autobiographische Skizze," xx-xxi, xxv.

149. Haeckel, *The Riddle of the Universe*, 233.

150. Ibid., 231; Johannes Müller, "Beobachtungen über die Gesetze und Zahlenverhältnisse der Bewegung in den verschiedenen Thierclassen mit besonderer Rücksicht auf die Bewegung der Insekten und Polymerien," *Isis* (1822): 62.

151. Haeckel, *Über die Biologie in Jena während des 19. Jahrhunderts*, 8.

152. Haeckel, *Der Kampf um den Entwickelungsgedanken*, 20, and *The Riddle of the Universe*, 269.

153. Haeckel, *The Riddle of the Universe*, 263.

154. Haeckel, *Entwicklungsgeschichte einer Jugend*, 56, letter of 1 June 1853.

155. Haeckel, *The History of Creation*, 1: 9.

156. Haeckel, "Autobiographische Skizze," xxiv.

157. Krausse, *Ernst Haeckel*, 79–80.

158. Elizabeth Pennisi, "Haeckel's Embryos: Fraud Rediscovered," *Science* 277 (1997): 1435. See also Stephen Jay Gould's detailed study *Ontogeny and Phylogeny* (Cambridge, MA: Harvard University Press, 1977).

159. Quoted in Krausse, *Ernst Haeckel*, 92.

Afterword

1. Emil du Bois-Reymond and Hermann von Helmholtz, *Dokumente einer Freundschaft: Briefwechsel zwischen Hermann von Helmholtz und Emil du Bois-Reymond 1846–1894* (Berlin: Akademie Verlag, 1986), 185, letter of 28 April 1858, and 186, letter of 29 May 1858.

2. Emil du Bois-Reymond, "Gedächtnisrede auf Johannes Müller," in his *Reden* (Leipzig: Veit, 1887), 2: 297–298.

3. Rudolf Virchow, *Johannes Müller: Eine Gedächtnisrede* (Berlin: August Hirschwald, 1858), 38.

4. Letter of Gustav Magnus to Jakob Henle, 29 April 1858, NL Hermann Hoepke, KE 74, UH. I am grateful to Christoph Gradmann for bringing this letter to my attention and to Soraya Chadarevian for assisting me with the translation.

5. Quoted in Wilhelm Haberling, *Johannes Müller: Das Leben des rheinischen Naturforschers* (Leipzig: Akademische Verlagsgesellschaft, 1924), 450–451. The dashes indicate that Haberling has intentionally omitted a passage.

6. Quoted in Gottfried Koller, *Das Leben des Biologen Johannes Müller 1801–1858* (Stuttgart: Wissenschaftsverlaggesellschaft, 1958), 234, letter of 22 January 1900.

7. Albert E. Gunther, *A Century of Zoology at the British Museum through the Lives of Two Keepers, 1815–1914* (London: Dawsons of Pall Mall, 1975), 239.

8. Ibid., 236–239.

9. Ibid., 242.

10. Du Bois-Reymond and Helmholtz, *Dokumente einer Freundschaft*, 143, letter of 30 May 1853. "Marienbader" may be a brandy from the Czech town of Marienbad, where Germans went to use the curative baths.

11. By reading du Bois-Reymond's memorial address as a consciously constructed, self-promoting narrative, Gabriel Finkelstein and Nicholas Jardine have made an important contribution to the history of science as an interpretive process. See Gabriel Ward Finkelstein, "Emil du Bois-Reymond: The Making of a Liberal German Scientist (1818–1851)," (Ph.D. diss., Princeton University, 1996), 173–174; and Nicholas Jardine, "The Mantle of Müller and the Ghost of Goethe: Interactions between the Sciences and Their Histories," in *History and the Disciplines: The Reclassification of Knowledge in Early Modern Europe*, ed. Donald R. Kelley (Rochester, NY: University of Rochester Press, 1997), 301–303.

12. Emil du Bois-Reymond and Carl Ludwig, *Two Great Scientists of the Nineteenth Century: Correspondence of Emil du Bois-Reymond and Carl Ludwig*, ed. Paul F. Cranefield, trans. Sabine Lichtner-Ayed (Baltimore: Johns Hopkins University Press, 1982), 64, letter of 31 March 1851.

13. Finkelstein, "Emil du Bois-Reymond," 154–155.

14. Peter Loewenberg, *Decoding the Past: The Psychohistorical Approach* (New York: Alfred A. Knopf, 1983), 21.

15. Loewenberg argues that an understanding of a person's social context begins with an exploration of his or her relationships, writing that "the unique person can best be comprehended in his full cultural context, meaning in relation to the lives of others with whom he has deeply rooted emotional affinities." Loewenberg, *Decoding the Past*, 29.

16. Donna Haraway, "Animal Conversations: Braiding Natures and Cultures," paper presented at Society for Literature and Science conference, 16 October 2004, Durham, NC.

17. Arleen Tuchman, *Science, Medicine, and the State in Germany: The Case of Baden 1815–1871* (New York: Oxford University Press, 1993), 57.

18. I am grateful to Jan C. Frich for leading me to this idea in the discussion following my lecture, "The Lab as a Literary Creation," in Oslo on 6 September 2005.

Bibliography

Ackerknecht, Erwin H. *Rudolf Virchow: Doctor, Statesman, Anthropologist.* Madison: University of Wisconsin Press, 1953.

Adams, Henry. *The Education of Henry Adams*, ed. Ira B. Nadel. New York: Oxford University Press, 1999.

Anderson, C. Thomas. "Robert Remak and the Multinucleated Cell: Eliminating a Barrier to the Acceptance of Cell Division." *Bulletin of the History of Medicine* 60 (1986): 523–543.

Anderton, Keith M. "The Limits of Science: A Social, Political, and Moral Agenda for Epistemology in Nineteenth-Century Germany." Ph.D. diss., Harvard University, 1993.

Andree, Christian. *Rudolf Virchow: Leben und Ethos eines grossen Arztes.* Munich: Langen Müller, 2002.

———. "Welches Verhältnis hatte Rudolf Virchow zu zeitgenössischen Dichtern, Künstlern, Verlegern und Editoren? Versuch einer Annäherung über die Korrespondenzpartner." *Würzburger medizinhistorische Mitteilungen* 12 (1994): 259–286.

Anonymous. Announcement. *Allgemeine medizinsiche Zentral-Zeitung* 20 (1851): 46.

Anonymous. "Nachruf." *Deutsche Klinik* no. 44 (4 November 1865): 413–414.

Balzac, Honoré de. "Avant-propos." In his *La Comédie humaine. Oeuvres complètes de Balzac.* 24 vols. Paris: Club de L'Honnête Homme, 1968.

Belloni, Luigi. "Haeckel als Schüler und Assistent von Virchow und sein Atlas der pathologischen Histologie bei Prof. Rudolf Virchow, Würzburg, Winter 1855–1856." *Physis* 15 (1973): 5–33.

Ben-David, Joseph. "Scientific Productivity and Academic Organization in Nineteenth-Century Medicine." *American Sociological Review* 25 (1960): 828–843.

———. *The Scientist's Role in Society.* Englewood Cliffs, NJ: Prentice-Hall, 1971.

———, and Awraham Zloczower. "Universities and Academic Systems in Modern Societies." *European Journal of Sociology* 3 (1962): 45–84.

Benaroyo, Lazare. "Rudolf Virchow and the Scientific Approach to Medicine." *Endeavour* 22 (1998): 114–116.

Bidder, Friedrich. *Aus dem Leben eines Dorpater Universitätslehrers: Erinnerungen des Mediziners Prof. Dr. Friedrich v. Bidder 1810–1894.* Würzburg: Holzner-Verlag, 1959.

———. "Vor Hundert Jahren im Laboratorium Johannes Müllers." *Münchener medizinische Wochenschrift* (January 1934): 60–64.

Biermann, Kurt B. "Friedrich Wilhelm Heinrich Alexander von Humboldt." In *Dictionary of Scientific Biography,* 6: 549–555.

Bischoff, Theodor. *Ueber Johannes Müller und sein Verhältnis zum jetzigen Standpunkt der Physiologie.* Munich: Academy of Sciences, 1858.

Bloom, Harold. *The Anxiety of Influence: A Theory of Poetry.* New York: Oxford University Press, 1997.

Boyd, Byron Albert. "Rudolf Virchow: The Scientist as Citizen." Ph.D. diss., University of North Carolina at Chapel Hill, 1981.

Bradbury, Savile. *Basic Measurement Techniques for Ligh Microscopy.* Oxford: Oxford University Press, 1991.

Brock, Thomas D. *Robert Koch: A Life in Medicine and Bacteriology.* Madison, WI: Science Tech Publishers, 1988.

"Burschenschaft." In *Brockhaus Enzyklopädie.* Mannheim: F. A. Brockhaus, 1987. 6: 228.

Cahan, David. "Introduction." In *Letters of Hermann von Helmholtz to His Parents: The Medical Education of a German Scientist 1837–1846,* ed. David Cahan. Stuttgart: Franz Steiner, 1993.

———. "Introduction." In Hermann von Helmholtz, *Science and Culture: Popular and Philosophical Essays,* ed. David Cahan. Chicago: University of Chicago Press, 1995.

Clark, William. "On the Ministerial Archive of Academic Acts." *Science in Context* 9 (1996): 421–486.

Clarke, Edwin, and L. S. Jacyna. *Nineteenth-Century Origins of Neuroscientific Concepts.* Berkeley: University of California Press, 1987.

The Complete Parallel Bible (New York: Oxford University Press, 1993.

Cranefield, Paul F. "Carl Ludwig and Emil du Bois-Reymond: A Study in Contrasts." *Gesnerus* 45 (1988): 271–282.

Coleman, William. "Carl Gegenbaur." In *Dictionary of Scientific Biography,* 5: 165–166.

Diepken, Paul. "Foreword." In Emil du Bois-Reymond and Carl Ludwig, *Two Great Scientists of the Nineteenth Century: Correspondence of Emil du Bois-Reymond and Carl Ludwig,* ed. Paul F. Cranefield, trans. Sabine Lichtner-Ayed. Baltimore: Johns Hopkins University Press, 1982.

Dierig, Sven. "'Die Instrumente waren noch theuer und selten': Schiek-Mikroskope im Umfeld von Johannes Müller." In *Unsichtbar—Sichtbar—Durchschaut: Das Mikroskop als Werkzeug des Lebenswissenschaftelers,* ed. Helmut Kettenmann, Jörg Zaun, and Stefanie Korthals. Berlin: Museumspädagogischer Dienst, 2001.

———. "'Die Kunst des Versuchens': Emil du Bois-Reymonds *Untersuchungen über thierische Elektricität.*" In *Kultur im Experiment,* ed. Henning Schmidgen, Peter Geimer, and Sven Dierig, 123–146. Berlin: Kulturverlag Kadmos, 2004.

———. "Rudolf Virchow und das Nervensystem: Zur Begründung der zellulären Neurobiologie." In *Das Gehirn—Organ oder Seele?* ed. Ernst Florey and Olaf Breidbach, 55–80. Berlin: Akademie Verlag, 1993.

Du Bois-Reymond, Emil. "Gedächtnisrede auf Johannes Müller." In his *Reden.* Leipzig: Veit, 1887. 2: 142–335.

———. "Ueber Geschichte der Wissenschaft." In his *Reden*. Leipzig: Veit, 1887. 2: 349–358.

———. *Jugendbriefe von Emil du Bois-Reymond an Eduard Hallmann*, ed. Estelle du Bois-Reymond. Berlin: Dietrich Reimer, 1918.

———. "Ueber die Lebenskraft." In his *Reden*. Leipzig: Veit, 1887. 2: 1–28.

———. Letters to Henry Bence Jones. SBPK, HA, 3 k 1852 (4).

———. "Naturwissenschaft und bildende Kunst." In his *Reden*. 2nd ed. 2 vols. Leipzig: Veit, 1912. 2: 390–425

———. "Der physiologische Unterricht sonst und jetzt." In his *Reden*. Leipzig: Veit, 1887. 2: 359–383.

———. *Untersuchungen über thierische Elektricität*. 2 vols. Berlin: Reimer, 1848–1884

———. and Hermann von Helmholtz. *Dokumente einer Freundschaft: Briefwechsel zwischen Hermann von Helmholtz und Emil du Bois-Reymond 1846–1894*. Berlin: Akademie Verlag, 1986.

———. and Alexander von Humboldt. *Briefwechsel zwischen Alexander von Humboldt und Emil du Bois-Reymond*, ed. Ingo Schwarz and Klaus Wenig. Berlin: Akademie Verlag, 1997.

———. and Carl Ludwig. *Two Great Scientists of the Nineteenth Century: Correspondence of Emil du Bois-Reymond and Carl Ludwig*, ed. Paul F. Cranefield, trans. Sabine Lichtner-Ayed. Baltimore: Johns Hopkins University Press, 1982.

———. *Zwei grosse Naturforscher des 19. Jahrhunderts: Ein Briefwechsel zwischen Emil du Bois-Reymond und Karl Ludwig*, ed. Estelle du Bois-Reymond. Leipzig: Johann Ambrosius Barth, 1927.

Du Bois-Reymond, Estelle. "Einleitung." In *Jugendbriefe von Emil du Bois-Reymond an Eduard Hallmann*, ed. Estelle du Bois-Reymond. Berlin: Dietrich Reimer, 1918.

Duchesneau, François. "Kölliker and Schwann's Cell Theory." In *La storia della medicina e della scienza tra archivo e laboratorio*, ed. Guido Cimino and Carlo Maccagni. Florence: L. S. Olschki, 1994.

Ebbecke, Ulrich. *Johannes Müller: Der grosse rheinische Physiologe*. Hannover: Schmorl und von Seefeld, 1951.

Eckenstein, Jean. *Der akademische Mentor für die Studirenden der Friedrich-Wilhelms-Universität zu Berlin*. Berlin: Wilhelm Schüppel, 1835.

Engelhardt, D. V. "Virchow et Haeckel et la liberté de la science." In *Médecine et philosophie à la fin du XIXème siècle*. Paris: Université Paris—Val de Marne, 1981.

Finkelstein, Gabriel Ward. "The Ascent of Man? Emil du Bois-Reymond's Reflections on Scientific Progress." *Endeavour* 24 (2000): 129–132.

———. "Emil du Bois-Reymond: The Making of a Liberal German Scientist (1818–1851)." Ph.D. diss., Princeton University, November 1996.

———. "M. du Bois-Reymond Goes to Paris." *British Journal for the History of Science* 36 (2003): 261–300.

Florkin, Marcel. *Naissance et déviation de la théorie cellulaire dans l'oeuvre de Théodore Schwann*. Paris: Hermann, 1960.

———. "Theodor Ambrose Hubert Schwann." In *Dictionary of Scientific Biography*, 12: 240–245.

Ford, Brian J. *Single Lens: The Story of the Simple Microscope*. New York: Harper & Row, 1985.

Freud, Sigmund. "Family Romances." In *Sigmund Freud: Collected Papers*, ed. James Strachey. 5 vols. New York: Basic Books, 1959.

———. *Totem and Taboo: Some Points of Agreement between the Mental Lives of Savages and Neurotics*. In *The Standard Edition of the Complete Psychological Works of Sigmund Freud*, trans. James Strachey. 24 vols. London: Hogarth Press, 1955.

Frey, Manfred. *Friedrich Miescher-His (1811–1887) und sein Beitrag zur Histopathologie des Knochens.* Basel: Benno Schwabe, 1962.

Fürbringer, M. "Wie ich Ernst Haeckel kennen lernte und mit ihm verkehrte und wie er mein Führer in den grössten Stunden meines Lebens wurde." In *Was Wir Ernst Haeckel Verdanken*, ed. Heinrich Schmidt. Leipzig: Unesma, 1914. 2: 335–350.

Geiger, Ludwig. "Briefe Alexander von Humboldts an Dr. Robert Remak 1839 bis 1855." *Jahrbuch für jüdische Geschichte und Literatur* (1916): 112–134.

Geppert, Carl Eduard. *Chronik von Berlin.* 3 vols. Berlin: Ferdinand Rubach, 1839–1841.

Gilman, Sander. *Jewish Self-Hatred: Anti-Semitism and the Hidden Language of the Jews.* Baltimore: Johns Hopkins University Press, 1986.

Goethe, Johann Wolfgang von. "Erster Entwurf einer allgemeinen Einleitung in die vergleichende Anatomie, ausgehend von der Osteologie." In *Goethe's Werke.* Abteilung 2: *Naturwissenschaftliche Schriften.* 13 vols. Weimar: Hermann Böhlau, 1893.

Goschler, Constantin. *Rudolf Virchow: Mediziner, Anthropologe, Politiker.* Cologne: Böhlau, 2002.

Gould, Stephen Jay. *Ontogeny and Phylogeny.* Cambridge, MA: Harvard University Press, 1977.

Gradmann, Christoph. *Krankheit im Labor: Robert Koch und die medizinische Bakteriologie.* Göttingen: Wallstein, 2005.

Günther, Klaus. "Die Gesellschaft Naturforschender Freunde zu Berlin: Johannes Müller und die Frage nach der Urzeugung." *Sitzungsberichte der Gesellschaft Naturforschender Freunde zu Berlin* 14 (1974): 26–36.

Gunther, Albert E. *A Century of Zoology at the British Museum through the Lives of Two Keepers, 1815–1914.* London: Dawsons of Pall Mall, 1975.

Haberling, Wilhelm. *Johannes Müller: Das Leben des rheinischen Naturforschers.* Leipzig: Akademische Verlagsgesellschaft, 1924.

Haeckel, Ernst. *Anthropogenie.* Leipzig: Wihelm Engelmann, 1874.

———. Autobiographical fragment. Manuscript 311. Ernst Haeckel Haus, Jena.

———. *Über die Biologie in Jena während des 19. Jahrhunderts.* Jena: Gustav Fischer, 1905.

———. "An Carl Gegenbaur." In his *Generelle Morphologie der Organismen.* 2 vols. Berlin: Georg Reimer, 1866.

———. *Entwicklungsgeschichte einer Jugend: Briefe an die Eltern 1852–1856.* Leipzig: K. F. Koehler, 1921.

———. "Ernst Haeckel: Eine Autobiographische Skizze." In *Ernst Haeckel: Gemeinverständliche Werke.* Leipzig: Alfred Kröner, 1924.

———. "Freie Wissenschaft und freie Lehre. Eine Entgegnung auf Rudolf Virchows Münchener Rede über 'Die Freiheit der Wissenschaft in der modernen Staat.'" 1878. Leipzig: A. Kröner, 1908.

———. *The History of Creation.* 2 vols. New York: Appleton, 1880.

———. *Der Kampf um den Entwickelungsgedanken.* Berlin: Georg Reimer, 1905.

———. *Die Perigenesis der Plastudule.* Berlin: Georg Reimer, 1876.

———. *Die Radiolarien.* Berlin: Georg Reimer, 1862.

———. *The Riddle of the Universe*, trans. Joseph McCabe. 1899. Buffalo, NY: Prometheus Books, 1992.

Hagner, Michael. "Sieben Briefe von Johannes Müller an Karl Ernst von Baer." *Medizinhistorisches Journal* 27 (1992): 138–154.

Haight, Gordon S. *George Eliot: A Biography.* London: Penguin, 1992.

Haraway, Donna. "Animal Conversations: Braiding Natures and Cultures." Paper presented at Society for Literature and Science conference, 16 October 2004, Durham, NC.

Hardy, Thomas. *The Well-Beloved: A Sketch of a Temperament.* London: Macmillan, 1952.
Harting, P. *Theorie und allgemeine Beschreibung des Mikroskopes.* 3 vols. Brunswick: Friedrich Vieweg, 1866.
Haynes, Roslynn D. *From Faust to Strangelove: Representations of the Scientist in Western Literature.* Baltimore: Johns Hopkins University Press, 1994.
Helmholtz, Hermann von. "An Autobiographical Sketch." In *Science and Culture: Popular and Philosophical Essays*, ed. David Cahan, 381–392. Chicago: University of Chicago Press, 1995.
———. "Das Denken in der Medicin." In *Vorträge und Reden von Hermann von Helmholtz*, 4th ed. Brunswick: Friedrich Vieweg, 1896. 2: 167–190.
———. "Emil du Bois-Reymond's fünfzigjähriges Doktor-Jubiläum." *Neue Freie Presse* (Vienna), 23 September 1893.
———. "Erinnerungen." In *Vorträge und Reden von Hermann von Helmholtz*, 4th ed. Brunswick: Friedrich Vieweg, 1896. 1: 3–21.
———. *Helmholtz's Treatise on Physiological Optics*, ed. James P. C. Southall. 3 vols. Bristol, UK: Thoemmes Press, 2000.
———. Lecture Notes. ABBAW, NL Helmholtz, nr. 538.
———. *Letters of Hermann von Helmholtz to His Parents: The Medical Education of a German Scientist 1837–1846*, ed. David Cahan. Stuttgart: Franz Steiner, 1993.
———. "Ueber die Natur der menschlichen Sinnesempfindungen." In his *Gesammelte Schriften*, ed. Jochen Brüning. Hildesheim: Olms-Weidmann, 2003. I.2.2: 591–609.
———. "On Thought in Medicine." In *Science and Culture: Popular and Philosophical Essays*, ed. David Cahan, 310–327. Chicago: University of Chicago Press, 1995.
Hemleben, Johannes. *Ernst Haeckel in Selbstzeugnissen und Bilddokumenten.* Reinbek bei Hamburg: Rowohlt, 1964.
Henle, Jakob. *Allgemeine Anatomie: Lehre von den Mischungs- und Formbestandtheilen des menschlichen Körpers.* Leipzig: Leopold Voss, 1841.
———. "Ueber die Ausbreitung des Epithelium im menschlichen Körper." *Müllers Archiv* (1838): 107–128.
———. *Handbuch der rationellen Pathologie.* 3 vols. Brunswick: Friedrich Vieweg, 1846.
———. *Von den Miasmen und Kontagien und von den miasmatisch-kontagiösen Krankheiten.* 1840. Leipzig: Barth, 1910.
———. "Ueber Schleim- und Eiterbildung und ihr Verhältnis zur Oberhaut." *Hufelands Journal der praktischen Heilkunde* 86 (1838): 3–62.
———. *Theodor Schwann: Nachruf.* Bonn: Max Cohen, 1882.
———. *Vergleichend-anatomische Beschreibung des Kehlkopfs mit besonderer Berücksichtigung des Kehlkopfs der Reptilien.* Leipzig: Leopold Voss, 1839.
Hintzsche, Erich. "Friedrich Gustav Jacob Henle." In *Dictionary of Scientific Biography*, 6: 268–270.
———. "Robert Remak." In *Dictionary of Scientific Biography*, 11: 367–370.
———. "Rudolf Albert von Koelliker." In *Dictionary of Scientific Biography*, 7: 437–438.
Hirsch, Gottwald Christian. "Die Forscherspersönlichkeit des Biologen Johannes Müller." *Sudhoffs Archiv für Geschichte der Medizin* 26 (1933): 166–190.
His, Wilhelm der Ältere. *Lebenserinnerungen und ausgewählte Schriften.* Bern: H. Huber, 1965.
Hoche, Alfred E. *Jahresringe: Innenansicht eines Menschenlebens.* Munich: J. F. Lehmann, 1940.
Hoepke, Hermann. "Der Bonner Student Jakob Henle in seinem Verhältnis zu Johannes Müller." *Sudhoffs Archiv* 53 (1969): 193–216.

———. "Jakob Henles Briefe aus Berlin 1834–1840." *Heidelberger Jahrbücher* 8 (1964): 57–86.
———. "Jakob Henle's Briefe aus seiner Heidelberger Studentenzeit (26. April 1830–Januar 1831)." *Heidelberger Jahrbücher* 11 (1967): 40–56.
———. "Jakob Henle's Gutachten zur Besetzung des Lehrstuhls für Anatomie an der Universität Berlin 1883." *Anatomischer Anzeiger* 120 (1967): 221—232.
Holborn, Hajo. *A History of Modern Germany.* 3 vols. New York: Alfred A. Knopf, 1959–1969.
Holmes, Frederic L. "The Role of Johannes Müller in the Formation of Helmholtz's Physiological Career." In *Universalgenie Helmholtz: Rückblick nach 100 Jahren, ed.* Lorenz Krüger. Berlin: Akademie Verlag, 1994.
Jacyna, L. S. "Images of John Hunter in the Nineteenth Century." *History of Science* 21 (1983): 85–108.
Jardine, Nicholas. "The Mantle of Müller and the Ghost of Goethe: Interactions between the Sciences and their Histories." In *History and the Disciplines: The Reclassification of Knowledge in Early Modern Europe*, ed. Donald R. Kelley. Rochester, NY: University of Rochester Press, 1997.
Kalisch, M. *Die Judenfrage in ihrer wahren Bedeutung für Preussen.* Leipzig: Veit, 1860.
Kant, Immanuel. *Kritik der Urteilskraft*, ed. Heiner F. Klemme and Piero Giordanetti. Hamburg: Felix Meiner, 2001.
Kisch, Bruno. "Robert Remak, 1815–1865." *Transactions of the American Philosophical Society* 44 (1954): 227–296.
Klein, Marc. "Jacob Mathias Schleiden." In *Dictionary of Scientific Biography*, 12: 173–176.
———. "Lorenz Oken." In *Dictionary of Scientific Biography*, 10: 194–196.
Kohut, Thomas A. "Psychohistory as History." *American Historical Review* 91 (1986): 336–354.
Kolkenbrock-Netz, Jutta. "Wissenschaft als nationaler Mythos: Anmerkungen zur Haeckel-Virchow-Kontroverse auf der 50. Jahresversammlung deutscher Naturforscher und Ärzte in München (1877)." In *Nationale Mythen und Symbole in der zweiten Hälfte des 19. Jahrhunderts: Strukturen und Funktionen von Konzepten nationaler Identität*, ed. Jürgen Link and Wulf Wülfing, 212–236. Stuttgart: Klett-Kotta, 1991.
Koller, Gottfried. *Das Leben des Biologen Johannes Müller 1801–58.* Stuttgart: Wissenschaftliche Verlagsgesellschaft, 1958.
Kölliker, A. *Erinnerungen aus meinem Leben.* Leipzig: Wilhelm Engelmann, 1899.
Koenigsberger, Leo. *Hermann von Helmholtz*, trans. Frances A. Welby. 1906. New York: Dover, 1965.
Krausse, Erika. *Ernst Haeckel.* Leipzig: B. G. Teubner, 1984.
———. "Johannes Müller und Ernst Haeckel: Erfahrung und Erkenntnis." In *Johannes Müller und die Philosophie*, ed. Michael Hagner and Bettina Wahrig-Schmidt, 223–237. Berlin: Akademie Verlag, 1992.
———. "Zum Verhältnis von Carl Gegenbauer (1826–1903) und Ernst Haeckel (1834–1919): Generelle und spezielle Morphologie." In *Miscellen zur Geschichte der Biologie*, ed. Armin Geus, Wolfgang Friedrich Gutmann, and Michael Weingarten, 84–97. Frankfurt am Main: Waldemar Kramer, 1994.
Krietsch, P. "Zur Geschichte der Prosektur der Charité Berlin." *Zentralblatt für Pathologie* 137 (1991): 531–541.
Kushner, Howard I. "Taking Erikson's Identity Seriously: Psychoanalyzing the Psychohistorian." *The Psychohistory Review* 22 (1993–1994): 7–34.
Laclos, Choderlos de. *Les Liaisons dangereuses.* Paris: Librairie Générale Française, 1987.
Lambert, Carole J. "Postmodern Biography: Lively Hypotheses and Dead Certainties." *Biography* 18 (1995): 305–327.

Lenoir, Timothy. "The Eye as Mathematician: Clinical Practice, Instrumentation, and Helmholtz's Construction of an Empiricist Theory of Vision." In *Hermann Helmholtz and the Foundations of Nineteenth-Century Science*, ed. David Cahan, 109–153. Berkeley: University of California Press, 1993.

———. "Helmholtz, Müller, und die Erziehung der Sinne." In *Johannes Müller und die Philosophie*, ed. Michael Hagner and Bettina Wahrig-Schmidt, 207–222. Berlin: Akademie-Verlag, 1992.

———. *Instituting Science: The Cultural Production of Scientific Disciplines*. Stanford, CA: Stanford University Press, 1997.

———. *The Strategy of Life: Teleology and Mechanics in Nineteenth-Century German Biology*. Chicago: University of Chicago Press, 1989.

Lenz, Max. *Geschichte der königlichen Friedrich-Wilhelms-Universität zu Berlin*. 5 vols. Halle: Buchhandlung des Waisenhauses, 1910.

Lesky, Erna. "Ernst Wilhelm von Brücke." In *Dictionary of Scientific Biography*, 2: 530–532.

Loewenberg, Peter. *Decoding the Past: The Psychohistorical Approach*. New York: Alfred A. Knopf, 1983.

———. *Fantasy and Reality in History*. New York: Oxford University Press, 1995.

Lohff, Brigitte. "Johannes Müller: Von der Nervenwissenschaft zur Nervenphysiologie." In *Das Gehirn—Organ oder Seele? Zur Ideengeschichte der Neurobiologie*, ed. Ernst Florey and Olaf Breidbach. Berlin: Akademie Verlag, 1993.

———. "Johannes Müllers Rezeption der Zellenlehre in seinem *Handbuch der Physiologie des Menschen*." *Medizinhistorisches Journal* 13 (1978): 247–258.

———. "Hat die Rhetorik Einfluss auf die Entstehung einer experimentellen Biologie in Deutschland gehabt?" In *Disciplinae Novae: Zur Entstehung neuer Denk- und Arbeitsrichtungen in der Naturwissenschaft*, ed. Christoph J. Scriba. Göttingen: Vandenhoeck und Ruprecht, 1979.

Machetanz, Hella. "Trichinen und die Duell-Forderung Bismarcks an Virchow im Jahre 1865." *Medizinhistorisches Journal* 13 (1978): 297–306.

Magnus, Gustav. Letters to Jakob Henle. UH, NL Hermann Hoepke, KE 74.

Marx, Karl, and Friedrich Engels. "Manifest der Kommunistischen Partei." *Ausgewählte Schriften*. 2 vols. Berlin: Dietz, 1970. 1: 25–57.

Menke, Richard. "Fiction as Vivisection: G. H. Lewes and George Eliot." *English Literary History* 67 (2000): 617–653.

Merkel, Friedrich. *Jacob Henle: Ein deutsches Gelehrtenleben*. Brunswick: Friedrich Vieweg, 1891.

———. *Jakob Henle: Gedächtnisrede*. Brunswick: Friedrich Vieweg, 1909.

Müller, Johannes. "Über den allgemeinen Plan in der Entwickelung der Echinodermen." *Physikalische Abhandlungen der könliglichen Akademie der Wissenschaften zu Berlin* (1852): 25–65.

———. "Von dem Bedürfnis der Physiologie nach einer philosophischen Naturbetrachtung." In his *Zur vergleichenden Physiologie des Gesichtssinnes des Menschen und der Thiere*. Leipzig: K. Knobloch, 1826.

———. "Beobachtung über die Gesetze und Zahlenverhältnisse der Bewegung in den verschiedenen Thierclassen mit besonderer Rücksicht auf die Bewegung der Insekten und Polymerien." *Isis* 1 (1822): 61–62.

———. "Bericht über die Fortschritte der vergleichenden Anatomie der Wirbelthiere im Jahre 1843." *Müllers Archiv* (1844): 50.

———. "Bestätigung des Bell'schen Lehrsatzes, dass die doppelten Wurzeln der Rückenmarksnerven verschiedene Functionen haben, durch neue und entscheidende Experimente." *Notizen aus dem Gebiete der Natur- und Heilkunde* 30 (1831): 113–117.

———. *Bildungsgeschichte der Genitalien aus anatomischen Untersuchungen an Embryonen des Menschen und der Thiere.* Düsseldorf: Arnz, 1830.

———. *Briefe von Johannes Müller an Anders Retzius von dem Jahre 1830 bis 1857.* Stockholm: Aftonbladets Aktiebolags Tryckeri, 1900.

———. *Über die Compensation der physischen Kräfte am menschlichen Stimmorgan.* Berlin: Hirschwald, 1839.

———. *Elements of Physiology*, trans. William Baly. Philadelphia: Lea and Blanchard, 1842.

———. "Über die Erzeugung von Schnecken in Holothurien." *Bericht über die Bekanntmachung geeigneten Verhandlungen der Königlichen Preussischen Akademie der Wissenschaften zu Berlin* (October 1851): 628–649.

———. "Ueber die Erzeugung von Schnecken in Holothurien." *Müllers Archiv* (1852): 1–36.

———. "Über den feinern Bau der krankhaften Geschwülste." *Monatsberichte der königlichen Preussischen Akademie der Wissenschaften zu Berlin* (December 1836): 107–113.

———. *Über den feineren Bau und die Formen der krankhaften Geschwülste.* Berlin: G. Reimer, 1838.

———. *Handbuch der Physiologie* des Menschen, 3rd ed. 2 vols. Koblenz: J. Hölscher, 1838–1840.

———. "Über die Larven und die Metamorphose der Echinodermen." *Monatsberichte der Königlichen Akademie der Wissenschaften zu Berlin* (1848): 75–109.

———. "Ueber die Metamorphose des Nervensystems in der Thierwelt." *J. F. Meckel's Archiv für Anatomie und Physiologie* (1828): 1–22.

———. "Ueber parasitische Bildungen." *Müllers Archiv* (1842): 193–212.

———. "Ueber die phantastischen Gesichtserscheinungen: Eine physiologische Untersuchung." 1826. In *Johannes Müller: Der grosse rheinische Physiologe,* by Ulrich Ebbecke. Hannover: Schmorl & Seefled, 1951.

———. "Zur Physiologie des Fötus." *Zeitschrift für die Anthropologie* (1824): 423–483.

———. *Zur vergleichenden Physiologie des Gesichtssinnes.* Leipzig: C. Knobloch, 1826.

———. "Ueber verschiedene Formen der Seethiere." *Müllers Archiv* (1854): 69–98.

———. and Jacob Henle. "On the Generic Characters of Cartilaginous Fishes, with Descriptions of New Genera." *The Magazine of Natural History* 2 (1838): 33–37, 88–91.

———. and Jakob Henle. "Über die Gattungen der Plagiostomen." *Wiegmann's Archiv* 1 (1837): 394–401.

———. and Theodor Schwann. "Versuche über die künstliche Verdauung des geronnenen Eiweisses." *Müllers Archiv* (1836): 66–89.

Nipperdey, Thomas. *Deutsche Geschichte 1800–1866: Bürgerwelt und starker Staat.* Munich: C. H. Beck, 1994.

Nyhart, Lynn K. *Biology Takes Form: Animal Morphology and the German Universities 1800–1900.* Chicago: University of Chicago Press, 1995.

Oken, Lorenz. *Lehrbuch der Naturphilosophie,* 3rd ed. Zurich: Friedrich Schulthess, 1843.

———. *Über das Universum als Fortsetzung des Sinnensystems: Ein pythagorisches Fragment.* Jena: Friedrich Frommann, 1808.

Olesko, Kathryn M., and Frederic L. Holmes. "Experiment, Quantification, and Discovery: Helmholtz's Early Physiological Researches." In *Hermann von Helmholtz and the Foundations of Nineteenth-Century Science,* ed. David Cahan. Berkeley: University of California Press, 1993.

Outram, Dorinda. "The Language of Natural Power: The 'Eloges' of Georges Cuvier and the Public Language of Nineteenth-Century Science." *History of Science* 16 (1978): 153–178.

Parnes, Ohad. "The Envisioning of Cells." *Science in Context* 13 (2000): 71–92.

Pearce, J. M. S. "Remak, Father and Son." *The Lancet* 347 (1996): 1669–1670.

Pennisi, Elizabeth. "Haeckel's Embryos: Fraud Rediscovered." *Science* 277 (5 September 1997): 1435.

Prüll, Cay-Rüdiger. "Zwischen Krankenversorgung und Forschungsprimat: Die Pathologie an der Berliner Charité im 19. Jahrhundert." *Jahrbuch für Universitätsgeschichte* 3 (2000): 87–109.

Rather, L. J., Patricia Rather, and John B. Frerichs. *Johannes Müller and the Nineteenth-Century Origins of Tumor Cell Theory.* Canton, MA: Science History Publications, 1986.

Rehkämpfer, Gerd. "Zur frühen Rezeption von Darwins Selektionstheorie und deren Folgen für die vergleichende Morphologie heute." *Sudhoffs Archiv* 81 (1997): 171–192.

Remak, Robert. "Anatomische Beobachtungen über das Gehirn, das Rückenmark und die Nervenwürzeln." *Müllers Archiv* (1841): 506–522.

———. "Ueber die Anzeigen zur Ausrottung krankhafter Geschwülste." *Medicinische Zeitung* 10 (1841): 131–132.

———. "Ein Beitrag zur Entwickelungsgeschichte der krebshaften Geschwülste." *Deutsche Klinik* 6 (1854): 170–174.

———. "Ueber die embryologische Grundlage der Zellenlehre." *Müllers Archiv* (1862): 230–241.

———. "Ueber die Enden der Nerven im elektrischen Organ der Zitterrochen." *Müllers Archiv* (1856): 467–472.

———. "Ueber die Entstehung des Bindegewebes und des Knorpels." *Müllers Archiv* (1852): 63–72.

———. "Ueber die Entwickelung des Hühnchens im Ei." *Müllers Archiv* (1843): 478–484.

———. "Ueber extracellulare Entstehung thierischer Zellen und über Vermehrung derselben durch Theilung." *Müllers Archiv* (1852): 47–57.

———. *Galvanotherapie der Nerven- und Muskelkrankheiten.* Berlin: August Hirschwald, 1858.

———. "Ueber die Ganglien der Herznerven des Menschen und deren physiologische Bedeutung." *Wochenschrift für die gesammte Heilkunde* no. 10 (9 March 1839): 149–154.

———. "Über gangliöse Nervenfasern beim Menschen und bei den Wirbelthieren." *Monatsbericht der königlichen Akademie der Wissenschaften zu Berlin* (1853): 293–298.

———. "Ueber die genetische Bedeutung des oberen Keimblattes im Eie der Wirbelthiere." *Müllers Archiv* (1849): 209–210.

———. "Ueber den Inhalt der Nervenprimitivröhren." *Müllers Archiv* (1843): 197–201.

———. "Neue Beiträge zur Kenntnis vom organischen Nervensystem." *Medicinische Zeitung* 9 (1840): 7–8.

———. "Neurologische Erläuterungen." *Müllers Archiv* (1844): 463–472.

———. "Ueber peripherische Ganglien in den Nerven des Nahrungsrohrs." *Müllers Archiv* (1858): 189–192.

———. "Ueber die physiologische Bedeutung des organischen Nervensystems, besonders nach anatomischen Thatsachen." *Monatschrift für Medizin, Augenheilkunde und Chirurgie* 3 (1840): 225–265.

———. "Ueber runde Blutgerinnsel und über pigmentkugelhaltige Zellen." *Müllers Archiv* (1852): 115–162.

———. *Ueber ein selbständiges Darmnervensystem.* Berlin: G. Reimer, 1847.

———. "Ueber die soganannten Blutkörperchen haltenden Zellen." *Müllers Archiv* (1851): 480–483.

———. "Ueber die Structur des Nervensystems." *Neue Notizen aus dem Gebiete der Natur- und Heilkunde* 7 (1838): 342–346.

———. "Ueber die Theilung der Blutzellen beim Embryo." *Müllers Archiv* (1858): 178–188.
———. "Ueber Theilung thierischer Zellen." *Müllers Archiv* (1854): 376.
———. *Untersuchungen über die Entwickelung der Wirbelthiere.* Berlin: G. Reimer, 1855.
———. "Ueber die Verrichtungen des organischen Nervensystems." *Neue Notizen aus dem Gebiete der Natur- und Heilkunde* 7 (1838): 65–70.
———. "Ueber vielkernige Zellen der Leber." *Müllers Archiv* (1854): 99–102.
———. "Vorläufige Mittheilung microscopischer Beobachtungen über den innern Bau der Cerebrospinalnerven und über die Entwickelung ihrer Formelemente." *Müllers Archiv* (1836): 145–159.
———. "Weitere mikroscopische Beobachtungen über die Primitivfasern des Nervensystems der Wirbelthiere." *Neue Notizen aus dem Gebiete der Natur- und Heilkunde* 3 (1837): 36–41.
Rheinberger, Hans-Jörg. "From the 'Originary Phenomenon' to the 'System of Pelagic Fishery': Johannes Müller (1801–1858) and the Relation between Physiology and Philosophy." In *From Physico-Theology to Bio-Technology: Essays in the Social and Cultural History of Biosciences. A Festschrift for Mikulás Teich,* ed. Kurt Bayertz and Roy Porter. Atlanta, GA: Rodopi, 1998.
Richards, Robert J. *The Romantic Conception of Life: Science and Philosophy in the Age of Goethe.* Chicago: University of Chicago Press, 2002.
Rosen, George. "Carl Friedrich Wilhelm Ludwig." In *Dictionary of Scientific Biography,* 8: 540–542.
Rosenberger, Eugenie. *Félix du Bois-Reymond, 1782–1865.* Berlin: Meyer und Jessen, 1912.
Rothfield, Lawrence. *Vital Signs: Medical Realism in Nineteenth-Century Fiction.* Princeton, NJ: Princeton University Press, 1992.
Rothschuh, Karl E. "Emil Heinrich du Bois-Reymond." In *Dictionary of Scientific Biography,* 4: 200–205.
———. "Emil du Bois-Reymond (1818–1896) und die Elektrophysiologie der Nerven." In *Von Boerhaave bis Berger: Die Entwicklung der kontinentalen Physiologie im 18. und 19. Jahrhundert mit besonderer Berücksichtigung der Neurophysiologie, ed. Karl E. Rothschuh.* Stuttgart: Gustav Fischer, 1964.
———. *History of Physiology,* ed. and trans. Guenter B. Risse. Huntington, NY: Robert E. Krieger, 1973.
Ruff, Peter W. *Emil du Bois-Reymond.* Leipzig: B. G. Teubner, 1981.
Schelling, Friedrich Wilhelm Joseph. *Einleitung zu seinem Entwurf eines Systems der Naturphilosophie: Historisch-Kritische Ausgabe.* Reihe I: *Werke,* ed. Manfred Durner and Wilhelm G. Jacobs. Stuttgart: Frommann-Holzboog, 2004.
Schleich, Carl Ludwig. *Besonnte Vergangenheit: Lebenserinnerungen 1859–1919.* Berlin: Ernst Rowohlt, 1921.
Schmiedebach, Heinz-Peter. *Robert Remak (1815–1865): Ein jüdischer Arzt im Spannungsfeld von Wissenschaft und Politik.* Stuttgart: Gustav Fischer, 1995.
Schott, Dieter. "'Die Medizin ist eine sociale Wissenschaft'—Rudolf Virchow und die 'Medicinische Reform' in der Revolution 1848/49." In *Die Revolution hat Konjunktur: Soziale Bewegung, Alltag und Politik in der Revolution 1848/49,* 87–108. Münster: Westfälisches Dampfboot, 1999.
Schulz, Kurd. "Rudolf Virchow und die Oberschlesische Typhusepidemie von 1848." *Jahrbuch des Schlesischen Friedrich-Wilhelms-Universität zu Breslau* 19 (1978): 107–120.
Schwann, Theodor. Letter of 6 December 1838. GSPK, Rep. 76Va, Sekt. 2, Tit. X, Nr. 11, Bd. VII, "Das anatomische Museum der Universität zu Berlin 1836–1858."
———. *Lettres de Théodore Schwann,* ed. Marcel Florkin. Liège: Société Royale des Sciences de Liège, 1961.

———. *Microscopical Researches into the Accordance in the Structure and Growth of Animals and Plants*, trans. Henry Smith. London: Sydenham Society, 1847.

———. "Ueber die Nothwendigkeit der atmosphärischen Luft zur Entwickelung des Hühnchens in dem bebrüteten Ei." *Notizen aus dem Gebiete der Natur- und Heilkunde* 41 (1834): 241–245.

———. "Versuche, um auszumitteln, ob die Galle im Organismus eine für das Leben wesentliche Rolle spielt." *Müllers Archiv* (1844): 127–159.

———. "Vorläufige Mittheilung, betreffend Versuche über die Weingährung und Fäulniss." *Annalen der Physik und Chemie* 41 (1837): 184–193.

———. "Ueber das Wesen des Verdauungsprocesses." *Müllers Archiv* (1836): 90–139.

Selow, Edith. "Frederick Wilhelm Joseph Schelling." In *Dictionary of Scientific Biography*, 12: 153–157.

Shakespeare, William. *Hamlet*. Harmondsworth, UK: Penguin Popular Classics, 1994.

Sheehan, James J. *German History 1770–1866*. Oxford: Clarendon Press, 1989.

Smith, Woodruff D. *Politics and the Sciences of Culture in Germany 1840–1920*. Oxford: Oxford University Press, 1991.

Sperber, Jonathan. *The European Revolutions, 1848–1851*. Cambridge: Cambridge University Press, 1994.

Steudel, Johannes. "Johannes Peter Müller." In *Dictionary of Scientific Biography*, 9: 567–574.

———. *Le Physiologiste Johannes Müller*. Paris: Université de Paris, 1963.

———. "Wissenschaftslehre und Forschungsmethodik Johannes Müllers." *Deutsche medizinsche Wochenschrift* 77(1952): 115–118.

Stürzbecher, Manfred. "Zur Berufung Johannes Müllers an die Berliner Universität." *Jahrbuch für die Geschichte Mittel- und Ostdeutschlands* 21 (1972): 184–226.

———. "Aus dem Briefwechsel des Physiologen Johannes Müller mit dem Preussischen Kultusministerium." *Janus* 49 (1960): 273–284.

———. "Zur Geschichte der Instruktionen des Prosektors der Charité 1831–1833." *Janus* 52 (1965): 40–75.

———. "Die Prosektur der Berliner Charité im Briefwechsel zwischen Robert Froriep und Rudolf Virchow." In *Beiträge zur Berliner Medizingeschichte: Quellen und Studien zur Geschichte des Gesundheitswesens vom 17. bis zum 19. Jahrhundert*, 156–220. Berlin: Walter de Gruyter, 1966.

Thomas, Roger. "Rudolf Virchow: Pathologist, Public Health Physician, Liberal Politician, Anthropologist, and Opponent of Anti-Semitism." *Journal of Medical Biography* 7 (1999): 200–207.

Tsouyopoulos, Nelly. "Schellings Naturphilosophie: Sünde oder Inspiration für den Reformer der Physiologie Johannes Müller?" In *Johannes Müller und die Philosophie*, ed. Michael Hagner and Bettina Wahrig-Schmidt. Berlin: Akademie Verlag, 1992.

Tuchman, Arleen Marcia. *Science, Medicine, and the State in Germany: The Case of Baden 1815–1871*. Oxford: Oxford University Press, 1993.

Turner, R. Steven. "Hermann von Helmholtz." In *Dictionary of Scientific Biography*, 6: 241–253.

Uschmann, Georg. *Ernst Haeckel: Biographie in Briefen*. Leipzig: Urania, 1983.

Virchow, Rudolf. *Cellular Pathology*, trans. Frank Chance. 1858. Ann Arbor, MI: Edwards Brothers, 1940.

———. *Disease, Life and Man: Selected Essays by Rudolf Virchow*, trans. Lelland J. Rather. Stanford, CA: Stanford University Press, 1958.

———. "Die endogene Zellenbildung beim Krebs." *Virchows Archiv* 3 (1851): 197–227.

———. "Erinnerungsblätter." *Virchows Archiv* 4 (1852): 541–548.

———. *Johannes Müller: Eine Gedächtnisrede*. Berlin: August Hirschwald, 1858.

———. Lecture notes on Johannes Müller's comparative anatomy course. ABBAW, NL R, Virchow nr. 2804.

———. Lecture notes on Johannes Müller's pathological anatomy course. ABBAW, NL R, Virchow nr. 2805.

———. Lecture notes on Johannes Müller's physiology course. ABBAW, NL R., Virchow nr. 2803.

———. Notebooks. ABBAW, NL R., Virchow nr. 2824.

———. "Zur pathologischen Physiologie des Bluts. Die Bedeutung der Milz- und Lymphdrüsen-Krankheiten für die Blutmischung (Leukaemie)." *Virchows Archiv* 5 (1853): 43–128.

———. "Ueber Perlgeschwülste (Cholesteatoma Joh. Müllers)." *Virchows Archiv* 8 (1855): 371–418.

———. "Prospectus." *Virchows Archiv* 1 (1847): n.p.

———. "Über die Reform der pathologischen und therapeutischen Anschauungen durch die mikroskopischen Untersuchungen." *Virchows Archiv* 1 (1847): 207–255.

———. *Rudolf Virchow: Briefe an seine Eltern 1839 bis 1864*, ed. Marie Rabl. Leipzig: Wilhelm Engelmann, 1906.

———. *Rudolf Virchow: Letters to His Parents 1839 to 1864*, ed. and trans. L. J. Rather. Canton, MA: Science History Publications, 1990.

———. *Sämtliche Werke.* 59:4, *Briefe: Der Briefwechsel mit den Eltern 1839–1864.* Berlin: Blackwell Wissenschafts-Verlag, 2001.

Waldeyer, Wilhelm. "J. Henle: Nachruf." *Archiv für mikroskopische Anatomie* 26 (1886): 1–32.

———. "Der Unterricht in den anatomischen Wissenschaften an der Universität Berlin im ersten Jahrhundert ihres Bestehens." *Berliner klinische Wochenschrift* (10 October 1910): 1863–1866.

Watermann, Rembert. *Theodor Schwann: Leben und Werk.* Düsseldorf: L. Schwann, 1960.

Wells, George A. "Johann Wolfgang von Goethe." In *Dictionary of Scientific Biography,* 5: 442–446.

Winau, Rolf. *Medizin in Berlin.* Berlin: Walter de Gruyter, 1987.

—. "Rudolf Virchow und die Berliner Medizin." *Notabene medici* 12 (1994): 433–440.

Wise, Norton. "What's in a Line? The 3 D's: Dürer, Dirichlet, du Bois." Rothschild Lecture, Harvard University, April 2001.

Zedlitz, Leopold Freiherr von. *Neustes Conversations—Handbuch für Berlin und Potsdam.* Berlin: A. W. Eisersdorff, 1834.

Zigman, Peter. "Ernst Haeckel und Rudolf Virchow: Der Streit um den Charakter der Wissenschaft in der Auseinandersetzung um den Darwinismus." *Medizinhistorisches Journal* 35 (2000): 263–302.

Zimmerman, Andrew. "Anti-Semitism as Skill: Rudolf Virchow's Schulstatistik and the Racial Composition of Germany." *Central European History* 32 (1999): 409–429.

Zloczower, Awraham. *Career Opportunities and the Growth of Scientific Discovery in Nineteenth-Century Germany.* New York: Arno Press, 1981.

Zwick, Jochen. "Akademische Erinnerungskultur, Wissenschaftsgeschichte und Rhetorik im 19. Jahrhundert: Über Emil du Bois-Reymond als Festredner." *Scientia Poetica: Jahrbuch für Geschichte der Literatur und der Wissenschaften* 1 (1997): 120–139.

Index

Academy of Sciences, Berlin, 16, 76, 92, 101, 124, 158
  attempts to elect Remak to, 167, 174, 179
  du Bois-Reymond and, 218–219
Academy, Art, of Berlin, 51, 79, 101, 118, 141
acoustics, 125
action, science as, 77, 110
Adams, Henry, 14
aesthetics, of experimentation, 107
alcoholism, 32, 147
Alexander the Great, 289 n.115
Altenhausen, Franziska von, 226
Altenstein, Karl von, 6
  and Jakob Henle, 55
  as a patron of Müller, 12, 14
  and Theodor Schwann, 64
  opposition to hiring Jewish professors, 167, 173
analogy, 86, 184
analysis, chemical, 22, 183
Anatomical Institute. *See* Institute, Berlin Anatomical
Anatomical Museum. *See* Museum, Berlin Anatomical
Anatomical Theater. *See* Institute, Berlin Anatomical
anatomy
  emergence of, 12
  pathological, 137–138, 141–142, 207
  relation to physiology, 22, 184
anti-Semitism, 153, 166
Archive for Anatomy and Physiology and for Scientific Medicine. See Müller's Archive
Archive for Pathological Anatomy and Physiology and for Clinical Medicine. See Virchow's Archive
Aristotle, 136
Assembly, Prussian, 34–35, 153, 176
auscultation, 145
autopsies, 140–141, 154–155, 224, 227
axis cylinder, 188
axon, 115, 170, 187

Baden, 5
Baer, Karl Ernst von, 10, 13, 33, 201, 218, 222
Baly, William, 109, 243 n.44
Balzac, Honoré de, 30, 39
Barez, Stephan, 146–147

barricades, 148
battery, invention of, 84
Bavaria, 151
Belgium, 65
Bell, Charles, 11, 50, 157
Bell's Law, 11, 50
Bence Jones, Henry, 85, 108–109
Berlin Anthropological Society, 155
Berlin Obstetrical Society, 151
Berlin Physical Society, 111, 120
Berlin
  cultural environment, 16
  daily life, 15–16
  demographics, 14
  economy, 15
  history, 14
Bernard, Claude, 22
Bethmann-Hollweg, Moritz August von, 180
Bichat, Xavier, 138
Bidder, Friedrich, 19, 25, 52–54, 227, 234, 245 n.104, 257 n.135
Bildung, 6
Bildungsbürgertum, 82, 264 n.109
biogenetic law, xii, 219–220, 222
biography
  du Bois-Reymond's, of Müller, 104–105, 108–109, 158, 224–226
  postmodern, xv
Bismarck, Otto von, xii, 147, 153
blood clotting, 162
blood poisoning, 143
Böhm, Dr. 225–226
Borsig, August, 5
Boumann, Johann, 16
Bronn, Heinrich Georg, 213, 289 n.112
Brücke, Ernst, xvi, 101, 111, 118, 179
Bunsen, Robert, 119–120
burner, Bunsen, 119
Burschenschaften, 5, 46, 51, 55–56, 121, 167, 252 n.20

Cabinet, Berlin Mineralogical, 27
cadavers, 18, 125
  conflict between the Charité and Berlin University over, 140–141, 152
Cagniard-Latour, Charles, 62
career traffic, 179
Caro, Frederica, 166
Carus, Carl Gustav, 136
Catholicism, Roman, 21, 146–147, 158–159, 193
cell bodies, 115, 170
cell theory, 62–64, 67, 73, 202, 233, 256 n.108
  Virchow and, 138, 170–171, 207
  Remak and, 170–171
Charité Hospital, 117, 139–145
  anatomical collection, 142–145
  and cadavers, 19, 140–141
Chevalier, Charles, 57
Chladni, Ernst von, 187
Chodowiecki, Daniel, 78, 80
cholesteatoma, 157
cholesterol, 157
chorda dorsalis, 63
class, social, xiii, 3, 73–74, 80, 252 n.17
  influence on science, 132, 232
  and the revolution of 1848, 149
cocaine, 228
Communist Manifesto, 86–87
companionship, need for scientific, 33, 144, 200, 233
comparative anatomy, xiii, 45, 74, 198, 201, 203, 211
  Haeckel's transformation of, 214, 220
competition
  among German universities, 12
  among scientists, 17, 51, 127, 203–204, 227
Confederation, Germanic, 5
conflicts, laboratory, xix
connective tissue, 157
cornea, 140
crayfish, 186, 212–213
creativity
  in history-writing, xiv
  literary, 96, 110
  of scientists, xv, xvii, 37, 110, 251 n.4
Cruveilhier, Jean, 143
Cultural Ministry, Prussian, 6, 122, 141, 145, 150, 181
  opposition to hiring Jewish professors, 167, 176
culture, influence on science, 232
current, 94
curves, mathematical, xiv, xviii

Cuvier, Georges, 8, 30, 49, 137, 214, 267 n.176
cyclostomes, 23
cysts, 195, 220
cytoblastema, 64, 162, 170
Czapski, Count, 180

daemonization, 110
Dangerous Liaisons, xv, 78, 96, 124
Darwin, Charles, 28, 163
  Haeckel's use of, 190, 200–201, 209–210, 213–214, 217, 221–222, 289 n.112
depression, 10, 33–34, 37, 227–228
Derrida, Jacques, 235
Detectives
  historians as, xiv, 228
  scientists as, 24
development, embryonic, 183, 185
dichotomies, 235
Diogenes, 120
Doktorvater, xvi, 42
drawing, 88, 196
Du Bois-Reymond, Emil, xii
  in 1848, 83–84
  academic career, 102, 126–127
  admiration of Schwann, 72
  Animal Electricity, 77, 81, 84–88, 120
  and anti-Semitism, 166
  and the Berlin Anatomical Museum, 30
  collaboration with Berlin machinists, 80, 83, 92–94, 109, 229
  and comparative anatomy, 101–102
  education, 80
  election to the Academy of Sciences, 92, 101
  electrophysiological experiments, 90–91
  family and cultural background, 79–83
  and the history of science, 77
  income and finances, 81, 101
  influence of Müller on, 85, 103, 109–110
  letters to Hallmann, xv, 95-99
  on the limits of scientific knowledge, 218
  living situation in Berlin, 89
  memorial address for Müller, 72, 76, 104–109, 158, 224–226, 229–230
  meeting with Helmholtz, 80, 116, 270 n.24
  as mentor to Helmholtz and Ludwig, 120–127
  and microscopy, 80
  on nature philosophy, 6
  as a physiology lecturer, 102
  promotion to associate professor, 179–180
  as a public speaker, 103
  relationship with Alexander von Humboldt, 91
  relationship with Haeckel, 217–218
  relationship with his father, 80–81
  relationship with Ludwig, 120, 126
  relationship with Müller, 48, 77, 109–110, 131, 229
  relationship with Virchow, 84
  response to Müller's death, 103
  scientific identity of, 77, 103–104
  support from Müller, 101
  wit, 192, 229
Du Bois-Reymond, Félix Henri, 79–82
dualism, 214
duelling, 46, 153, 167

echinoderms, 24, 107
Eck, Gottlieb, 145
Edict of Emancipation, Prussian, 167
Edict of Nantes, 78
eels, electric, 88
Egloff, Elise, 71
Ehrenberg, Christian Gottfried, 167, 172, 175, 183, 213
Eichhorn, Johann, 29, 34, 36, 145–147
  opposition to appointing Jewish professors, 167, 174, 177, 179
electricity, animal, 7, 77, 84–88, 90–91, 103
electrophysiology, 84–85, 90–91
electrotherapy, 172, 182, 188, 232
Eliot, George, 108
embryology, 136, 171, 178, 182, 185
embryos, 170
Engels, Friedrich, 86–87
Entwicklungsjahre, 191
erections, 10
evolution, 21, 28, 163
  Haeckel on, 190, 198, 200, 209–210, 213–214, 220, 289 n.112
experimentation, 22, 107
experiments, physiological, xiii, 11, 13, 20, 22, 90–91, 102, 122

fatherhood, xvii
Faulkner, William, xiv
fermentation, 62, 117–118
fibers
  primitive, 173
  of Remak, xii
fiction, xi, xiv–xv
Flourens, Marie Jean Pierre, 267 n.176
force, conservation of, 118
forms, animal, 20, 24, 29, 201, 220, 229, 233
fossils, 107
Frantzius, Alexander, 154
Frederick the Great, 4, 167
Freud, Sigmund, xvi–xvii
  and cocaine, 228
  "The Family Romance", xvii
  Totem and Taboo, xvi–xvii, 235
Friedländer, Nathan, 283 n.73
Friedrich Wilhelm III, 167
Friedrich Wilhelm IV, 34–35, 167, 176, 180
Friedrichs-Wilhelms-Institut, 112–113, 133–135, 139, 146
frogs, as experimental animals, 11, 85, 90, 93, 106, 182, 233
Froriep, Robert, 51
  as Charité prosector, 141–145
  as mentor to Virchow, 143–144
  rivalry with Müller, 142, 162
functionality, 87–88, 93

Galvani, Luigi, 84, 106
galvanometer, 77, 90–91, 93–94, 233
galvanotherapy, 172
ganglia, 170
  sympathetic, 170
ganglion cells, 114–115, 170
ganoid fishes, 23
Gegenbaur, Carl, 202–204
General Association of Berlin Doctors, 168
genitalia, 10
germ layers, embryonic, xii, 171
German Progressive Party, 153
Germany, history of, 4
Goethe, Johann Wolfgang von
  anatomical studies, 7
  Elective Affinities, 38, 219
  and evolution, 7, 201, 214, 221
  "Hope," 78
  as an inspiration for Haeckel, 201
  as an inspiration for Müller, 38
  Lewes's biography of, 108–109
  Poetry and Truth, 191
  Wilhelm Meister, 259 n.7
Görcke, Johann, 146
Grimm, Heinrich, 144
Grimm, Jakob, 38
growth
  cell, 170
  embryonic, 157, 170
  pathological, 157, 170
Gunther, Albert, 37, 227
Gymnasien, 6
gynecology, 151

Habilitation, 17, 51, 176, 203
Haeckel, Ernst, xii
  academic career, 204
  conflict with Virchow over teaching evolution, 163, 209–210
  doctoral thesis on the crayfish, 212–213
  education, 191, 202
  and evolutionary theory, 190, 198, 200–201, 209–210, 213–214, 217, 220, 222
  family and cultural background, 191–193
  General Morphology, 204, 214
  and microscopy, 196–198
  on Müller's death, 226–227
  Müller's influence on, 213, 223
  Perigenesis of the Plastidules, 218, 290 n.133
  relationship with du Bois-Reymond, 217–218
  relationship with Müller, 200, 209–210, 223, 230–231
  relationship with Virchow, 208–210, 220, 288 n.96
  Riddle of the Universe, 214–215, 217
  scientific identity, 190
Haller, Albrecht von, 135, 157, 160
Hallmann, Eduard, xv, 68, 95–99
Hamlet, 76–77, 84, 110
Handwerker, 241 n.3
Haraway, Donna, 233
Hardy, Thomas, 24
heart, 170
heat, animal, 118

Hecker, Friedrich, 145
Hegel, Georg Wilhelm Friedrich, 20, 51, 87
Helgoland, 31, 191, 196, 211–212, 215–217, 219
Helmholtz, Ferdinand, 112
Helmholtz, Hermann von, xii
 acoustical studies, 125
 on anatomy, 117, 124–125
 and anti-Semitism, 166
 and the Berlin Physical Society, 111
 candidacy for the Berlin physics chair, 104, 126
 clinical rotations, 117
 doctoral research on ganglion cells, 115-116, 270 n.22
 education, 112–117
 family and cultural background, 112–113
 fermentation studies, 117–118
 measurement of nerve impulse velocity, 123
 meeting with du Bois-Reymond, 80, 116, 270 n.24
 as a mentor to du Bois-Reymond, 126
 and microscopy, 114–115
 muscle chemistry studies, 118
 as museum helper, 101
 notes on Müller's lectures, 113–114
 Physiological Optics, 125, 127-129
 relationship with his father, 269 n.4
 relationship with Müller, 131, 230, 270 n.22, 271 n.38
 representation of Müller's physiology, 127–131
 Sensations of Tone, 125
 as staff surgeon in Potsdam, 117
 visual studies, 125, 127–130, 273 n.94
hemorrhages, 155
Henle, Jakob, xii
 and comparative anatomy, 56, 87
 conflict with Virchow, 71
 education, 45–47
 efforts to teach microscopy, 58, 68–69, 257 n.141, 258 n.151
 family background, 43–44
 General Anatomy, 71
 Habilitation on the intestine, 59
 incarceration, 55–56
 income and finances, 45, 51
 influence of Müller on, 66, 74
 Jewish background, 43–44, 252 n.11
 and the *Journal for Rational Medicine*, 71
 laryngeal studies, 56
 living situation in Berlin, 20, 51, 53
 marriages, 71
 membrane and epithelial studies, 58–59
 move to Zurich, 34
 participation in *Burschenschaften* and duelling, 46, 51, 55–56
 on pathology, 71
 personality, 45, 49, 52, 55
 pus and mucous studies, 59
 proposal that microorganisms cause disease, 59, 164
 and the pupil membrane, 47–48
 relationship with Müller, 44, 48–52, 67–71, 74, 229, 258 n.151
 relationship with Virchow, 71, 163–164
 scientific ambitions, 47
 support of experimental physiology, 125
 travel with Müller, 48–49
heroes, scientific, 209, 211, 214
Hesse, Ludwig Ferdinand, 133
His, Wilhelm, 179
histology, 177
history-writing
 agenda in, xiv
 motives for, xv, xvii, xix
 and self-promotion, 20, 159, 190, 228, 259 n.1, 291 n.11
 sources in, xiv
Hohenzollern dynasty, 34
holothuria, 23, 182
Hölscher, Jakob, 22
Holtzendorff, Franz von, 156
horses, 88–89
Huguenots, 78
Humboldt, Alexander von, 11
 electrophysiological experiments, 7, 84, 91
 promotion of Berlin science, 13
 support of du Bois-Reymond, 81, 91, 96, 120–121
 support of Henle, 49, 55
 support of Helmholtz, 124
 support of Müller, 36
 support of Remak, 173–177, 180
 Views of Nature, 88

Humboldt, Wilhelm von, 16, 81
Huxley, Thomas Henry, 201
hypochondria, 10, 243 n.50

identity, scientific, 190, 228, 233
images, 196
income, of university professors, 13
industrialization, German, 5
inflammation, 143, 156, 162
influence, anxiety of, xvii, 109–110
insomnia, 32–33, 224, 248 n.180
Institute, Berlin Anatomical 17–19, 54, 58, 140, 233
Institute, Berlin Physical 126
Institute, Berlin Physiological, 89, 92, 100, 127
Institute, Bonn Anatomical, 125
interpretation, xiv
intimacy, scientific, 47–50, 233

jealousy, among scientists, 123, 204, 229, 272 n.58
Jena, Battle of, 4
Judaism, 167, 169, 252 n.11
Jüngken, Johann Christian, 83, 138–139, 144
Junkers, 4, 132

Kant, Immanuel, 218
  and Helmholtz's visual studies, 128–129
  and inborn capacities for perception, 79, 112
  and *Zweckmässigkeit*, 7, 23, 87
Kierkegaard, Søren, 261 n.40
Koch, Robert, xii
Kölliker, Albert, 58
  as Haeckel's teacher, 191, 193–194, 196, 198, 203–206, 209, 212, 220
  as Henle's student, 68–69, 204
  as Remak's student, 178
Koreff, David Ferdinand, 167
Kulturkampf, 147
kymograph, 120

laboratories, scientific, xi, xiv, 194, 234, 239 n.1
Laboratorium, Physiologisches, 92
Lachmann, Johannes, 209
Laclos, Choderlos de, xv, 78, 96
Ladenberg, Adalbert von, 36, 81, 91
Lamarck, Jean Baptiste, 214, 222
language, 77, 154
larynx, 56
laudanum, 228
Law of Conservation of Force, 118
Law of Specific Sense Energies, xi, 9
  importance for Helmholtz's sensory physiology, 125, 127–130, 273 n.87
lectures, popular, 103–104, 126
Leipzig, Battle of, 4
Lenin, Vladimir Ilych, 214
letters, personal, xvi, 95–99, 134, 191–192
Leubuscher, Rudolf, 149, 155
leukaemia, 162
Lewes, George Henry, 108–109
Leydig, Franz, 196
Lichtenstein, Heinrich, 27
Liebig, Justus, 22, 118
life force, 62, 64–65, 73, 234
  du Bois-Reymond's rejection of, 76, 85–88, 93–94
  Haeckel's rejection of, 198, 214
  Helmholtz's rejection of, 130–131
  Schwann's rejection of, 229
life, great plan of, 20, 76, 124, 139, 161, 164, 184, 234
Liphardt, C. E. von, 58
literary techniques, xv
Lohmeyer, Johann, 145
loneliness, 33, 201
loop of Henle, xii
Louis XIV, 78
Ludwig, Carl, 90, 101, 119–122
  career after 1849, 126
  democratic activism, 121
  relationship with du Bois-Reymond, 120, 126
Luther, Martin, 106

Magendie, François, 11, 106, 135, 157, 160
Magnus, Gustav, 70, 126, 226, 258 n.159
  home laboratory, 80, 93, 111, 116
manifestos, 181, 262–263 n.71
Mannigfaltigkeit, 25, 198–199
marine organisms, 23, 107, 202
Marx, Karl, 44, 86–87
masturbation, 147
materialism, 164, 222, 231

Matteucci, Carlo, 84, 93, 101, 126, 233
Mayer, Carl, 143, 151
Mayer, Rose, 151
Meckel von Hemsbach, Heinrich, 151
Meckel, Johann Friedrich, 14
medical education, 112–117, 193–195, 245 n.100
Medical faculty, Berlin University, 177, 181, 283 n.73
medicine, rational, 163
memorial addresses, xvi
  du Bois-Reymond's, for Müller, xvi–xvii, 72, 76, 104–109, 158, 224–226, 229–230
  Virchow's, for Müller, 158–162, 225
memory, xviii, 77, 230, 273 n.101
Mendelssohn, Moses, 167
Mendelssohn-Bartholdy, Felix, 44, 252 n.11
mentorship, laissez-faire, xvi, 100, 234–235
Merseburg, 191
Messina, 213
metaphor, 187, 219
methods
  exact, 215
  scientific, 160, 164, 186
Metternich, Klemens Wenzel Lothar von, 4–5
Meyer, Eli Joachim, 169
Meyer, Feodore, 169
Mianowsky, Josef, 180
microscopes
  achromatic, 57–58, 62
  cost of, 58, 255 n.87
  Fraunhofer, 8, 57–58, 183, 255 n.87
  Schiek, 58, 183, 191, 196–197, 255 n.87
microscopy, 8, 22, 56–59, 68–69, 138, 140, 143, 152, 157, 160, 183
  joy of, 197–198
midwives, 169
Miescher-His, Friedrich, 255 n.87
Milton, John, 107
Mitscherlich, Carl Gustav, 83, 134, 177, 179
Mohr, Carl, 22
monism, 214
Morgagni, Giovanni Batista, 137–138
morphine, 226
morphology, 204, 214
motivation, of scientists, 24, 231
movement, animal, 8
Müller, Johannes
  in 1848, 35–37, 108, 161
  ambition of, 11–12
  and animal electricity, 84–85, 106
  application for Berlin anatomy and physiology chair, 13
  assistance from patrons, 6, 8, 11–12
  and the Berlin Anatomical Museum, 14, 26-32
  and the Berlin pathology chair, 151–152
  and *Burschenschaften*, 5
  and cell theory, 67–68, 73
  comparative anatomy lectures, 25, 113–114, 136–137, 198, 210–211, 216
  Comparative Physiology of Vision, 9
  death, 102–103, 203, 213, 222, 224–228
  depressive episodes, 10, 33–34, 37, 227
  doctoral thesis, 8, 242 n.38
  family background, 3, 8, 32
  on fantasy images, 9, 32, 250 n.220
  on genitalia, 23
  as gross anatomy lab supervisor, 19–20
  Handbook of Human Physiology, xi, 21, 96, 106, 109, 128, 172
  on the human voice, 20, 25
  income and finances, 32
  insomnia, 32–33
  on marine organisms, 23, 107
  and microscopy, 8, 22, 68–69, 157, 160, 183
  and nature philosophy, 6–8, 21
  on physiology, 8, 21, 26
  on physiological experiments, 11, 13, 20, 22, 90, 122
  political conservatism, 35, 108, 161, 221
  possible mental illness, 33, 249 n.189
  relationship with du Bois-Reymond, 101, 109–110, 131, 229
  relationship with Haeckel, 209–210, 213, 223, 230–231
  relationship with Helmholtz, 118, 131, 230, 271 n.38
  relationship with Henle, 44, 48-50, 66, 68-71, 74, 229, 258 n.151
  relationship with Schwann, 50–51, 67–68, 73–74, 229
  relationship with Virchow, 139, 143–144, 152, 160, 230

Müller, Johannes (*continued*)
and religion, 21, 158–159, 227
on the role of philosophy in science, 8, 23, 160
scientific style, 60–62
shipwreck, 37, 41, 101–102
on the soul, 21
spinal nerve root studies, xi, 11, 50, 90–91, 160, 162, 172
style as an adviser, 67, 100, 117, 234
teaching schedule, 17–18
on tumors, 22, 25–26, 138, 157, 162, 170, 182
on the visual system, 23
work habits, 31, 212
as a writer, 37–41, 107, 187
Müller, Maria, 10
Müller, Max, 10, 105, 212
Müller, Nanny, 10, 32, 44, 69, 105, 226, 228
Müller's Archive, 22, 51, 154–155
Mursinna, Christian Ludwig, 113
Muscle
chemistry, 118
contraction, 60
gastrocnemius, 90
tension, 60
Museum, Berlin Anatomical, 15, 26–31, 98, 107–108, 173, 233
admissions, 28
custodian Thiele, 28, 30, 83
funding, 29, 32
helpers, 51, 101, 118
history of, 142
renovations, 28
Museum, Berlin Zoological, 27
myograph, 102, 131
myxinoids, 23

Nachzeichnen, 196
Napoleon, 4, 44, 79
narrative, xiv–xv, 188
breaks in, xv
du Bois-Reymond's memorial address for Müller as, 105–110, 239 n.2, 267 n.177, 291 n.11
Haeckel's use of, 191, 194–195, 200, 211, 215, 221
Virchow's memorial address for Müller as, 158–162
nature philosophy, 6–8, 21, 45
du Bois-Reymond on, 86–88, 105
Haeckel and, 210, 221
Virchow on, 159
Naturforscher, 84, 130
Naturphilosophie. See nature philosophy
nerve fibers, 115–116, 184
ganglious, 184
sympathetic, 170, 187
nerve impulse, velocity of, 78, 123–124
nerve roots, spinal, xi, 11, 50, 90–91, 160, 162, 172, 184
nerve, sciatic, 90
nerves, myelinated, 170
nervous system, 169–170
as a medium for disease, 163
sympathetic, 170, 183
net, for gathering plankton, 24, 212
Neuchâtel, 79, 81
neuron theory, 115
neurosis, xvii
Nice, 191, 206, 212
Nietzsche, Friedrich, 261 n.40
Noah's ark, 29

objectivity, 220, 223
observation, 22
Oken, Lorenz, 7–8, 135, 221
ontogeny, xii, 222
ophthalmometer, 102
ophthalmoscope, 124
opium, 224, 228
order
in comparative anatomy, 195, 200, 214
in narrative, 219, 224, 290 n.135
in scientific knowledge, 195
Ørsted, Hans Christian, 93

Paris, 48–49
Pasteur, Louis, xii
pathology, 151–152, 154
rational, 163
value of microscopy for, 22, 138
patricide, xvi
pelagic fishery, 24, 31–32, 37, 213, 234
Pépinière. *See* Friedrichs-Wilhelms-Institut
perception, spatial, 129, 273 n.95
percussion, 145

personalities
influence on science, 232
interactions of, xviii, 233
perspectives, multiple, xv
Pfeufer, Karl, 71
Phoebus, Philipp, 140–141
phylogenetic tree, 199
phylogeny, xii, 198–199, 222
physiology
comparative, 215
emergence of, 12
relation to anatomy, 22, 184
Pistor, Philipp Heinrich, 57–58
plagiostomes, 23
plankton, 31, 38, 234
plexus chorioideus, 195, 220
Plössl, Simon, 57
Poland, 166, 169
polarity, 8
politics, influence on science, 132, 232
Pomerania, 132, 156
popularization, 193
Posen, 166, 175
primitive band, 170, 173, 189
Pringsheim, Nathanael, 218
projection, 265 n.136
prosector, 18
Charité, 140–145, 173–174
protein, digestion of, 60–62, 250 n.227
Prussia
anti-Catholic campaigns of, 65, 147
educational reforms in, 5
history of, 4
political reforms in, 4
psychoanalysis, xvi–xviii
psychohistory, xvii–xviii, 232, 240 n.16
pupil membrane, 47
Purkinje, Jan Evangelista, 13, 69, 187–188, 255 n.87
pus, in the blood, 143, 156
putrefaction, 118

radiolaria, 191, 196, 212–213
rank, academic, 13, 244 n. 68
Ranken arteries, 10
Rathke, Martin, 10, 55, 83, 124
Rätsel, 26, 217, 220
readers, awareness of, xv, 155–156, 187
realism, 39, 251 n.231
rebellion
against parents, 192, 195–196
against scientific advisers, 219, 251 n.4
record, historical, xvi
desire to influence, 104, 127
effect of current science on, xiii
effect of relationships on, xiii
red thread, 219, 223
Rehfues, Philipp Josef von, 8, 42
Reichert, Karl Bogislaw, 31, 70, 95, 101, 177, 265 n.128
Reichstag, 153
Reimer, Georg, 154
Reinhardt, Benno, 135, 144, 147, 154
relationships, student-adviser, xiii, xvii
influence on science, xviii–xix, 127, 233
Remak, Ernst Julius, 169
Remak, Robert, xii
in 1848, 35, 168–169
as candidate for the Berlin pathology chair, 181–182
as candidate for Charité prosector, 144–145, 173–174, 181
as candidate for the Königsberg physiology chair, 122, 174, 180, 184
and cell theory, 170–171
and comparative anatomy, 183–184
dedication to medicine, 185
education, 166–167, 172
family and cultural background, 165–168, 180
income and finances, 169, 178–179
marriage to Feodore Meyer, 169
and medical reform, 168–169
and microscopy, 177, 179, 183
Müller's influence on, 167, 172–174, 183
relationship with Henle, 188–189
relationship with Müller, 173, 175, 183, 230
relationship with Virchow, 181–182
representation of Müller's science, 185
as Schönlein's assistant, 173–174, 185
support from Alexander von Humboldt, 173–177, 180
and sympathetic ganglia, 116, 170
as a teacher, 177–180
on tumors, 170, 185
as a writer, 186–189
Remak, Salomon Meier, 166

resonators, 125
respiration, fetal, 8
retina, 129
return of the Dead, 110
Retzius, Anders, 10, 28, 48, 70
revolution of 1848
 causes of, 34
 liberal vs. radical positions in, 35, 168
 Müller's response to, 35–37, 161
 Remak's role in, 35, 168–169
 Virchow's role in, 35, 145–153
rhetoric, 86, 156, 250 n.214
Rhine, Confederation of, 4
Rhineland, 4–5
 stereotypes, 51–52
Richter, Marie, 71
rococo, 39, 93, 231, 250 n.225, 264 n.114
Rokitansky, Karl, 143, 162
Rosenthal, Isidor, 251 n.1
Royal Society, 126
Rudolphi, Carl Asmund, 8, 12, 159
Rütimeyer, Ludwig, 222

Saint-Hilaire, Geoffroy, 30
scabies, 171
Schelling, Friedrich Wilhelm Joseph von, 7–8
Schenk, August, 219
Scherer, Johann, 198
Schiek, F. W., 58, 197
Schleich, Karl Ludwig, 148, 155
Schleiden, Matthias, Jakob, 63, 66, 96, 170, 201–202, 206, 233
Schlemm, Friedrich, 13, 18, 177
Schmidt, Heinrich, 192
Schönlein, Johann Lukas, 50–51, 144–145, 173–174, 180
Schultz-Schultzenstein, Heinrich, 179
Schulze, Johannes, 6, 105, 121, 142, 226, 234, 250 n.214
Schwann cells, xii
Schwann, Theodor, xii, 37
 assessment of Müller's science, 72–74
 and cell theory, 62–64, 67, 73, 162, 170, 233, 256 n.108
 on cell growth, 64, 170
 education, 50
 and experimental physiology, 50, 60–62, 72–73
 family background, 50
 fermentation studies, 62, 118
 income and finances, 60, 64–65
 living situation in Berlin, 20, 51, 53
 Microscopical Researches, 63–65, 68
 move to Louvain, 34, 65
 muscle tension studies, 60
 personality, 50
 protein digestion studies, 60–62
 rejection of life force, 62, 64, 72–73
 relationship with Müller, 50–51, 62, 65, 67–68, 73–74, 229
 and Roman Catholicism, 50, 65–66
 scientific style, 60–62
 studies of embryonic need for oxygen, 60
 study of gall, 66
 support from Müller, 67
Schwerin, Count, 36
science, history of, xiv–xv
 and du Bois-Reymond, 77, 104, 108
 and Haeckel, 221
 and Virchow, 156, 159, 164
sea cucumbers, 37, 107
seduction, xv, 95–99
Semper, Carl, 222
serfs, liberation of, 4
Sethe, Anna, 204, 216, 222
sewer system, Berlin, 163
Shakespeare, William
 Hamlet, 76–77, 84, 110
 King Henry IV, Part I, 94
Sicily, 213
Siemens, Werner von, 5, 111
Silesia, 146–148, 161
Smith, Henry, 99
socialism, 161
Society of German Scientists and Physicians, 8, 10, 209
soul, 65
space
 for research, 19–20, 54, 89–92, 141, 233, 235
 for teaching, 178
spleen, 160
Steffens, Henrich, 86
stereotypes
 of Prussians and Rhinelanders, 51–52
 of scientists, 248 n.181

stethoscope, 50, 145
Stoecker, Adolf, 153
style, scientific, xv, 272 n.58
suicide, xii, 226, 228, 230
syphilis, 171
systems, of knowledge, 164

temporal bone, 96
Tiedemann, Friedrich, 14
transmission, nerve impulse, 85
trichinosis, 155, 163
Troschel, F. H., 227
truth, xiv–xv
tuberculosis, 208
tumors
  Müller on, 22, 25–26, 67, 138, 157, 162, 170, 182
  parasitic, 170
  Remak on, 170, 182
  Virchow on, 155, 157, 162–163, 170
typhus, 146–148, 161, 208

unification, German, 5, 79, 153
University, Berlin
  in 1848, 36, 149–150
  academic calendar, 17
  academic structure, 16–17
  admission of Jews to faculty, xii, 167, 174–177, 179–180
  founding of, 6
  main building, 16
  student enrollment at, 16, 176
University, Bonn, 5–6, 242 n.25
University, Dorpat, 12, 19, 54–55, 101
University, Göttingen, 71, 258 n.159
University, Heidelberg, 47, 125, 163, 272 n.74
University, Jena, 202–203
University, Königsberg, 118, 127, 174
University, Krakow, 169
University, Louvain, 34, 65
University, Marburg, 119–120
University, Würzburg, 50–51, 151, 163, 178, 191, 204–206, 287 n.70
University, Zurich, 34, 204, 206

Velten, Olga von, 118–119
Vilnius, 173, 175
Virchow, Johann Christoph, 133
Virchow, Rudolf, xii
  in 1848, 35, 145–153, 161, 168–169
  and anti-Semitism, 153, 166
  as candidate for the Berlin pathology chair, 151–152, 166, 181–182, 209, 230
  and cell theory, 138, 170–171, 207
  Cellular Pathology, 153, 155, 157, 163, 208
  as Charité prosector, 140–145, 230
  clinical rotations, 139–140
  conflict with Haeckel over teaching evolution, 163, 209–210
  dedication to medicine, 133, 139, 146
  dismissal from the Charité, 151
  doctoral thesis on the cornea, 140
  duel challenge, 153
  education, 133–139
  family and cultural background, 133
  as Haeckel's teacher, 191, 194–195, 200, 204, 206–209, 220
  income and finances, 134
  memorial address for Müller, 158
  Müller's influence on, 138, 152, 162–164, 230
  marriage to Rose Mayer, 151
  and medical reform, 146, 149, 151, 155–156, 168–169
  and microscopy, 138, 140, 143, 152, 157, 160
  notes on Müller's lectures, 135–138
  political service, 148, 153
  relationship with Haeckel, 208–210, 220, 288 n.96
  relationship with Henle, 163
  relationship with his father, 133–134, 157
  relationship with Müller, 139, 143–144, 152, 160, 230
  relationship with Remak, 181–182
  representation of Müller's science, 156–162
  and Roman Catholicism, 21, 147, 158–159
  and the Silesian typhus epidemic, 146–148
  on tumors, 155, 157, 162–163, 170
  as a writer, 153, 187
Virchow's Archive, 138, 147, 154–157
vision, 23
  binocular, 9, 125, 135, 273 n.95
  color, 127

visual learning, 196
vitalism, 21, 72, 104, 106, 130, 215, 229
  new, 164
Vogt, Karl, 222
voice, human, 20
Volta, Alessandro, 84

Waldeyer, Wilhelm, 69, 177
Walter, Johann Gottlieb, 27
Walther, Philipp von, 10
War, Franco-Prussian, 79
Weiss, Christian Samuel, 13, 27
Westphal, Carl, 169
Wilhelm I, 180
Windischman, Karl, 34
Wolff, Caspar Friedrich, 136
Wolffian bodies, 10
words, favorite, xv, 25–26, 184
workaholism, 33
Wundt, Wilhelm, 218

Young, Thomas, 125

Zeiller, Nanny. *See* Müller, Nanny
zeitgeist, 232
Zweckmässigkeit, 7, 44, 106, 199
  du Bois-Reymond's redefinition of, 78, 87–88, 109